AF844227

IISS

Library & Information Department
Arundel House, 13-15 Arundel Street
London WC2R 3DX
tel +44 (0)20 7395 9122 *e-mail* library@iiss.org

This book is to be returned on or before the last date below.
Renewals (max. 2) may be made by telephone and e-mail.

1 2 APR 2016

THE VIRTUAL BATTLEFIELD: PERSPECTIVES ON CYBER WARFARE

Cryptology and Information Security Series

The Cryptology & Information Security Series (CISS) presents the latest research results in the theory and practice, analysis and design, implementation, application and experience of cryptology and information security techniques. It covers all aspects of cryptology and information security for an audience of information security researchers with specialized technical backgrounds.

Coordinating Series Editors: Raphael C.-W. Phan and Jianying Zhou

Series editors

Feng Bao, *Institute for Infocomm Research, Singapore*
Kefei Chen, *Shanghai Jiaotong University, China*
Robert Deng, *SMU, Singapore*
Yevgeniy Dodis, *New York University, USA*
Dieter Gollmann, *TU Hamburg-Harburg, Germany*
Markus Jakobsson, *Indiana University, USA*
Marc Joye, *Thomson R&D, France*
Javier Lopez, *University of Malaga, Spain*
Nasir Memon, *Polytech University, USA*
Chris Mitchell, *RHUL, United Kingdom*
David Naccache, *École Normale Supérieure, France*
Gregory Neven, *IBM Research, Switzerland*
Phong Nguyen, *CNRS / École Normale Supérieure, France*
Andrew Odlyzko, *University of Minnesota, USA*
Adam Young, *MITRE Corporation, USA*
Moti Yung, *Columbia University, USA*

Volume 3

Recently published in this series

Vol. 2. M. Joye and G. Neven (Eds.), Identity-Based Cryptography
Vol. 1. J. Lopez and J. Zhou (Eds.), Wireless Sensor Network Security

ISSN 1871-6431

The Virtual Battlefield: Perspectives on Cyber Warfare

Edited by

Christian Czosseck

Cooperative Cyber Defence Centre of Excellence (CCD CoE), Tallinn, Estonia

and

Kenneth Geers

Naval Criminal Investigative Service (NCIS), and Cooperative Cyber Defence Centre of Excellence (CCD CoE), Tallinn, Estonia

THE INTERNATIONAL INSTITUTE
FOR STRATEGIC STUDIES
ARUNDEL HOUSE
13-15 ARUNDEL STREET
LONDON WC2R 3DX
WITHDRAWN

IOS
Press

Amsterdam • Berlin • Tokyo • Washington, DC

© 2009 The authors and IOS Press.

All rights reserved. No part of this book may be reproduced, stored in a retrieval system, or transmitted, in any form or by any means, without prior written permission from the publisher.

ISBN 978-1-60750-060-5
Library of Congress Control Number: 2009937744

Publisher
IOS Press BV
Nieuwe Hemweg 6B
1013 BG Amsterdam
The Netherlands
fax: +31 20 687 0019
e-mail: order@iospress.nl

Distributor in the USA and Canada
IOS Press, Inc.
4502 Rachael Manor Drive
Fairfax, VA 22032
USA
fax: +1 703 323 3668
e-mail: iosbooks@iospress.com

LEGAL NOTICE
The publisher is not responsible for the use which might be made of the following information.

PRINTED IN THE NETHERLANDS

The Virtual Battlefield: Perspectives on Cyber Warfare
C. Czosseck and K. Geers (Eds.)
IOS Press, 2009
© 2009 The authors and IOS Press. All rights reserved.

Preface

On January 14, 2009, I posted a Call for Papers (CFP) to Bugtraq for a Conference on Cyber Warfare. Within hours, I received an email from n3td3v, an infamous computer security commentator [1]:

> How can you have a security conference on "cyber warfare" when it doesn't exist and has never taken place.

n3td3v has a point. Estimating the threat posed by cyber attacks is not easy. Case studies are few in number, much information lies outside the public domain, and there have been no wars – yet – between modern, cyber-capable militaries. While the era of cyber espionage is already here [2], a possible era of broad-scale cyber warfare still lies in the future [3].

Nevertheless, an examination of international affairs over the past two decades suggests that cyber battles of great consequence are easy to find. Since the earliest days of the World Wide Web, Chechen guerilla fighters, armed not only with rifles but with digital cameras and HTML, have clearly demonstrated the power of Internet-enabled propaganda. During the 1999 war over Kosovo, likely non-state actors tried to disrupt NATO military operations through computer hacking, and were able to claim minor victories [4]. In 2007, Syrian air defense was reportedly disabled by a cyber attack moments before the Israeli air force demolished an alleged Syrian nuclear reactor [5]. In 2009, the entire nation-state of Kyrgyzstan was knocked offline during a time of political crisis [6].

What military officers call the 'battlespace' grows more difficult to define – and to defend – over time. Advances in technology are normally evolutionary, but they can be revolutionary: artillery reached over the front lines of battle; rockets and airplanes crossed national boundaries; and today, cyber attacks can target political leadership, military systems, and average citizens anywhere in the world, during peacetime or war, with the added benefit of attacker anonymity.

Information Technology (IT) now pervades our lives. In 1965, Gordon Moore correctly predicted that the number of transistors on a computer chip would double every two years [7]. There has been concomitant growth in almost all aspects of IT, including the widespread availability of practical encryption, user-friendly hacker tools, and Web-enabled open source intelligence (OSINT). It should therefore no longer be surprising that political and military strategists use and abuse computers, databases, and the networks that connect them in order to achieve their objectives [8]. In the early 1980s, this concept was already known in the Soviet Union as the Military Technological Revolution (MTR); following the U.S. victory in the 1991 Gulf War, the Pentagon's Revolution in Military Affairs was almost a household term [9].

Cyberspace, narrowly defined, is a collection of networked computers. But the extent to which humans (and other computers) obtain their information and marching orders from somewhere in cyberspace grows by the day. This is in part what hackers call the expanding 'attack surface'. In national security terms, the concepts of attack, defense, and security remain unchanged, as do the threats posed by adversary propaganda, espionage, and attacks on critical infrastructure. The difference is that traditional threats are now Internet-enabled; they employ a new delivery mechanism that can increase the speed, diffusion, and even the power of an attack. The cyber skirmishes we witness today likely foreshadow a long march that cyber warfare will make from a corollary of real-world disputes to a lead role in conflicts of the future.

This book consists of the research papers presented at the Cooperative Cyber Defense Centre of Excellence (CCD CoE) Conference on Cyber Warfare, which took place in Tallinn, Estonia, in June 2009. Individually and collectively, they explore the relationship between computer security and national security. Unsurprisingly, the devil is found in the details: the challenge of attribution, the calculation of damages, the security of critical infrastructure, ethics, jurisdiction, responsibility, and much more. This book is divided into two sections: Strategic Viewpoints and Technical Challenges and Solutions.

Strategic Viewpoints

Chapter 1: "Cyber Wars: A paradigm shift from Means to Ends." Amit Sharma, from the Indian Ministry of Defence, argues that cyber warfare is different from other types of conflict, and merits its own set of rules. He warns against trying to fit cyber warfare into the traditional definitions found in the Law of Armed Conflict (LOAC). Sharma believes that it is easy to underestimate the strategic potential of cyber warfare, stating that cyber attacks alone are powerful enough to achieve political goals. In his view, cyber warfare will cease to be merely a force multiplier for conventional warfare; rather, conventional warfare will be used to support the objectives of cyber warfare. Sharma examines the legal treatment of nuclear weapons, and deterrence theory, for possible application in the cyber domain.

Chapter 2: "Towards an Evolving Theory of Cyberpower." Dr. Stuart H. Starr, from the Center for Technology and National Security Policy (CTNSP) at the National Defense University (NDU), discusses the development of a theory of 'cyberpower' in research that systematically addresses five key areas: it defines the key terms that are associated with cyber issues; it categorizes the elements, constituent parts, and factors that yield a framework for thinking about cyberpower; it explains the major factors that are driving the evolution of cyberspace and cyberpower; it connects the various elements of cyberstrategy so that a policy maker can place issues in proper context; and it anticipates key changes in cyberspace that are likely to affect decision making.

Chapter 3: "*Sub Rosa* Cyber War." Martin C. Libicki of the RAND Corporation explains that the battlefield terrain of cyberspace allows not only for stealthy attack and defense, but even for the existence of a stealthy war. *Sub rosa* cyber war is a conflict in which the warring parties do not publicly acknowledge battlefield victories and defeats, or even the existence of an ongoing war. Two reasons such a conflict is possible include the difficulty of good cyber battle damage assessment and the diabolical challenge of cyber attack attribution. Further, opponents may desire to keep a cyber conflict *sub rosa* in order to preserve freedom of action; public awareness and scrutiny could complicate negotiations or lead to unwanted escalation. Libicki cautions that *sub rosa* cyber war carries serious risks, such as insufficient operational oversight and a dubious assumption that the conflict is truly *sub rosa* to third parties.

Chapter 4: "Warfare and the Continuum of Cyber Risks: A Policy Perspective." Andrew Cutts, the Director of Cyber Security Policy at the U.S. Department of Homeland Security, explains that nation-states are beginning to appreciate the potential benefits and costs of employing cyber attacks as a means of projecting national power. He offers a framework for evaluating national cyber security policy that includes prioritizing competing missions and balancing short- and long-term objectives. The focus of this chapter is on the long-term, strategic threat. To get ahead of the most serious cyber risks, national security leadership must strive to find the appropriate balance of resources, energy, and focus, specifically to distinguish between the most frequent threats and those that are the most consequential.

Chapter 5: "Cyber Terrorism: A New Dimension in Battlespace." Major J. P. I. A. G. Charvat, from the Centre of Excellence Defence against Terrorism in Ankara, Turkey, examines the emerging threat of cyber terrorism. First, he discusses the phenomenon of terrorism and the motivations of terrorist organizations. Next, he explores the way terrorists now use IT to disseminate propaganda, to recruit new members, and to support conventional attacks. Finally, he considers the possibility that a terrorist organization might adopt a pure cyber attack strategy in an attempt to inflict electronic or physical damage.

Chapter 6: "Borders in Cyberspace: Can Sovereignty Adapt to the Challenges of Cyber Security?" Forrest Hare, from the School of Public Policy at George Mason University, writes that while cyberspace has no borders, nation-states do. Therefore, governments must consider how they will define and defend their sovereignty in this new domain. Crucially, Hare argues that governments should realize that to some degree they will be forced to coordinate and integrate their efforts on the international level. To help explore the nature of boundaries in cyberspace, this research makes use of two very different frameworks: the first is a comparison of the challenges of cyber security to international drug trafficking; the second employs game theory to support the cyber security decision-making process.

Chapter 7: "Towards a Global Regime for Cyber Warfare." Dr. Rex Hughes, Co-Director of the Cyber Security Project, Chatham House, London, explains that when novel, disruptive technologies dramatically alter the nature of warfare, history shows that the likelihood of a new arms race is high. As more nations aspire to project national power in cyberspace, a digital arms race may be around the corner. Therefore, diplomats should find a coherent set of principles, rules, and norms to govern state security and military operations in cyberspace. Hughes argues that the most important cyber challenge facing national security thinkers today concerns how to prevent a major arms race in this arena. This chapter introduces readers to the Law of Armed Conflict, examines how it might apply to cyber warfare, and outlines the steps required to create a global regime for cyber warfare.

Chapter 8: "What Analogies Can Tell Us About the Future of Cybersecurity." David Sulek and Ned Moran of Booz Allen Hamilton examine the benefits and drawbacks of using historical analogies to understand cyber warfare. When such analogies are appropriately chosen and systematically applied, they can clarify the present situation and offer decision-makers strategic insight; vice versa, poor analogies obscure objectives, unnecessarily complicate choices, and create blind spots. In every case, analogies are bound to fail unless they incorporate objective analysis and their hand is not overplayed. The authors consider the well-known Electronic Pearl Harbor, and explore a range of new ideas: Cyber Katrina, Cyber Sputnik, Cyber Balkanization, Cyber Tribes, Cyber Conquistadors, and Cybernization.

Chapter 9: "The Information Sphere Domain – Increasing Understanding and Cooperation." In this chapter, Dr. Patrick D. Allen (Johns Hopkins University, Applied

Physics Lab) and Dennis P. Gilbert, Jr (Booz Allen Hamilton) write that a great advantage always accrues to a competitor who understands and operates within a domain better than their opponent. First, the authors define what constitutes a domain, and describe how new domains are created over time. Their research outlines the 'Information Sphere' domain, which has features that are both similar to and different from the four traditional, physical domains: air, land, sea, and space.

Chapter 10: "Sun Tzu was a Hacker: An Examination of the Tactics and Operations from a Real World Cyber Attack." Billy K. Rios, of GreyLogic, LLC, examines the nuts and bolts of a real world, international cyber attack. Rios was in a unique position to witness the "communications, execution, and responses" of the cyber attackers and defenders in the 2008 war between Russia and Georgia. The author concludes that the availability and effectiveness of cyber attack tools ensure that data packets, fired purely on the battlefield terrain of computer networks, are likely to play a role in all future international conflicts. This research paper considers cyber attacks in the light of traditional concepts of maneuver warfare, as described in Marine Corps Doctrinal Publication 1 (MCDP-1).

Chapter 11: "Belarus in the Context of European Cyber Security." Fyodor Pavlyuchenko, of www.charter97.org, examines the use of Internet censorship as a government tool that can be used to suppress political dissent within a nation-state. The author, representing one of the most popular – and hacked – online news sites in Belarus, catalogues a decade of cyber attacks that the site has suffered (usually during times of political tension). The author believes that the use of political DoS attacks threatens not only freedom of expression in Belarus, but the integrity of Internet resources in other European countries as well. Finally, he contends that the cyber conflict in Belarus is analogous to the ongoing struggle between state and non-state actors in Russia.

Chapter 12: "Politically Motivated Denial of Service Attacks." Jose Nazario of Arbor Networks dissects DDoS attacks that have been used in support of political or ideological goals. He explains how DDoS attacks have evolved from a platform used to inflict "punitive damage" on a target to a sophisticated means of Internet censorship. Nazario uses extraordinary access to a wide range of data, including Internet backbone traffic, border gateway protocol (BGP) routing data, botnet communications, and community chatter, to create a compelling narrative. He concludes that while most Internet attackers appear to be non-state actors, they are nonetheless capable of using botnets of significant size and power to launch effective DDoS attacks.

Chapter 13: "A Brief Examination of Media Coverage of Cyberattacks (2007-Present)." Cyrus Farivar, a freelance technology journalist, critiques the media coverage of three politically-oriented cyber attacks: Kyrgyzstan (2009), Georgia (2008), and Estonia (2007). He reflects on where journalistic analysis was correct and where it was off the mark. On balance, he argues that there is enormous room for improvement, and that it is in the interest of the cyber security community, the media, and policymakers to improve their understanding of and their ability to write on the subject of cyber attacks, so that public understanding and appreciation of this new threat will increase.

Technical Challenges and Solutions

Chapter 14: "Behavioral Analysis of Zombie Armies." Olivier Thonnard, Wim Mees (both from the Royal Military Academy, Belgium) and Marc Dacier (Symantec

Research Labs, France) present ground-breaking research on the behavior of 'zombie armies'. They characterize the long-term behavior, global characteristics, strategic evolution, size, lifespan, and resilience of botnets. Their research highlights the uneven spatial distribution of infected computers across the Internet, specifically on a limited number of "unclean" or "zombie-friendly" networks. Most of the botnets they studied have an impressive attack capability in terms of bandwidth and the number of ways they are able to probe and exploit other computers; further, like real-world armies, they can coordinate their efforts with other botnets. Their analysis employed data from the European Union (EU)-funded WOMBAT project (Worldwide Observatory of Malicious Behaviors and Attack Threats).

Chapter 15: "Proactive Botnet Countermeasures – An Offensive Approach." Felix Leder, Tillmann Werner, and Peter Martini, from the Institute of Computer Science IV, University of Bonn, Germany, contend that cyber defenders are at a disadvantage vis-à-vis attackers in part because computer science evolves too quickly for cyber law to keep up. In this chapter, the authors describe a technically feasible, proactive way to detect and defeat computer botnets, based on the assumption that reactive measures alone are insufficient. Their research formalizes botnet topologies, via real-world examples, and derives effective strategies for attacking them. However, they explain that their approach employs tactics that go well beyond what current cyber law encompasses, and argue that "controversial discussions" are needed to explore the political, legal, ethical, and liability-based ramifications of a proactive, counter-botnet approach – sooner rather than later.

Chapter 16: "When Not to Pull the Plug – The Need for Network Counter-Surveillance Operations." Scott Knight and Sylvain Leblanc of the Royal Military College of Canada argue that one traditional response to a cyber attack – to immediately remove compromised machines from the network – has two primary drawbacks: it warns the intruder that he or she has been discovered, and it can prevent the collection of crucial evidence to help determine motive and estimate damage. The authors delineate a Network Counter-Surveillance Operation designed to maintain quiet contact with a hacker while information is gathered on intentions, techniques and more.

Chapter 17: "Autonomic Computer Network Defence Using Risk State and Reinforcement Learning." Luc Beaudoin (Defense Research and Development Canada), Nathalie Japkowicz and Stan Matwin (both from the University of Ottawa) write that humans are simply not able to handle the complexity and speed of many forms of cyber attack, and that it is essential to automate certain aspects of traffic analysis and attack mitigation. They argue that a level of autonomic computer network defense can be achieved using reinforcement learning and dynamic risk assessment, with a view toward determining optimal action sequences (or policies) to recover from computer network risk situations. In their view, this approach will benefit commercial network management and security products by aiding in the selection of automatic mitigation actions, as risk states are sensed.

Chapter 18: "Enhancing Graph-based Automated DoS Attack Response." Gabriel Klein, Marko Jahnke, Jens Tölle (all from FGAN-FKIE, Germany), and Peter Martini (University of Bonn, Germany) argue that timely and appropriate responses to DoS attacks are critical in both civilian and military settings. While intrusion detection systems (IDS) are capable of detecting DoS attacks, the growing sophistication and speed of such attacks increase the need for automated countermeasures to support computer network defense. Per force, it is necessary to quickly evaluate the potential effects of proposed countermeasures on network resources. This chapter discusses GrADAR, an intuitive, graph-based approach for automatically assessing the likely effects of DoS countermeasures on a network. Further, it proposes an enhancement which takes into account the effects of the workload on resource availability.

Chapter 19: "On n^{th} Order Attacks." Daniel Bilar, from the University of New Orleans, USA, explores a class of cyber attack designed to subvert "mission-sustaining ancillary systems". In technical terms, ancillary systems could be throughput control, visualization environments, memory resource allocation, manufacturing, and the supply chain; the purpose of such systems – and the real target of the attacker – is its political, military, or economic mission. The attacker's goal could be to disrupt power management, logistics, elections, or even the social welfare of the target. Bilar discusses historical, current and forward-looking examples, with special emphasis on attacks against computerized, open societies.

Chapter 20: "Business and Social Evaluation of Denial of Service Attacks in View of Scaling Economic Countermeasures." Louis-Francois Pau, from the Copenhagen Business School and Rotterdam School of Management, writes that DoS attacks not only affect computer network resources; they can have a direct, negative impact on the bottom line of a business in the real world. This chapter proposes a method to determine the direct and indirect costs associated with DoS attacks, which is a necessary step in determining countermeasures aimed at legal- or policy-driven dissuasion, retaliation, compensation, and restoration. Dr. Pau's method relies on time-preference dynamics applied to monetary mass for the restoration of capabilities, on long-term investments to rebuild capabilities, and on the usability level of capabilities after an attack. A real-world example of a DoS attack on a corporate data centre is provided. In conclusion, the author gives specific policy recommendations and suggests information exchange requirements.

Chapter 21: "Virtual Plots, Real Revolution." Roelof Temmingh (Paterva) and Kenneth Geers (NCIS/CCD CoE) investigate whether computer botnets could evolve from spam and Distributed Denial of Service (DDoS) generators to semantic creatures that could voice opinions, arguments, and even threats via the Internet. Key to their argument is the assumption that only a small percentage of information on the Web is truly unique. In theory, a malicious actor could create a virtual population of fraudulent identities from stolen and/or randomized biographies, pictures, and histories of Internet activity, which could be used to support a criminal, political, military or terrorist agenda. The increasingly impersonal nature of Internet communications will make timely threat evaluation difficult.

Many thanks to all who were involved in organizing the 2009 Conference on Cyber Warfare, and especially to Christian Czosseck, for his diligence in helping to edit this book.

Kenneth Geers
Tallinn, Estonia
August 7, 2009

References

[1] "Who is "n3td3v"?", Hacker Factor Solutions, White Paper, Release 1.4.1, 12-October-2006, www.hackerfactor.com/papers/who_is_n3td3v.pdf.

[2] "Espionage report: Merkel's China visit marred by hacking allegations". *Spiegel Online*, August 27, 2007, http://www.spiegel.de/international/world/0,1518,502169,00.html; Cody, E. "Chinese official accuses nations of hacking". *Washington Post*, September 13, 2007, http://www.washingtonpost.com/wp-dyn/content/article/2007/09/12/AR2007091200791_pf.html#.

[3] However, this is already a point on which reasonable people can disagree: "How robot drones revolutionized the face of warfare", Cable News Network (CNN), July 24, 2009. http://edition.cnn.com/2009/WORLD/americas/07/23/wus.warfare.remote.uav/index.html.

[4] Geers, Kenneth. "Cyberspace and the Changing Nature of Warfare," *SC Magazine*, August 27, 2008, http://www.scmagazineus.com/Cyberspace-and-the-changing-nature-of-warfare/article/115929/.

[5] Fulghum, David A., Wall, Robert, and Butler, Amy. "Cyber-Combat's First Shot," *Aviation Week & Space Technology*. November 26, 2007. Vol. 167, Iss. 21, p. 28.

[6] Keizer, Gregg. "Russian 'cyber militia' knocks Kyrgyzstan offline", *Computerworld*, 01/28/2009, www.networkworld.com/news/2009/012809-russian-cyber-militia-knocks-kyrgyzstan.html.

[7] "Moore's Law", Intel Corporation, www.intel.com/technology/mooreslaw/.

[8] Adams, James. "Virtual Defense", *Foreign Affairs*, May/June 2001, 80, 3, p. 98.

[9] Mishra, Shitanshu. "Network Centric Warfare in the Context of 'Operation Iraqi Freedom'" Strategic Analysis, Vol. 27, No. 4, Oct-Dec 2003, Institute for Defence Studies and Analyses, http://www.idsa.in/publications/strategic-analysis/2003/oct/Shitanshu.pdf.

About the Contributors

Editors

Christian Czosseck is a Scientist at the Cooperative Cyber Defence Centre of Excellence (CCD CoE) in Tallinn, Estonia, and the German representative there since 2008. He has been a soldier in the German Armed Forces (Bundeswehr) for more than 12 years. Christian graduated first in his class in Computer Science at the Universität der Bundeswehr in Munich, and for the past six years he has held numerous Information Assurance positions in the German military.

Kenneth Geers, Naval Criminal Investigative Service (NCIS), is a Scientist and the U.S. representative to the CCD CoE. He has served as an intelligence analyst, a French and Russian linguist, and computer programmer in support of strategic arms control initiatives. Kenneth has been a student in six countries, earned the Defense Language Institute Provost's Award for Outstanding Scholastic Achievement, wrote his Master's thesis on nuclear proliferation at the Jackson School of International Studies, University of Washington, Seattle, and is a Certified Information Systems Security Professional (CISSP).

Authors

Luc Beaudoin earned a degree in electrical engineering with honors from the Royal Military College of Canada, a Masters in Business Administration from the University of Québec, and is now a MSc student in Systems Sciences at the University of Ottawa. Luc served ten years in the Canadian Forces as a telecommunications officer, where he was the first Watch Officer at the Canadian Forces Network Operations Centre. He joined the Defence Research and Development Canada- Network Information Operations section in 2003, where he leads network security projects associated with situational awareness, dynamic risk response, decision making and automated defense.

Daniel Bilar is an Assistant Professor of Computer Science at the University of New Orleans, and holds a PhD in Engineering Sciences from Dartmouth College. Daniel was a founding member of the Institute for Security and Technology Studies at Dartmouth, conducting counter-terrorism technology research for the U.S. Departments of Justice and Homeland Security. Active research topics include the detection, classification and containment of highly-evolved malicious software, quantitative risk analysis/management of networks, and the optimization of business and innovation processes.

J. P. I. A. G. Charvat graduated from Anglia Ruskin University with a degree in Modern History in 1992. After a year at the Royal Military Academy Sandhurst, he was commissioned into the Royal Military Police in 1993. He has served in the UK, Cyprus, Germany, Bosnia, and in the U.S. at the Army Military Police School, Fort Leonard Wood. After company command, he was posted to the Centre of Excellence

Defence against Terrorism in Ankara, Turkey. Julian also holds a Certificate in Terrorism Studies from St Andrew's University.

Andrew Cutts joined the Department of Homeland Security in March 2008, in the Office of Policy as Director for Cyber Security. Prior to joining DHS, he directed the Cyber Conflict Research Institute at Norwich University. From 2002–2004, he was a research program manager at the Institute for Security Technology Studies at Dartmouth College. He was in the private sector from 1995 to 2002, where he became Vice President for Information Technology at a high-tech engineering services firm based in Connecticut. Before that, he served as a Naval Intelligence Officer for nine years.

Cyrus Farivar is a freelance technology journalist and radio reporter/producer from Oakland, California (USA). He has lived in Lyon (France), Saint-Louis (Senegal), Melbourne (Australia), and Geneva (Switzerland). He is currently writing a book tentatively called "The Internet of Elsewhere," due out from Rutgers University Press in 2010. He reports for National Public Radio, The World (WGBH/PRI/BBC), Canadian Broadcasting Corporation, and freelances for The Economist, Foreign Policy, Slate, The New York Times, Popular Mechanics, and Wired.

Kenneth Geers, Naval Criminal Investigative Service (NCIS), is a Scientist and the U.S. representative to the CCD CoE. He has served as an intelligence analyst, a French and Russian linguist, and computer programmer in support of strategic arms control initiatives. Kenneth has been a student in six countries, earned the Defense Language Institute Provost's Award for Outstanding Scholastic Achievement, wrote his Master's thesis on nuclear proliferation at the Jackson School of International Studies, University of Washington, Seattle, and is a Certified Information Systems Security Professional (CISSP).

Dennis P. Gilbert, Jr. is a Cybersecurity Strategist and Information Operations (IO) Practitioner with Booz Allen Hamilton. Mr. Gilbert was Vice President in an IT company, and served in the U.S. Air Force for 21 years, where he held leadership positions in Information Assurance (IA), satellite communications, space control, and electronic warfare. He now serves as a trusted advisor to organizations within the U.S. Department of Defense and Intelligence Community.

Forrest Hare is a Lieutenant Colonel in the United States, assigned to the Office of the Secretary of Defence. Most recently, he was responsible for developing the United States Air Force cyberspace strategy as part of the Military Strategy for Cyberspace Operations. In addition, he has served in numerous information operations positions world-wide. Lt Col Hare is currently a PhD student at the George Mason School of Public Policy, studying national security policy for cyberspace. He has taught Economics and Geography at the United States Air Force Academy and at the University of Maryland, Asian Division. He received his Bachelor of Science degree from the United States Air Force Academy, and a Master of Arts from the University of Illinois. He also conducted post-graduate studies at the University of Fribourg, Switzerland, under a Swiss University Grant.

Rex Hughes is a Senior Resident Member at Wolfson College, Cambridge and Research Associate of The Cambridge-MIT Institute. Dr. Hughes' current research examines the global governance of cyber security. With Dr. Paul Cornish, he established the Cyber Security Project at Chatham House which published in March 2009 its first major report, *Cyberspace and the National Security of the United Kingdom*. Prior to his Cambridge years Hughes founded and directed the world's first multidisciplinary Internet Studies program at the University of Washington in Seattle, where in partnership with IBM-Lotus he led the development of *iEnvoy*, the first secure diplomat-to-diplomat Internet communications platform deployed by the U.S. Department of State.

Marko Jahnke is a senior researcher at the Research Institute for Communication, Information Processing, and Ergonomics (FKIE) in Wachtberg, Germany. He has several years of experience in applied information and communications security. He currently leads a multi-participant research project that focuses on intrusion detection in tactical MANETs.

Gabriel Klein is a researcher at the Research Institute for Communication, Information Processing, and Ergonomics (FKIE) in Wachtberg, Germany. His research focuses on intrusion detection in tactical MANETs. He is currently spending a 4-month visit as a guest researcher at the Cooperative Cyber Defence Centre of Excellence (CCD-CoE) in Tallinn, Estonia.

Scott Knight is an Associate Professor in the Department of Electrical and Computer Engineering at the Royal Military College (RMC) of Canada. Dr. Knight was appointed to the academic faculty at RMC in 2000 on retirement from 21 years of service in the Canadian Air Force. He founded the RMC Computer Security Laboratory, and continues to lead this research group in his present appointment.

Sylvain P. Leblanc is an Assistant Professor and PhD candidate in Software Engineering with the Department of Electrical and Computer Engineering at the Royal Military College of Canada. Sylvain (Sly) served for 23 years as a Signals Officer with the Canadian Forces. He is a member of the RMC Computer Security Laboratory, and works closely with the Defence Research and Development Centre – Ottawa. His specific research interests are in active network defence and computer network defence processes.

Felix Leder is a PhD student at the University of Bonn. After working for Nokia, he turned his attention fully to his favourite field of research: IT security. His current research interests are botnet mitigation tactics and new methodologies for executable file and malware analysis. A lot of his spare time is spent working on the Honeynet Project.

Martin Libicki (PhD, U.C. Berkeley) is a senior management scientist at RAND, focusing on the impact of IT on national security. He is the author of *Conquest in Cyberspace: National Security and Information Warfare*, *Information Technology Standards: Quest for the Common Byte*, *Cyber-Deterrence and Cyber-War*, *What is Information Warfare*, *The Mesh and the Net: Speculations on Armed Conflict in a Time of Free Silicon*, and *Who Runs What in the Global Information Grid*. His most recent assignments included helping to organize the Air Force for cyber-war, exploiting cell phones in counter-insurgency, developing a post-9/11 IT strategy for the U.S. Department of Justice, using biometrics for identity management, assessing DARPA's Terrorist Information Awareness program, conducting information security analysis for the FBI, and evaluating In-Q-Tel.

Ned Moran is an Adjunct Professor in the Computer Science Department at Georgetown University, and also works as an Intelligence Analyst with the Terrorism Research Center. One of his particular areas of interest is how modern terrorist groups use the Internet to organize and operate. In this capacity, Mr. Moran conducts cyber investigations and forensics on suspected terrorist websites. Ned received an M.A. in Security Studies with honors from Georgetown University's School of Foreign Service where he focused on Technology and National Security Policy.

Jose Nazario is the manager for security research at Arbor Networks, responsible for analyzing Internet security threats, reverse engineering malicious code, and software development. Dr. Nazario develops security mechanisms for Arbor's Peakflow platforms via the Active Threat Feed (ATF). His research interests include large-scale Internet trends such as reachability and topology measurement, Internet-scale events such as DDoS attacks, botnets and worms, source code analysis, and data mining. He is the author of *Defense and Detection Strategies against Internet Worms* and *Secure Architectures with OpenBSD* and maintains WormBlog.com, a site devoted to worm detection and defense research.

Louis-Francois Pau is Professor of Mobile communications and media at the Copenhagen business school and at the Rotterdam school of management. Previously, he was CTO of L.M. Ericsson's Network Systems division and CTO for Digital Equipment/Hewlett Packard Europe. He has been on the faculties of the Danish Technical University, Ecole Nationale Supérieure des Télécommunications (Paris), M.I.T., and University of Tokyo. He is a Fellow of IEEE (USA), BCS (UK), JSPS (Japan).

Fyodor Pavlyuchenko was born in Minsk, Belarus in 1976. He graduated from Belarus State University in 1999. Since 1997, he has been involved in a number of high-profile Belarusian Internet projects. Since 2000, he has developed, maintained and promoted www.charter97.org, the leading independent news website in Belarus. Fyodor created the IT-resource bybanner.com for the dissemination of information on Belarusian Internet news, and has participated in the creation and development of Belarusian newspapers and online publications.

Billy K. Rios is a Security Engineer for Microsoft, where he studies emerging risks and cutting edge attacks and defenses. Previously, Billy was a Senior Security Consultant for VeriSign, a penetration tester for the Ernst and Young Advanced Security Center, a U.S. Department of Defense Intrusion Detection Analyst and Marine Corps Officer. He has presented research at Blackhat, RSA, Bluehat, DEFCON, PacSec, HITB, and ASIA.

Amit Sharma is Deputy Director/Scientist C in the Institute for System Studies and Analysis, Defence Research and Development Organization, Ministry of Defence, Government of India. He has worked in the field of Information Security, Strategic Information Dissemination Systems, Net Centric Warfare, C4I2SR systems and Secure and survivable networks. He did his B Tech (honors) in Computer Science and Technology at the National Institute of Technology, Hamirpur India, and is currently pursuing his Masters in Global Security from Defence College of Management Technology, UK Defence Academy, Cranfield University, United Kingdom.

Stuart H. Starr is President of the Barcroft Research Institute (BRI). Prior to founding BRI, he was Director of Plans, MITRE; Assistant Vice President, C2 and Systems Assessment, M/A-COM Government Systems; Director, Long Range Planning and Systems Evaluation, OASD(C3I), OSD; and Senior Project Leader, Institute for Defense Analyses. He received his PhD in Electrical Engineering from the University of Illinois and was a Fellow at MIT's Seminar XXI. Dr. Starr is a Fellow, Military Operations Research Society (MORS); Associate Fellow, AIAA; Member of the Army Science Board; a Senior Research Fellow at the Center for Technology and National Security Policy, National Defense University; and a frequent participant in Blue Ribbon Panels of NATO, the National Research Council, and the Air Force Science Advisory Board.

David Sulek is a Principal with Booz Allen Hamilton's U.S. Security Team with 16 years of strategy, policy analysis, and general management consulting experience, and has worked with the President's National Security Telecommunications Advisory Committee. He leads a team of policy analysts focused on homeland security, national preparedness, information sharing, cyber security, and public-private partnership issues. He received a master's degree in national security studies from the Edmund A. Walsh School of Foreign Affairs at Georgetown University, and a bachelor's degree in political science from Syracuse University.

Roelof Temmingh has been in the computer security industry for 15 years. In 2000, he co-founded SensePost as Technical Director and later was in charge of Research and Development. At SensePost, he developed many successful security assessment tools, including Wikto and Suru, and contributed to several books (*Aggressive Network Self-Defense*, *How to Own a Continent*, *Nessus Network Auditing*). He is a regular speaker at many of the top international security conferences (Blackhat, Defcon, FIRST, CansecWest, RSA, etc.). At the start of 2007, he left the company to start Paterva.

Olivier Thonnard is a PhD student at EURECOM (France), researching honeypot traffic analysis under the supervision of Marc Dacier (Symantec Research). As a military officer, he teaches at the Royal Military Academy, Polytechnic Faculty, Brussels. His research interests include intrusion detection and network traffic analysis, specifically relating to the global analysis of Internet threats. Olivier is developing data mining techniques based on clustering, correlation methods, and clique algorithms.

Tillmann Werner is a computer scientist at the University of Bonn. Previously, he worked as an incident handler at the German national CERT. He is a member of the Honeynet Project and has been doing research in the area of network-based attacks for more than five years.

Contents

Part II. Technical Challenges and Solutions

Part I

Strategic Viewpoints

The Virtual Battlefield: Perspectives on Cyber Warfare
C. Czosseck and K. Geers (Eds.)
IOS Press, 2009
© 2009 The authors and IOS Press. All rights reserved.
doi:10.3233/978-1-60750-060-5-3

Cyber Wars: A Paradigm Shift from Means to Ends

Amit SHARMA[a,1]
[a] *Institute for System Studies and Analysis (I.S.S.A), Defence Research and Development Organization (D.R.D.O), Ministry of Defence, India*

Abstract. The last couple of decades have seen a colossal change in terms of the influence that computers have on the battle field, to an extent that defence pundits claim it to be a dawn of a new era in warfare. The use of computers and information in defence has manifested into various force multipliers such as Information Operations, C4I2SR Systems, Network Centric Warfare, to the extent that commentators are terming this information age as a Revolution in Military Affairs (RMA). These advances have not only revolutionized the way in which wars are fought, but have also initiated a new battle for the control of a new dimension in the current contemporary world: The Cyber Space.

Over time cyber warfare has assumed the shape of an elephant assessed by a group of blind people, with every one drawing different meanings based upon their perceptions. Under these circumstances there was a gradual paradigm shift in military thinking and strategies, from the strategic aspect to the tactical aspect of cyber warfare laying more emphasis on cyber attacks and counter measures. This resulted in the formation of a notion that cyber warfare or information warfare is a potent force multiplier, which in a sense downgraded the strategic aspects of cyber war to a low grade tactical warfare used primarily for a force enhancement effect. The author believes this is wrong, cyber war is a new form of warfare and, rather than cyber war merely being an enhancement of traditional operations, traditional operations will be force multipliers of cyber war.

This paper tries to shatter myths woven around cyber warfare so as to illuminate the strategic aspects of this relatively misinterpreted notion. This paper will elucidate the scenarios and mechanisms illuminating the process of using the strategies of cyber war, so as to achieve conventional objectives. The paper will also analyze the doctrine and strategies including first and second strike capabilities with regard to cyber war. This paper identifies a paradigm shift from the conventional belief of cyber warfare acting as a force multiplier for conventional warfare to the recognition, that conventional warfare will be acting as a force multiplier around cyber war and hence making cyber war as the primary means of achieving grand strategic objectives in the contemporary world order.

Keywords: Cyber wars; cyber warfare; information warfare; strategy and doctrine.

The opinions expressed or implied in this paper are solely those of the author, and are based on open sources. Under no circumstances these may be correlated or perceived as the views of Government of India in general and the Author's organization in particular.

[1] The Author is Deputy Director/Scientist 'C' in Information Security Division of Strategic Information Dissemination Systems of I.S.S.A., D.R.D.O., Ministry of Defence, Government of India. The Author is a Chevening Scholar from India and is currently pursuing M Sc. Global Security at UK Defence Academy. Email: amitsharma.drdo@gmail.com and asharma.dcmt@defenceacadamy.mod.uk .

Introduction

"One hundred victories in one hundred battles is not the most skillful. Seizing the enemy without fighting is the most skillful."

Sun Tzu Sixth Century B.C [1]

Sun Tzu in sixth century B.C. eloquently referred to the fact that, the best form of warfare is to take down the enemy without fighting with him.[2] Over time, as the warfare has evolved, this notion has gained impetus, especially with the genesis of cyberspace and cyber warfare. It was for the first time, that Sun Tzu's notion of, "Seizing the enemy without fighting is the most skillful" , could be imagined as happening in its entirety, using this potent new weapon which, in current contemporary world has no limits, no boundaries and to a surprise no visible restrictions or legislations. Although over time the notion of information warfare has matured and manifested into a form which has a colossal impact on how the contemporary wars are fought, but this has also resulted in the downgrading of strategic side of information warfare or cyber warfare to a decisive tactical force multiplier capable of turning the tides in war. Whilst this force enhancer aspect of cyber warfare is an important and decisive component of conventional warfare, but against the conventional wisdom this is not the end, but merely a beginning of the strategic aspect of cyber warfare.

In order to analyze the strategic aspect of cyber warfare, Luttwak's criteria of integration of a strategic warfare across all spectrum of affairs right from the tactical to the grand strategic level,[3] provides an important criterion for postulating the strategic framework for cyber warfare; Or in terms of Liddell-Hart, the coordination and assessment of means to achieve ends at all levels plays [4] a dominant role in casting a cyber warfare strategy. In light of these considerations, the author will elucidate the framework in which cyber warfare will have a strategic effect by acting as primary means to achieve conventional ends, hence will induce a paradigm shift from the conventional notion of cyber warfare as a tactical force multiplier to the notion of strategic cyber warfare acting as primary means of achieving grand strategic objectives in the contemporary world order. The author will accomplish this objective by deriving the elixir of Clausewitz's Trinitarian warfare and applying the concepts of Rapid dominance and Parallel warfare in cyber space so as to generate the strategic paralytic effect envisaged in effect based warfare. The author will conclude by shattering the conventional dictum of cyber defence, based on the notion of "defence in layers" and legal aspects of Law of Armed Conflict; by providing the only feasible and viable cyber defence strategy relying on the application of Rational Deterrence Theory (RDT) in general and on the idea of Mutually Assured Destruction (MAD) in particular so as to maintain the strategic status quo.

1. Cyber Irony- The Revolution in Military Affairs

Over last couple of decades, Information assets have had an irrefutable impact on the way, in which conventional wars are fought, to an extent that military theorists have termed it to be a Revolution in Military Affairs (RMA).[5] This extensive reliance of conventional warfare on information in contemporary conflicts is often misrepresented as information warfare rather than information-enabled warfare or information-enhanced warfare. The sudden significance ushered to this relatively new paradigm of information-enabled warfare, where information warfare is acting as a decisive force multiplier, has also raised certain existential questions for its survival predominantly that; whether this new paradigm of Information enabled warfare is really a strategic information warfare which will be the primary means of achieving ends or is it just a misinterpreted notion created by its loyal supporters, only to pacify the appetite of the change-hungry military world?

The answer to this existential question lies in the debate revolving around the notion, that information warfare is a revolution in military affairs. Alvin and Heidi Toffler have termed this information-enabled warfare as a third wave[6]; similarly most of the contemporary military theorists have termed this misinterpreted information warfare as a revolution in military affairs. At this juncture an important argument looms around the relation of a revolution in military affairs and its strategic effect. Throughout the history whenever there has been an occurrence of revolution in military affairs, it has always been followed with a strategic effect; for example revolutions in military affairs such as guns, artillery, airpower, nuclear weapons and so on, have always been accompanied with their strategic impact in creating a new world order. This important relationship between revolution in military affairs and its strategic effect is clearly missing in case of information warfare.

Hence if information warfare is really a revolution in military affairs then ideally it should have a strategic effect and since that effect is clearly missing, it can be concluded that something somewhere is missing. This gap is due to the misinterpretation of information warfare as mere decisive tactical information-enabled warfare acting as a force multiplier for conventional warfare. The Author believes that this notion is a fallacy; information warfare is more than just information-enabled warfare, which albeit represents an important aspect of information or cyber warfare, but not in totality. Cyber warfare is a strategic warfare which can be used as a principle means to achieve strategic ends and as required by Luttwak's criterion for strategic warfare [7], the framework for the strategic cyber warfare is to be defined across all spectrum of affairs right from the grand strategic to the tactical level.

2. The Grand strategic cyber warfare – the triad theory of cyber warfare

"War is thus an act of force to compel our enemy to do our will"
Clausewitz [8]

Clausewitz in his book *On War* clearly elucidated the fact that the end of the war is to compel the enemy to do your will [9] and Sun Tzu argues that the best form of warfare is the one in which the enemy is seized without a fight [10].Cyber warfare derives the essence of both of these great military theorists as it is a warfare which is capable of compelling the enemy to your will by inducing strategic paralysis to achieve desired ends and this seizing of enemy is done almost without any application of physical force.

Clausewitz formulated the theory of nature of war based upon the conception of Trinity. This elusive Trinitarian warfare according to Clausewitz held the key to victory in a war. Clausewitz predominantly constructed this trinity around three dominant tendencies, the blind force composed of primordial violence, hatred and enmity; the play of uncertainty and chance in which the creative spirit roams; and the reason for violence or the political instrument.[11] The tendencies are abstracted as, the people or the will to fight a war in terms of finances, manpower and support; the military or the means; and the government or the effort, the leadership and the direction which is essential for a nation.

These three tendencies extensively interact with each other and have a continuously changing relationship. Till the time they are present and are interacting, the nation will sustain even when hit with the worse case scenarios. It is only when all of the three components are destroyed together or in conventional terms are subjected to parallel warfare; it is only then a 'cascade effect' will be generated to induce a strategic paralytic effect onto the Nation and the Nation as a system will crumble resulting in chaos and mayhem. In the current contemporary world in general and the developed countries in particular, the reliance on modern technology is not treated as a luxury, but a necessity where all the three tendencies are extensively dependent upon cyber space in one form or the other (Figure 1).

The modern militaries extensively depend on information assets and cyber space especially in scenarios where the deployment is across the globe. These information assets are used extensively in command and control systems especially in joint or coalition operations; in net-centric warfare operations involving global information grids; for logistic; for surveillance right from the information gathering which requires data links with satellites to information dissemination involving strategic information dissemination system; for communication right from the tactical field or theater data link operations and networks to strategic command and control networks involving communication satellites; global positioning and navigation satellites and networks for not only navigation, but for precision targeting; and so on(Figure 1).

The scenario is the same with the tendency involving the people or the will to fight a war in terms of finances, manpower and support. In almost all the developed countries and in some developing countries, people rely extensively on computers and cyber assets for almost all of their daily chores. Whether utilities such as gas and electricity; or health, transportation and banking facilities, all rely on cyber or network assets for their functioning. The scenario is even worse in case of banking and economic institutions like stock exchanges where money, which is nothing but numbers or information, travels across the national boundaries to the remotest corner of the world due to globalization of financial and banking instruments of economy. These technological advantages have not only eased the life of people, but have also made them vulnerable; as day by day people are becoming hopelessly dependent on these facilities to an extent that they take them for granted. The media and communication networks have almost become the vital sensory organs of this technologically dependent society.

This western society according to Bill Durodie is becoming more and more individualistic. It is becoming a society where people are socially disconnected; politically disengaged; and are in scientific disbelief.[12] In this society where perceptions overweight the reality, the people are becoming more and more 'risk averse' and are constantly living in an environment of fear. This state of society is classically defined by Ulrich Beck as a Risk Society.[13]

These societies are socially disconnected; politically disengaged; in scientific disbelief; and are constantly living in an environment of fear. The sudden disappearance of almost all of their facilities on which they are hopelessly dependent upon, will result in catastrophic outcomes where chaos, fear, bedlam, anarchy and basic animal instincts of man will prevail resulting in mayhem and complete destruction of nation as a system (Figure 2).

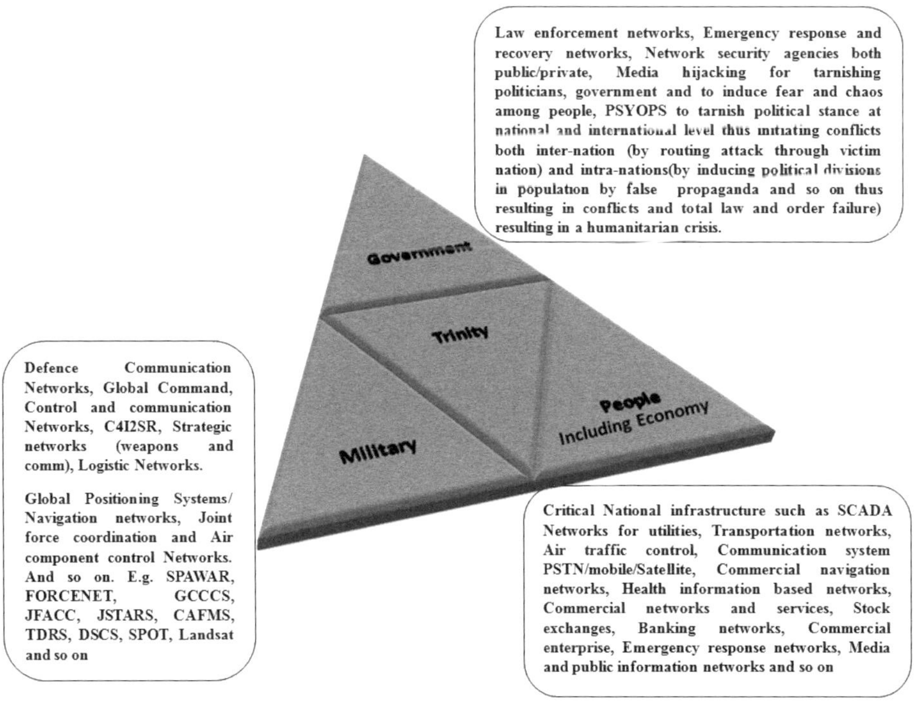

Figure 1: The notion of trinity in terms of strategic cyber warfare. (Source: Author)

The third and the most dominant tendency constitute the government, the political instrument and the leadership. Government plays an important role in providing leadership to the nation and in combining efforts and means to achieve political aims; hence for the failure of nation as a system, it is important that along with the other two tendencies the political instrument should also be destroyed.

In current contemporary world, governments play as the political instrument in the trinity by means of excising control and gaining the mandate of people. These objectives are achieved by using effective law enforcement and by providing a secure, secular and democratic environment to people to attain control and mandate over them. The law enforcement and security agencies rely extensively on criminal records and other coordination networks such as emergency response and recovery networks which although act as a force multiplier for them, but at the same time make them vulnerable to strategic cyber warfare. Another important aspect to gain mandate and control of people is the media. Media is an important tool that frames the perception and psychological frame of mind of the population. The 'CNN effect' is a potent tool to influence the mindsets of people.[14] Media around the world is extensively interlinked through networks which not only makes information to disseminate easily, but also make media more susceptible to strategic cyber warfare. These Media networks can be hijacked for tarnishing the image of politicians and the government of a victim nation; and can be used to induce fear and chaos among people. PSYOPS can be fully employed on these hijacked media networks to tarnish the political stance of the victim nation at national and international levels, thus initiating conflicts both at inter-nation level (by routing an attack through a victim nation) and intra-nation level (by inducing political divisions in the population by false propaganda and so on, thus resulting in conflicts and total law and order failure), hence creating the symptoms of a failed state which has anarchy, fear and chaos, resulting in a humanitarian crisis and failure of the state (Figure 2).

As described above, these three tendencies form the core of a nation or constitute the nation as a system of systems. All of these components are quite resilient in nature unless and until they are simultaneously attacked and destroyed, they are quite capable of reviving one another. Thus in order to achieve a strategic effect and to gain rapid dominance, parallel warfare should be initiated in cyber space so as to destroy all the three tendencies simultaneously (Figure 2). The Author believes that the cyber attacks such as *Titan Rain; attacks on Estonia and Georgia;* and so on were not successful due to the fact that they were tactical in nature and were targeting the individual components of the trinity at a time. Since these components are resilient when they are together, so even if one of the tendencies is fully destroyed in a cyber attack, the other two tendencies will tend to rescue it, hence a strategic paralytic effect will not be achieved. The author believes that it is for this reason the cyber attacks conducted so far could not achieve a strategic effect and were ineffective.

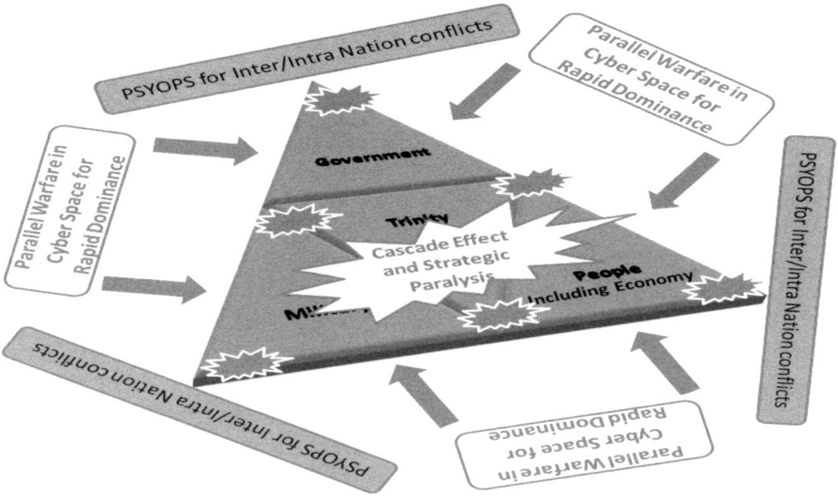

Figure 2: Cyber Trinity based parallel cyber warfare attack to induce strategic paralytic effect on a victim nation (Source: Author)

In order to achieve a strategic paralytic effect via the application of cyber warfare, all the three components of the trinity should be attacked simultaneously. These cyber attacks should be performed by using the paradigm of parallel warfare[15], which relies on gaining rapid dominance by producing the desired effect of paralyzing the control of the enemy by performing rapid decisive operations at all levels i.e. the strategic, operational and tactical; across the spectrum involving related assets and critical components of all the three tendencies of the cyber trinity; and with rapid succession so as to reduce the chances of counter attack or of the "defensive phenomena" of *pulling the plug.*

When the art of simultaneous parallel warfare to achieve rapid dominance is combined with the strategy to simultaneously attack the three components of the cyber trinity, it will generate a *cascade effect* rendering the victim in a paralytic state with the loss of control over the state and failure of the state-as-a-system, thus generating ramifications which are way ahead than mere arithmetic benefits generated by a successful attack. This failure of Nation as a system-of-systems and the disruption or paralysis of the state will destroy the victim's will and capability to fight thus 'compelling it to submit to your will'.

This strategy of strategic cyber warfare against the Trinity in cyber space to achieve strategic paralysis provides for an alternative warfare or means to achieve strategic effect of rendering the enemy ineffective to operate as per its wishes, eventually is more important than the conventional paradigm of destruction-based warfare to annihilate the forces it depends upon for its defence, hence generating not only a strategic victory, but also a constructive conflict termination. This constructive conflict termination is not only desired, but is imperative especially in the contemporary world, with examples of conflicts with flawed exit strategies resulting in victories turning into protracted wars such as the Iraq war.

3. Campaign planning- The orchestration of strategy

Once the strategy for conducting cyber warfare to achieve the strategic end of compelling the enemy to submit to your will by rendering the enemy ineffective is defined, the next important task is to integrate that strategy to a campaign plan which spreads across all levels of warfare. This process of accessing the assets and means to achieve the desired effect and ends by orchestrating the strategy across all levels of warfare is termed as the campaign planning and the scenario is the same in pursuing a strategic cyber warfare campaign.

As done in conventional warfare the campaign planning for strategic cyber warfare is also based on phasing, but against the conventional dictum of executing these phases in sequential or near sequential manner, the strategic cyber warfare tends to use these phases in almost near parallel manner. At any temporal instance, each of the phases will have a substantial effect over the entire theater of operations in general and individual zones in particular in conformance with the parallel warfare dictum. When applied in order to orchestrate the strategic cyber warfare strategy of generating a paralytic effect on the adversary by initiating parallel cyber attacks on all the three components of the Trinity, all phases of the campaign will overlap extensively and the campaign will be in the form of simultaneous waves of these overlapping phases tailored according to the theater conditions at all levels.

The campaign for strategic cyber warfare in line with conventional dictum[16] will consist of five broad stages; Shape, Deter, Seize initiative, Dominate and Exit (Figure 3). Out of these the Shape and Deter are part of the pre conflict phases; Seize initiative and Dominate are usually the conflict phases; and the Exit is the post conflict phase. Although these phases are lucidly categorized as pre-conflict; conflict; and post-conflict phases, these categorizations tend to overlap extensively. For example even though there would be a conflict in progress in certain parts of the theater, but still in certain other parts, the Shape and Deter phases may be exercised to limit the conflict from further escalation.

The initial phase or the Shape phase revolves around shaping the conflict.[17] This is done using extensive peacetime cyber reconnaissance, such as mapping enemy's or potential adversary's cyber assets, network design, layout, vulnerabilities, critical components and dependence; assessing enemy's cyber defence capabilities both offensive and defensive; boosting national cyber defence capabilities not only in military, but in all the components of the cyber trinity; identification of critical assets/targets which would initiate a 'cascade effect'. In order to identify these components the 'Critical component theory' which was utilized by the allied air force for strategic bombing of Germany during the Second World War[18] can prove an important asset, And performing all the above operations on your own assets also, can be useful to identify potential susceptibility and vulnerabilities.

Apart from cyber reconnaissance, an important component included in the peacetime operations involves the process of inducing vulnerabilities in enemy's cyber assets. This is achieved either by using cyber means such as installing covert malware such as trojans, rootkits, dormant stealth malware and so on; or by using covert means such as exporting bugged firmware by using front door companies, thus making enemy systems susceptible to Permanent Denial Of Service (PDOS) attacks and by covertly gaining information of enemy's critical system software such as operating systems, by gaining access to the skeleton keys for the backdoors, usually channeled through vendor influence; and so on.

Deter like the Shape phase is also a pre-conflict phase which extensively revolves around shaping the future conflict by gaining a credible and known deterrence. This phase capitalizes on the information gained during the cyber reconnaissance of the adversary's and of personal cyber infrastructure and their susceptibility to strategic cyber attack. Based upon this information, relevant steps are initiated to harden personal infrastructure in terms of 'layers of defence' and redundancy; and to prepare for exploiting the enemy's vulnerabilities and then testing them in simulated environments by conducting cyber war games.

An important part of this phase is to develop a cyber deterrence which is credible and is made known to the enemy. The credibility of the cyber deterrence can be achieved by creating a *Cyber Triad capability*, equivalent to a *Nuclear Triad*[19] which will have capability for orchestrating a second strike in case of failure of the deterrence. Cyber Triad capability can consist of Regular defence/military assets and networks as forming the first section of the triad; the second section of the triad can consist of an isolated conglomerate of air-gapped networks situated across the friendly nations as part of cooperative defence, which can be initiated as credible second strike option; and the third section of the triad can consist of a loosely connected network of cyber militia involving patriotic hackers, commercial *white hats* and private contractors which can be initiated after the initial strike or in case of early warning of a potential strike. This Cyber Triad creates a scenario of a credible and undisputable cyber deterrence and second strike capability thus assuring a *Mutually Assured Destruction (MAD)* in cyber space.

The later feature of the cyber deterrence involving the policy of making the cyber deterrence known to the enemy, can be achieved by following a 'cyber countervailing' strategy in line with the 'countervailing' nuclear strategy followed by NATO forces during the cold war.[20] This strategy revolved around making known to the potential adversary that the implication of a nuclear strike would be far greater than the potential gains an adversary can achieve by initiating the first strike. The scenario will remain the same in the strategic cyber warfare where the potential enemy should be made known of the potential risk it might be facing in light of initiating a first strike. This can be achieved by means of media coverage; extensive war games; and to some extent by covert instantiation of limited cyber warfare attacks on the adversary. The author believes that the recent attacks such as the Titan Rain; the attacks on Estonia and Georgia; and attacks on various other countries around the world such as UK, France Germany, India and so on; can be a part of a *cyber countervailing strategy.*

Once the conflict has been initiated there can be two possibilities; either the cyber deterrence has failed and the enemy has initiated the first strike or due to certain unforeseen events, friendly forces needed to initiate the first strike. In the former situation it should be always assumed that the adversary's first strike will have, if not an all out decapitating effect, a certain degree of effect on friendly cyber infrastructure, but an important aspect here is to detect the attack before it could generate any strategic paralytic effect. Once this detection is achieved, it is required that necessary steps should be taken so as to activate the cyber triad to initiate the second strike capability. This can be achieved by taking defensive counter measures and securing cyber infrastructure; initiating the second triad involving a coordinated and real-time integration of an isolated conglomerate of air gapped networks situated across friendly nations as part of cooperative defence; and in a worst case scenario, where there is a total loss of offensive and defensive capabilities, the third component of the triad consisting of a loosely connected network of cyber militia involving patriotic hackers; commercial *white hats* and private contractors should be activated in order to initiate a

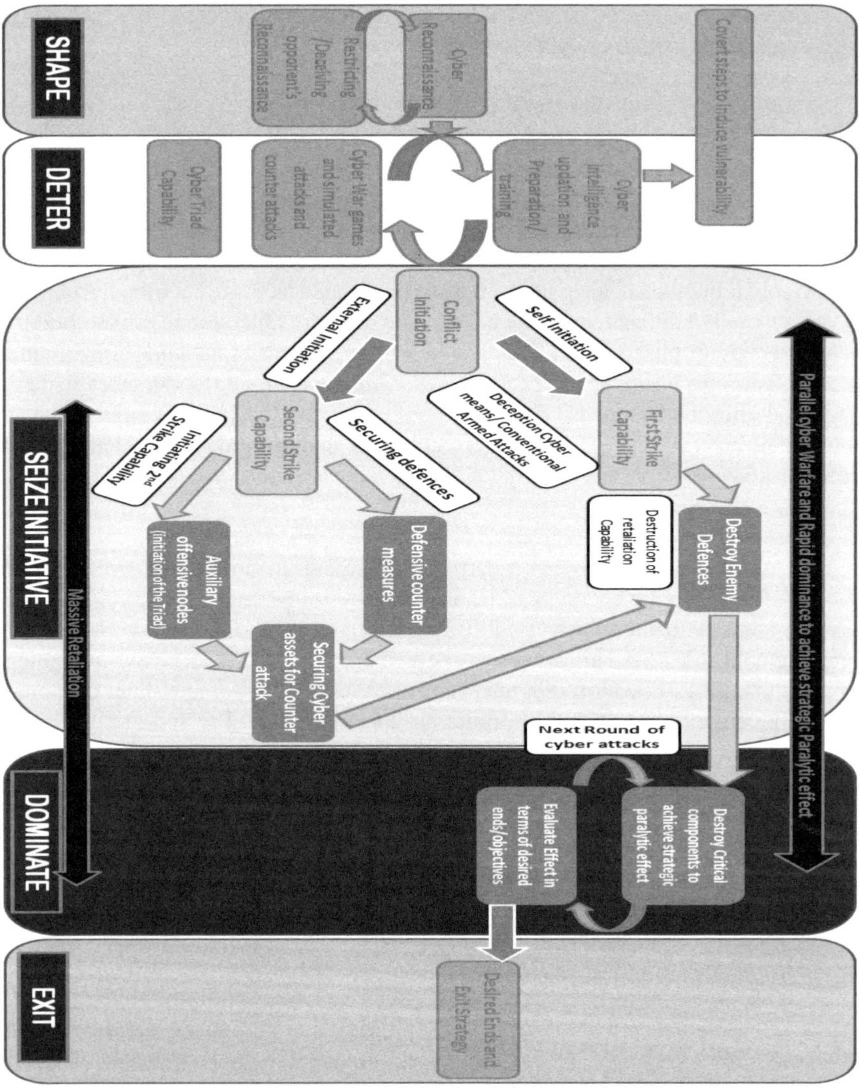

Figure 3: Cyber campaign planning for strategic cyber warfare. (Source Author)

protracted conflict. This protracted conflict is essential to gain time so as to revitalize your own capabilities in worst case scenarios.

Similarly if the first strike is to be initiated, the primary aim should be on the total destruction of the cyber triad. This is done so as to destroy enemy's capability of retaliation and to provide strategic freedom of operation in cyber space to friendly forces. This strategic freedom of operation in cyber space is essential in order to initiate the strategic cyber warfare on the critical components of the cyber trinity for achieving the strategic paralytic effect. An important consideration at this point should be on the fact that these phases should be conducted in an overlapping fashion to perform parallel warfare in cyber space.

Once the necessary objectives are achieved and the desired end-state is reached due care should be taken to stabilize the situation. Although the effects of strategic cyber warfare like the nuclear warfare are far reaching, devastating, and in a way cannot be measured, they still can be contained and the post-conflict situation can be stabilized by initiating external support in the form of reestablishing the cyber services and facilities and retrieving of data, if not fully, then to a decent level, so as to carve out an exit strategy. Like in conventional conflicts, the chances of a cyber insurgency consisting of patriotic hackers and of humanitarian crisis loom at large especially in a post-conflict scenario. Hence due care should be taken to reestablish the essential services and command and control infrastructure after the conflict. An aggressive media strategy to counter dissident feeling and anguish among the people should be utilized so as ease the stabilization process and assist with post conflict rehabilitation.

4. Cyber Defence – A conventional fallacy

The current conventional wisdom on cyber defence relies on the notion of 'defence in layers'[21] and on International legal regulations especially by drawing similarities between cyber attacks and armed conflicts and then applying the law of armed conflict [22] appropriately. The notion of 'defence in layers' is a tried and tested dictum which is extensively used to protect both the commercial and the defence networks. It relies on installing multiple layers of defences so as to make the penetration almost impossible. Even though this notion has extensively been used to protect cyber infrastructure, it is a known fact that such a system is as strong as its weakest link. No matter how much the system is hardened and no matter how many layers are used to secure the system, there is still no guarantee that the system security is foolproof. It is safe only up until the time when someone doesn't find any vulnerability or an exploitable construct in the system, which can be exploited to gain access in to the system. Yes, this notion of defence at least assures that the penetrator will require time to defeat multiple layers of security. It is this time that is crucial for defenders to take necessary action to thwart the threat. Hence this provides for a minimum deterrence, but nevertheless is not a complete and foolproof solution.

The other aspect of conventional wisdom on cyber defence relies on the legal framework of international law both *jus ad bellum* and *jus in bello*. This defensive strategy relies on the deterrence generated by the legal punitive aftermath of a cyber conflict on the erring side. The notion relies extensively on drawing similarities between the conventional international armed conflicts and cyber conflicts; and then applying the international laws which will fit the scenario. Most of the discussion in this realm revolves around self defence for *jus ad bellum* and Law of armed conflict for *jus in bello*. Extensive analysis is done in order to find the lines of similarity between the armed conflicts and cyber attacks, but these arguments in a sense degrade the strategic aspect of cyber warfare to mere tactical cyber attacks or are centered towards the means rather than the strategic ends.

The author believes that this notion of defence is a fallacy as the underlying assumption of treating strategic cyber warfare as mere tactical cyber attacks is in itself a fallacy. If cyber warfare is performed so as to achieve the strategic paralytic effect, then the consequences of such a warfare would be far reaching and to an extent not measurable in conventional terms. The *cascade effect* initiated as an aftermath of strategic cyber warfare would generate a chain of 'unintended consequences' that are almost impossible to tackle using the conventional framework of law of armed conflict. The only appropriate legal frame work to handle strategic cyber warfare would be based on the legal frame work of *jus ad bellum* and *jus in bello* for nuclear weapons, which unfortunately is a long debated notion and has an incoherent international opinion visible extensively in the ICJ's opinion [23] over the use of nuclear weapons.

In light of these circumstances, where the conventional dictum of cyber defence is a mere fallacy the only viable and achievable option for cyber defence would rely on the age old dictum of deterrence and to an extent on the cold war principle of Mutually Assured Destruction. This cyber deterrence can be guided by the Rational Deterrence Theory (RTD) which relies on the underlying assumption of actors to be rational and performing cost benefit analysis before reaching any logical conclusions; and the outcome variation depends upon the variations of opportunities which the antagonists have.[24] Ashen and Snidal argue that the key concepts for achieving deterrence based on RDT will be on the credibility of the deterrence capabilities and the rational actor assumption of decisions relying on cost benefit analysis.[25] In terms of cyber deterrence the credibility can be achieved by the creation of a cyber triad as part of the Deter phase of cyber campaign planning. This cyber triad capability can consist of Regular defence/military assets and networks as forming the first section of the triad; the second section of the triad can consist of an isolated conglomerate of air-gapped networks situated across the friendly nations as part of cooperative defence; and the third section of the triad can consist of a loosely connected network of cyber militia involving patriotic hackers, commercial *white hats* and private contractors

This credible second strike capability assures the dictum of Mutually Assured Destruction (MAD) in cyber space and hence an option for defence in terms of deterrence. This capability should be made known to the potential advisories as part of cyber *countervailing* strategy to warn them of undesired consequences and punitive costs they may bear in the event of a cyber conflict. This form of deterrence is generally classified as the *deterrence by punishment*; the other form of deterrence is classified as the *deterrence by denial*.[26] This deterrence by denial in cyber defence can be achieved by preemptive cyber strikes on the adversary's cyber offensive capabilities. However, in scenarios of state actors this policy may result in a further escalation of conflict; hence utmost care and thought process in regard to attribution of cyber attacks should be taken before initiating such a strike. In the case of non-state actors the *deterrence by denial* in the form of preemptive cyber strikes, offer a credible deterrence mechanism for thwarting any such threat. In both the scenarios, extensive reconnaissance and surveillance both by cyber and conventional means can act as suitable tools for attribution and target selection.

It follows that cyber deterrence can act as an important means for thwarting both the state and non-state threats by means of *deterrence by punishment* and *deterrence by denial*. Also it is clearly evident that none of the cyber defence notions can provide for a holistic cyber defence; hence a cyber defence strategy should be a combination of the notion of 'defence in layers'; the legal aspects of International law, although whether the Law of Armed conflict may be applicable in its entirety is debatable, but that is out of the scope for this article; and by generating a credible cyber deterrence based on the cyber triad, thus assuring a Mutually Assured Destruction in cyber space and hence a strategic status quo.

5. Cyber finale

The last couple of decades have seen a colossal change in the way in which conventional wars are being fought, with information being an integral component. Information and information assets have made such an indelible impact on warfare that military pundits often misinterpret this information-enabled warfare to be information warfare, where information assets act as decisive force multipliers, which are capable of changing the outcome of wars. Although it is a considerable feat, but in a sense it degrades the strategic aspect of information warfare to mere tactical cyber attacks. Over years this paradigm of considering information warfare as mere tactical force multiplier has gained impetus This paradigm has created an environment where "means are emphasized more than the desired ends". The author believes that this is a fallacy and calls for a paradigm shift from "means to ends". The Information warfare is a strategic warfare which derives the essence of both Sun Tzu and Clausewitz as it is a type of warfare which is capable of compelling the enemy to do your will by inducing strategic paralysis to achieve desired ends and this seizing of the enemy is done with virtually no application of physical force.

The strategic information warfare is capable of achieving desired strategic ends by inducing a strategic paralytic effect onto the nation; and the nation as a system will crumble resulting in chaos and mayhem. This strategic effect relies on the framework defined across all spectrum of affairs, right from the grand strategic to the tactical level, and is achieved by the strategy of strategic cyber warfare against the Clausewitz's Trinity in cyber space to achieve strategic paralysis and rapid dominance using parallel warfare. This strategy provides alternative means to achieve the strategic effect of rendering the enemy ineffective to operate as per its wishes, which eventually is more important than the conventional paradigm of destruction-based warfare to annihilate the forces, the enemy depends upon for its defence; hence generating not only a strategic victory, but also a constructive conflict termination.

In order to achieve these ends, a comprehensive orchestration of strategy in the form of campaign planning is required, which in line with conventional dictum will consist of five broad stages; *Shape, Deter, Seize initiative, Dominate and Exit*; categorized further as pre-conflict, conflict and post-conflict phases. Out of these the most crucial are the pre-conflict phases of Shape and Deter, which involve extensive cyber reconnaissance and the creation of a credible and known cyber deterrence based on *cyber triad* capability and *cyber countervailing strategy*. This creates a credible second strike option in cyber space, which assures a strategic status quo based on Mutually Assured Destruction in event of a cyber conflict.

In terms of cyber defence the conventional wisdom of treating of cyber warfare as mere tactical cyber attacks or force multipliers and relying on the legal framework of Law of Armed Conflict (LOAC) for defence is a fallacy; as it not only undermines the strategic aspect of cyber warfare in the form of generating strategic paralytic effect to achieve political ends, but it also relies on a legal framework for defence, which is contemplated on a false assumption of drawing similarities with conventional armed conflicts. The only warfare which matches cyber warfare in strategic terms is nuclear warfare; and as there is a gross division of International Law in terms of *Jus ad bellum and jus in Bello* for Nuclear weapons, the scenario unfortunately will remain the same for cyber warfare also.

Under these circumstances, the only feasible and viable cyber defence strategy will rely on the application of Rational Deterrence Theory in general and on the idea of Mutually Assured Destruction in particular so as to maintain the strategic status quo. Hence the author recommends that the nations should reconsider their cyber defence strategies, and should define a strategy based on a combination of the notion of "defence in layers"; legal instrument; and potent cyber triad based cyber deterrence. This would be the only viable and achievable option in current contemporary and futuristic conflicts, which the author predicts, will be dominated by strategic cyber warfare as the "primary means to achieve strategic ends".

References

[1] Sun Tzu translated by Samuel B. Griffith, *The Art of War*. (Oxford University Press, 1963).
[2] Sun Tzu translated by Samuel B. Griffith, *The Art of War*. (Oxford University Press, 1963).
[3] Luttwak Edward, Strategy: The Logic of War and Peace, (London: Harvard university press, 1987).
[4] Liddell Hart Basil Henry, *Strategy*, (New York: Meridian, 1991).
[5] Charles A. Ray, "Cyber war and Information Warfare: A Revolution in Military Affairs or Much Ado about Not Too Much?", National War College Report, 1997.
[6] Alvin and Heidi Toffler, The Third Wave- The Classic Study Of Tomorrow, (Bantam Books,1980)
[7] Luttwak Edward, Strategy: The Logic of War and Peace, (London: Harvard university press, 1987).
[8] Michael Howard and Peter Paret, eds., On War (Princeton University Press, 1984) Book I, Chapter 1, section 2, pp 75.
[9] Michael Howard and Peter Paret, eds., On War (Princeton University Press, 1984) Book I, Chapter 1, section 2, pp 75.
[10] Sun Tzu translated by Samuel B. Griffith, *The Art of War*. (Oxford University Press, 1963)
[11] Michael Howard and Peter Paret, eds., On War (Princeton University Press, 1989) pp 583.
[12] Bill Durodie, "The Limitation of Risk Management" , *TIDSSKRIFTET POLITIK* ,Vol. 8 No.1,2004.
[13] Ulrich Beck, *Risk Society: Towards a New Modernity*, (New Delhi: Sage, 1992).
[14] Belknap, Margaret H. *The CNN Effect: Strategic Enabler or Operational Risk?*, U.S. Army War College Strategy Research Project. 2001.

[15] Deptula David, *Firing for Effect: Change in the Nature of warfare,*(Arlington: Aerospace Education Foundation,1995).
[16] Joint Publication 3-0 *Doctrine for Joint Operation,* December 2005,IV-34.
[17] Joint Publication 3-0 *Doctrine for Joint Operation,* December 2005,IV-34.
[18] Pape Robert, *Bombing to Win: Airpower and coercion in war*,(Ithaca: Cornell university press,1996).
[19] Aldridge Robert C, First strike!: the Pentagon's strategy for nuclear war,(Cambridge: South End press,1983)
[20] Powell Robert, "Nuclear Deterrence and the Strategy of Limited Retaliation", *The American Political Science Review*, Vol. 83, No. 2 (Jun., 1989), pp. 503-519
[21] Harold F. Tipton and Micki Krause, *Information Security Management Handbook,* (Florida: CRC Press, 2006).
[22] Matthew E. Haber, Computer Network Attack and the Laws of Armed Conflict: Searching for Moral Beacons in Twenty-First-Century Cyber warfare, ARMY Command and general staff college, Fort Leavenworth, 2002.
[23] The 1996 ICJ's opinion over the Legality of Nuclear Weapons.
[24] Achen, Christopher and Snidal Ducan," Rational Deterrence Theory and comparative case studies", *World Politics*, Vol. XLI, No. 2, pp.139-169.
[25] Achen, Christopher and Snidal Ducan," Rational Deterrence Theory and comparative case studies", *World Politics*, 1998, Vol. XLI, No. 2, pp.139-169.
[26] Downs George," The Rational Deterrence Debate", *World Politics*, 1998, Vol. XLI, No. 2, pp.225-237.

THE INTERNATIONAL INSTITUTE
FOR STRATEGIC STUDIES
ARUNDEL HOUSE
13-15 ARUNDEL STREET
LONDON WC2R 3DX

The Virtual Battlefield: Perspectives on Cyber Warfare
C. Czosseck and K. Geers (Eds.)
IOS Press, 2009
© 2009 The authors and IOS Press. All rights reserved.
doi:10.3233/978-1-60750-060-5-18

Towards an Evolving Theory of Cyberpower

Dr. Stuart H. STARR[1,a]
[a] *Center for Technology and National Security Policy (CTNSP)*
National Defense University (NDU)

Abstract. In the 2006 Quadrennial Defense Review, a request was made to have the Center for Technology and National Security Policy (CTNSP), National Defense University (NDU), develop a theory of cyberpower. It was noted that there was a need to develop a holistic framework that would enable policy makers to address cyber issues in proper perspective.

To satisfy that tasking, CTNSP convened five workshops, drawing on experts from government, industry, academia, and think tanks. Those workshops addressed a broad set of issues related to the evolution of cyberspace, cyberpower, cyberstrategy, and institutional factors that influence those factors (e.g., governance, legal issues).

To develop the desired theory, this paper systematically addresses five key areas. First, the paper *defines* the key terms that are associated with cyber issues. Particular emphasis is placed on the terms "cyberspace", "cyberpower", and "cyberstrategy". Second, the paper *categorizes* the elements, constituent parts, and factors that yield a framework for thinking about cyberpower. Third, the paper *explains* the major factors that are driving the evolution of cyberspace and cyberpower. To support that effort, the paper presents strawman principles that characterize major trends. Fourth, the paper *connects* the various elements of cyberstrategy so that a policy maker can place issues in proper context. Finally, the theory *anticipates* key changes in cyberspace that are likely to affect decision making.

In view of the dramatic changes that are taking place in cyberspace, it is important to stress that this effort must be regarded as a preliminary effort. It is expected that the theory will continue to evolve as key technical, social, and informational trends begin to stabilize.

Keywords: cyberspace, cyberpower, cyberstrategy, cyber institutional factors

Introduction

This white paper represents a continuing effort to evolve a theory of cyberpower. The white paper begins by characterizing the components of a "theory of cyberpower". Consistent with that characterization, we identify key terms and put forth straw man definitions of those terms. We then identify the specific objectives that will be

[1] **Disclaimer:** The views expressed in this article are those of the author and do not reflect the official policy or position of the National Defense University, the Department of Defense or the U.S. Government. All information and sources for this paper were drawn from unclassified materials.

addressed in this theory. In accord with those objectives, we present a holistic framework to categorize and discuss key categories. Within this holistic framework, we discuss the intellectual capital required to address these issues.

Subsequently, we discuss theoretical dimensions of the key categories: cyberspace, cyberpower, cyberstrategy, and institutional factors. In addition, we discuss the challenges associated with connecting across these categories and anticipating the future cyber activities and issues of interest.

We conclude the white paper by summarizing major findings and identifying the next steps that should be taken to refine this evolving theory of cyberpower.

1. Context

To provide context for this white paper, this section discusses elements of a theory, objectives, approach, structure, key definitions, and required intellectual capital.

1.1. Elements of a Theory

A theory of warfare should address five key issues [1]. First, it should introduce and **define** the key terms that provide the foundation of the theory. Second, it should give structure to the discussion by **categorizing** the key elements of the theory. Third, it should **explain** the elements in these categories by summarizing relevant events and introducing key frameworks or models. Fourth, it should **connect** the various elements of the subject so that key issues can be treated comprehensively. Finally, it should seek to **anticipate** key trends and activities so that policy can be germane and useful.

This framework for a theory raises one immediate issue. There is interest in the ability to predict, rather than anticipate, key activities. However, as described below, the cyber problem is in the midst of explosive, exponential change. In the midst of this exceptional uncertainty, it is infeasible to make reliable predictions. Thus, we have adopted the less challenging task of "anticipating" key trends and activities.

Finally, it is important to stress the following caveat: since this is an evolving effort to develop a theory of cyberpower, the emerging theory will **not** be complete. Furthermore, as discussed below, early efforts to develop a theory for a discipline have inevitably been somewhat **wrong.**

To provide some context for theoretical developments, it is useful to note the challenges that the theories associated with physics have faced in its evolution. Contemporary physics theory has evolved over hundreds of years, dating back to the seminal contributions of Galileo and Newton. In this discipline, there is a common base of knowledge, although there are significant variants for specific sub-areas (e.g., quantum mechanics, classical dynamics, relativity). In addition, there are strong links to other "hard science" disciplines (e.g., math, chemistry, biology). Although the definitions of key terms and concepts are generally established, it should be noted that there were many false starts (e.g., a hundred years ago, physicists had (incorrectly) postulated the existence of an ether through which electromagnetic waves propagated as they traversed a vacuum). Even in contemporary times, discussions still persist about the fundamental definitions of matter (e.g., quarks with a variety of properties).

Within the sub-areas of physics, there is broad agreement about key categories (e.g., solid, liquid, and plasma physics). In these key sub-areas, mathematical models have generally been developed drawing on experiments and observations. Many of these mathematical models have proven to be extremely accurate and precise in explaining and predicting outcomes. However, there are still efforts underway to connect many of the key sub-areas of physics. For example, there is considerable work underway in the area of "string theory" to develop a unified understanding of basic phenomena, although some critics have argued that this is likely to be a dead end [2].

To highlight the challenges facing the "cyber theorist", it is useful to contrast the discipline of physics with that of cyberspace. The cyberspace of today has its roots back in the 1970s when the Internet was conceived by engineers sponsored by ARPA. Detailed analysis of cyberspace issues often requires even broader cross-disciplinary knowledge and skills than physics. These include, *inter alia,* computer scientists, military theorists, economists, and lawyers. Each of these disciplines has its own vocabulary and body of knowledge. Thus, it is quite challenging for these stakeholders to communicate effectively. This is manifested in debates about the most basic of terms (e.g., "cyberspace") where key definitions are still contentious. Consistent with the heterogeneous nature of the problem, it is not surprising that prior efforts to characterize this space have not been successful. At present, there is no agreed upon taxonomy to support a comprehensive theory.

As noted above, key attributes of a theory include its ability to explain and predict (or at least, to anticipate). There are many reasons why prior theoretical cyber efforts have foundered. These include the facts that key facets of the field are changing exponentially, there is little or no agreement on key frameworks, and the social science element of the discipline (e.g., understanding of cognition, human interactions in virtual societies) makes it very difficult to develop models that reliably explain or anticipate outcomes. Finally, we are unable to connect the disparate elements of the field because a holistic perspective of the discipline has not yet been created.

1.2. Objectives

This white paper addresses the five elements of a military theory: define, categorize, explain, connect, and anticipate. In the areas of "explain" and "anticipate", the focus is on identifying and characterizing key "rules of thumb" and principles for cyber elements.

The scope of the white paper is restricted in two key areas. First, we focus attention on the national security domain. Changes in cyberspace are having a major affect on social, cultural, and economic issues, but we address them only tangentially. Second, we limit attention to the key cyberpower issues that are confronting the national security policy maker. Thus, there is no attempt to generate a comprehensive theory of cyberpower that touches on broader issues.

1.3. Approach

The preliminary theory of cyberpower emerged from three initiatives. First, we drew insights from observations of cyber events, experiments and trends. Second, we extrapolated from prior national security methods, frameworks, theories, tools, data, and studies, which were germane to the problem. Finally, we formulated and hypothesized new methods, frameworks, theories, and tools to deal with unexplained trends and issues.

Subsequently, the theory has evolved based on two activities. First, over the past year, several conferences and workshops have been convened that focused on three key issues: cyber-deterrence, risk management, and international perspectives on cyber issues. Second, a number of papers have been generated that focused on assessing cyber policy issues and the US government's use of social networks. Accordingly, the preliminary theory has been evolved to reflect the insights that emerged from those activities.

1.4. Structure

This white paper has adopted the holistic cyber framework depicted in Figure 1. This framework is patterned after the triangular framework that the military operations research community has employed to decompose the dimensions of traditional warfare. In that framework, the base consists of systems models, upon which rests more complex, higher orders of interactions (e.g., engagements, tactical operations, campaigns).

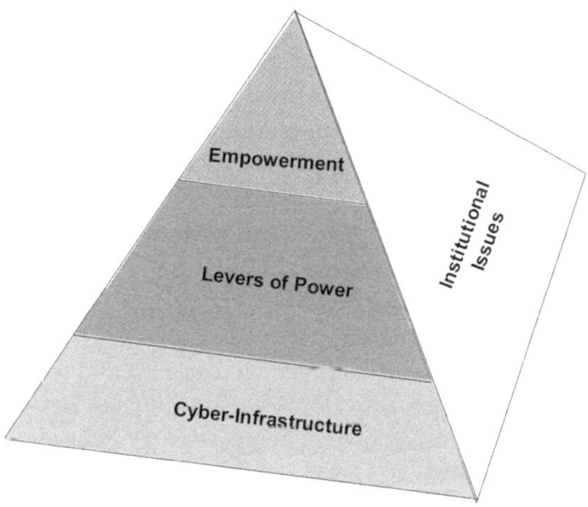

Figure 1. Broad Conceptual Framework

By analogy, the bottom of the pyramid consists of the components, systems, and systems-of-systems that comprise the cyber-infrastructure. The output from this cyber-infrastructure enhances the traditional levers of power: political/diplomatic, informational, military and economic (P/DIME). These levers of power, in turn, provide the basis for empowerment of the entities at the top of the pyramid. These entities include, *inter alia,* individuals, terrorists, trans-national criminals, corporations, nation states, and international organizations. Note that while nation states have access to all of these levers of power, the other entities generally have access to only a sub-set of them. In addition, initiatives, such as deterrence and treaties, may provide the basis for limiting the empowerment of key entities.

The pyramid suggests that each of these levels is affected by institutional factors. These include factors such as governance, legal considerations, regulation, sharing of information, and consideration of civil liberties.

It must be emphasized that this framework is merely one of many frameworks that could be constructed to conceptualize the cyber domain. However, it has proven useful in decomposing the problem and developing subordinate frameworks to address key cyber issues.

1.5. Key Definitions

As noted above, there is a continuing discussion about the appropriate definitions for key cyber terms. In the definition posed by William Gibson, in his 1984 book "Neuromancer" [3] cyberspace was characterized as: "A consensual hallucination… A graphic representation of data abstracted from banks of every computer in the human system."

For the purposes of this theory, this white paper has adopted the formal definition of cyberspace that the Deputy Secretary of Defense formulated: "…the interdependent network of information technology infrastructures, and includes the Internet, telecommunications networks, computer systems, and embedded processors and controllers in critical industries" [4]. This definition does not explicitly deal with the information and cognitive dimensions of the problem. To deal with those aspects explicitly, we have introduced two complementary terms: cyberpower and cyberstrategy.

This white paper has adopted the following definition for the term "Cyberpower". It is "the ability to use cyberspace to create advantages and influence events in the other operational environments and across the instruments of power." In this context, the instruments of power include the elements of the P/DIME paradigm. For the purposes of this evolving theory, primary emphasis will be placed on the military and informational levers of power.

Similarly, the term "Cyberstrategy" is defined as "the development and employment of capabilities to operate in cyberspace, integrated and coordinated with the other operational domains, to achieve or support the achievement of objectives across the elements of national power." Thus, one of the key issues associated with cyberstrategy deals with the challenge of devising "tailored deterrence" to affect the behavior of the key entities empowered by developments in cyberspace.

One of the major issues associated with cyberspace is the question of whether it is "…an operational domain…". To explore this issue, note that the term "domain" is not defined formally in key national security and military products. However, it is cited in selected policy documents. For example, the 2004 National Military Strategy [5] states that "The Armed Forces must have the ability to operate across the air, land, sea, space, and cyberspace domains of the battlespace". Furthermore, in the 2006 Quadrennial Defense Review (QDR), it notes that "The DoD will treat cyberspace as a domain of warfare".

Joint Publication 3-0 [6] identifies several key features of a domain: it can be described physically; there are distinctions in means, effects, and outcomes; and military and combat operations can be conducted in and through the domain.

One can make the argument that cyberspace is a domain through the following logic. It is widely accepted that (outer) space is a domain. In comparison to "space", "cyberspace" has the following bounding attributes that suggest that it is a military domain. It is subject to ongoing levels of combat; it is characterized by greater ease of access; and it is more difficult to identify and track military operations within it.

If cyberspace is a domain, it has significant practical implications. It will require the allocation of resources to support organization, training, and equipping of "cyber-forces". It implies the need to develop a culture that is consistent with cyber activities. Finally, it has implications in the development of a professional cadre and establishment of a structured career progression. Thus, for the purposes of this evolving theory, it will be assumed that cyberspace is "an operational domain".

Consistent with this white paper's definition, the elements of the holistic framework can be recast as depicted in Figure 2.

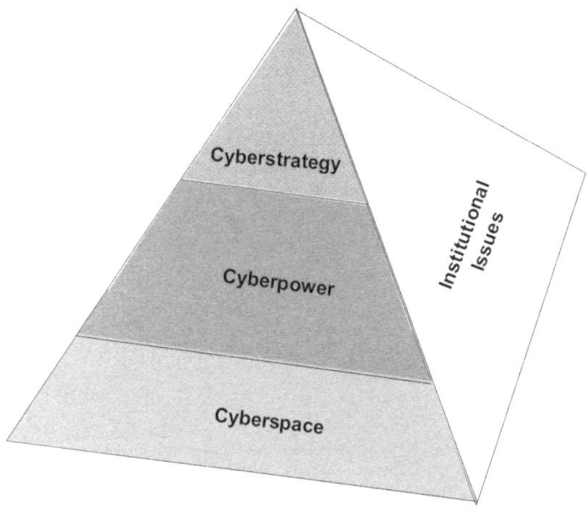

Figure 2. Cyberspace, Cyberpower, Cyberstrategy, and Institutional Factors

1.6. Required Intellectual Capital

To deal with the rich array of cyber policy issues that confront senior decision makers, it will require a diverse set of intellectual capital. Figure 3 suggests the differing types of knowledge that will be required to address issues within and across the categories of interest.

For example, in the realm of cyberspace, there is a need for physicists, electrical engineers, computer scientists, systems engineers, and system-of-system engineers. These professionals will play key roles in developing the hardware components (e.g., microprocessors, hard drives), software protocols and standards (e.g., implementing Internet Protocol version 6 (IPv6)), applications and services, and the systems that exploit this hardware and software (e.g., command, control, and communications systems).

In the realm of cyberpower, there is a need for subject matter experts (SMEs) that are qualified to deal with issues of Politics, Diplomacy, Information, Military, and Economics. This implies extensive reliance on economists (both micro- and macro-) and social scientists with training in such diverse fields as sociology, cultural anthropology, psychology, and demographics.Furthermore, in the area of military knowledge, there is a need for participation by military planners, operators, and analysts.

In the realm of cyberstrategy, there is a need for interdisciplinary experts who are able to deal with the full range of political, military, economic, social, informational,

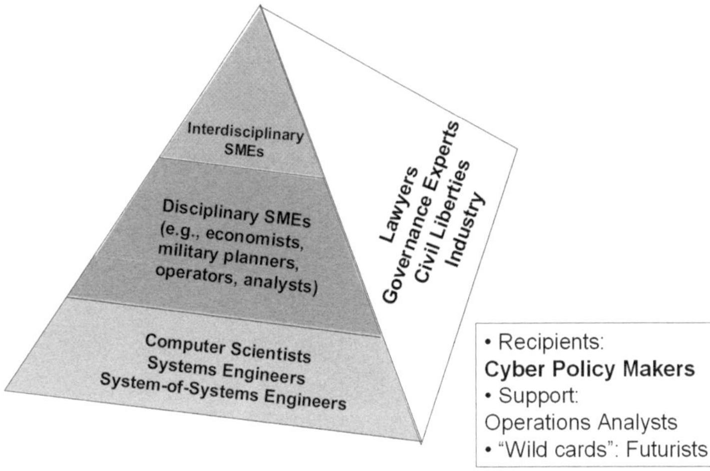

Figure 3. Required Intellectual Capital

and infrastructure (PMESII) issues associated with entities that are empowered by changes in cyberspace. In particular, analysts are needed who have had experience in addressing deterrence among these entities.

Finally, in the realm of institutional factors, the key skills required are legal, governance, civil liberties, and industrial experience.

It is anticipated that one of the main users of this intellectual capital will be cyber policy decision makers. They will also need operations analysts to help them orchestrate and harness this heterogeneous intellectual capital, and futurists to help them conceptualize possibilities that require unfettered imaginations.

2. Theoretical Perspectives

Three of the major objectives of a theory of cyber are to help **explain, connect,** and **anticipate** key aspects of the problem to the decision maker. To do so, it will require the formulation of conceptual models for the various categories introduced above. In formulating these conceptual models, it is useful to recall the famous epithet from the statistician George Box: "all models are wrong; some are useful" [7]. The challenge for the theorist is to suggest and apply appropriate models that are useful for the decision maker, and to delineate the range of their utility.

This section systematically introduces a variety of conceptual models that are germane to the many policy questions associated with cyber issues. Structurally, we will pursue a "bottom-up" approach and address cyberspace, cyberpower, cyberstrategy, and institutional factors. For each area, we will introduce a variety of models and frameworks that will help the decision maker explain key observables and conceptualize the issues of interest. This will be followed by articulating key "rules of thumb" and principles that highlight major issues of interest.

2.1. Theoretical Aspects of Cyberspace

This section of the white paper identifies key trends in cyberspace and discusses cyberspace "rules of thumb" and principles.

2.1.1. Key Trends

This section of the white paper briefly explains key trends in cyberspace. Trends are introduced in five key areas: growth in users, features of key components (e.g., microprocessors, hard drives), architectural features (e.g., Internet Protocols), and military systems-of-systems.

2.1.1.1. Growth in Users

The most remarkable aspect of the Internet has been the exponential growth in users, world-wide. Figure 4 illustrates that growth over a thirty-three year period. It can be seen that the user population increased from approximately 1M users in 1992 to 1,200M users in 2007. It is projected that the Internet will have 2B users by 2010. This number is projected to grow substantially if the One Laptop Per Child (OLPC) project is brought to fruition. That project aims to get many millions of low-cost laptops in the hands of children in under-developed countries.

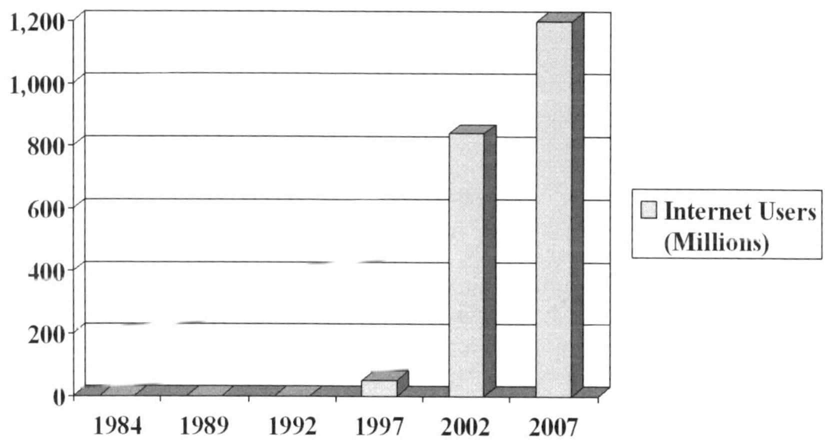

Figure 4. Number of Internet Users (Millions)

The Navy's Special Studies Group depicted this growth from another perspective. They used 50M users as a benchmark for penetration of a mass medium. That level was achieved by radio in 38 years, television in 13 years, and the Internet in 6 years (beginning with the introduction of the World Wide Web).

Another key element of cyberspace is cellular telephony. As a point of reference, the first cell phone call was made in 1973. It is estimated that today, thirty five years later, approximately 3.3B cell phones are in use, world-wide.

Two other benchmarks serve to calibrate the problem. It is estimated that around 210B e-mails were sent every day in 2008. That is equivalent to 2M e-mails sent every second. It is estimated that on the order of 70 percent of these e-mails may be spam or viruses.

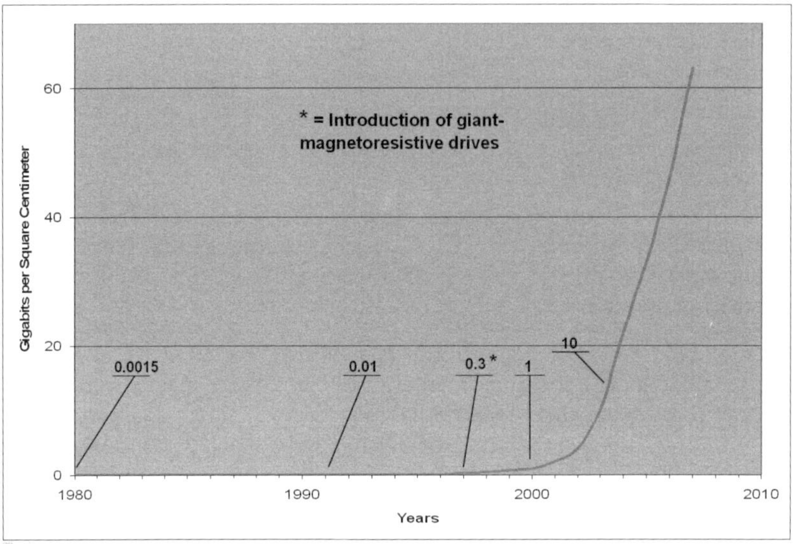

Figure 5. Hard Drive Capacity

2.1.1.2. Components

From a theoretical perspective, the physics of the hardware that supports cyberspace has a significant impact on its performance. This is particularly manifested in the design of microprocessors and hard drives.

2.1.1.2.1. Microprocessors

Clock cycles of modern microprocessors exceed 4 GHz. Therefore, under ideal circumstances, electrons can move a maximum of 0.075 meters in a single processor clock cycle, nearing the size of the chip itself. With clock cycles going even higher[2], electronic signals cannot propagate across a chip within one clock cycle, meaning elements of the chip cannot communicate with other elements on the other side of the same chip. Thus, this limitation maximizes the effective size of a single integrated microprocessor running at high clock speeds. Addressing this limitation is one of the reasons that various processor manufacturers have moved chip architectures toward multi-core processors, where multiple, semi-independent processors are etched on a single chip. Current chips typically have two or four cores although there are instances where 1000 to 4000 cores are a single die.

2.1.1.2.2. Hard Drives

Figure 5 depicts computer hard drive storage capability (in gigabits per square centimeter) over the last twenty five years. It is notable that the improvement in memory was marginal for the first twenty years until IBM engineers applied the phenomenon of giant magnetoresistance[3]. Currently, improvements in memory are

[2]As a bounding case, note that in 2008 the fastest US computer, Roadrunner (built by IBM and Los Alamos National Laboratory), was capable of more than 1 petaflop (i.e., 1 quadrillion floating point calculations per second) [8].

[3] The Nobel Prize in Physics for 2007 was awarded to Albert Fert, France, and Peter Grunberg, Germany, who independently discovered this phenomenon.

manifesting exponential improvement, making it feasible to create very portable devices, such as iPods, with extremely high storage capability.

These two examples suggest that a careful technology assessment is needed to assess if and when bottlenecks in technology will be overcome that limit current performance.

2.1.1.3. Architectural Features

Figure 6 schematically depicts the architecture of the existing Internet. The key innovations of this architecture revolve around the key protocols and standards instantiated in the Transmission Control Protocol/Internet Protocol (TCP/IP) stack and the use of a router to transmit packets from the sender to the user.

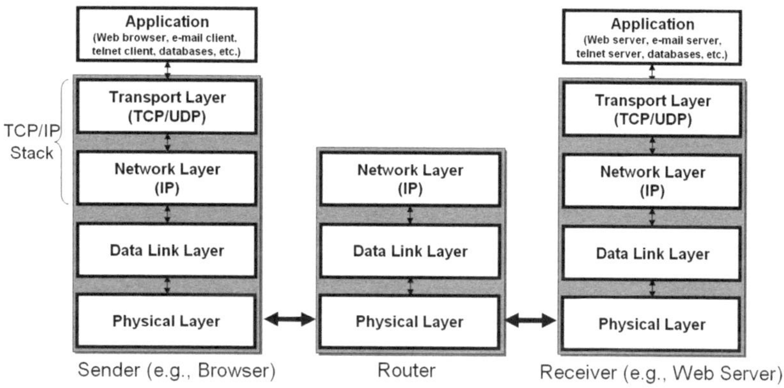

Figure 6. Protocol Layering and Routing Packets Across a Network

Originally, this architecture was devised by a group of colleagues for whom security was a secondary issue. Thus, the primary emphasis was to implement an architecture that facilitated the interoperability among heterogeneous networks. In addition, a decision was made to implement IP addresses that consisted of 32 bits (or approximately 4 billion addresses).

These two decisions have led to several major limitations in the current architecture. In light of the security shortfalls in the existing architecture, there is interest in alternative architectures that are designed around different priorities (e.g., highest priority: security; second priority: connectivity among highly mobile users). Consistent with those revised priorities, new architectural efforts are underway at the National Science Foundation and DARPA.

Second, the constraint on IP addresses (as well as concern about enhanced security and mobility) has led to the adoption of IPv6. Since it allocates 128 bits to IP addresses, it will give rise to an extraordinarily large number of IP addresses[4].

Both of these innovations pose a problem to the cyberspace community: how can one transition from the current architecture to an alternative architecture, efficiently and

[4] IPv6 will provide 2^{128} addresses. This would provide $5x10^{28}$ addresses for each of the 6.5B people alive today. Alternatively, our sun converts $3.4e^{38}$ hydrogen nuclei to helium nuclei each second. Each hydrogen atom could have its own IPv6 address.

effectively, without creating new security vulnerabilities? This is an on-going challenge that the computer science community must confront over the next decade.

2.1.1.4. Military Systems-of-Systems

The military community has embraced the underlying computer science principles associated with the Internet, although they have enhanced security for classified systems by developing "air gapped" networks (e.g., Secure Internet Protocol Router Net (SIPRnet), Joint Worldwide Intelligence Communications System (JWICS)). Figure 7 provides a cartoon of that implementation for the notional Global Information Grid (GIG).

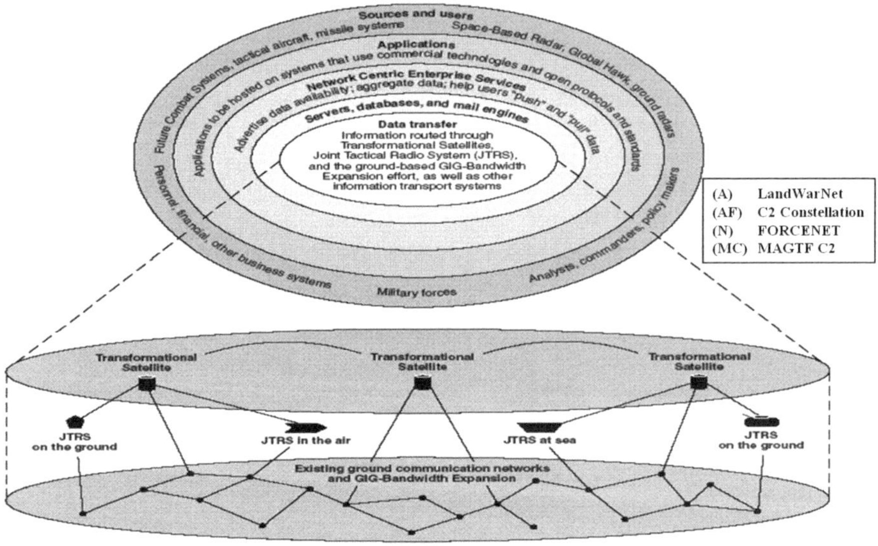

Figure 7. A Framework to Characterize the GIG

There are several distinctive aspects of the evolving GIG. First, for the transport layer, the plan is to employ a heterogeneous mix of satellite (e.g., Advanced Extremely High Frequency), airborne (e.g., selected Joint Tactical Radio Systems (JTRS)), and surface (e.g., fiber optic) telecommunications media. As a side note, the military is finding it difficult to develop many of these elements within acceptable levels of performance, schedule, and cost.

Second, there is interest in employing a Service Oriented Architecture (SOA) to provide loose coupling among key systems. Third, they have developed Communities of Interest to address the challenges associated with the data that will flow through the systems (e.g., specify metadata; deal with issues of pedigree). It has been articulated that they wish to transition from the principle of "need to know" to "need to share". Finally, they hope to assimilate the Services' visions of future systems into the GIG (e.g., USA LandWarNet; USN ForceNet; USAF C2 Constellation).

In order to achieve this vision it will require the concerted efforts of the military's system-of-systems engineers [9].

2.1.2. Cyberspace "Rules of Thumb" and Principles

To help explain the various trends in cyberspace, one can provide several "rules of thumb" and strawman principles. Several "rules of thumb" are employed in the community that are incorrectly characterized as "laws". For example Moore's "Law" indicates that the number of transistors on a chip approximately doubles every 18 months[5]. This has contributed to the production of devices that have decreased cost, enhanced computational power, and decreased size. Although this trend is generally representative of past behavior, there is concern that it may be extremely difficult to sustain that trend in the indefinite future without a fundamental, expensive change in the underlying technology (e.g., transition to nanotechnology). Second, as noted above in Figure 5, recent break-throughs in physics have put the growth in hard drive capacity on an exponential curve, vice a conservative linear curve.Ultimately, this curve will reach a level of saturation (an "S-curve") that is representative of a mature technology. Lastly, the current limitation in IP addresses will be dramatically overcome once the transition to IPv6 is implemented.

Based on prior cyber research activities, several strawman cyberspace principles can be articulated. First, the offensive has the advantage. This is due, in part, to the "target rich" environment that an adversary faces. This makes it difficult for the defense to prioritize and defend selected targets. In addition, the existing architecture makes it very challenging to attribute an attack if an adversary seeks to be anonymous. If cyberspace is to be more resistant to attack, it will require a new architecture that has "designed in" security. However, it will be a challenge to transition, effectively and efficiently, from the current legacy system to a more secure objective system.

2.2. Theoretical Aspects of Cyberpower

This section of the white paper briefly explains key trends in the military and information dimensions of cyberpower. It focuses on changes in environmental theories of power and risk, net-centric operations (NCO), and the mission-oriented approach to influence operations.

2.2.1. Environmental Theories of Warfare

In the discussions that led to this study, it was observed that the naval theories of Alfred Mahan played a major role in shaping the US perspectives and strategies on naval power. It was suggested that cyber power needed a comparable perspective to shape its strategy in cyberspace.

Consistent with that interest, this study re-evaluated the various environmental theories of power. These included analyses of land power (Mackinder [11]), naval power (Mahan [12]), airpower (Douhet [13]), and space power (Gray and Sloan [14]). Based on these analyses, four common features of environmental power theories were identified: technological advances; speed and scope of operations; control of key features; and national mobilization.

[5] To put this change in context, note that in 1971, processor speeds were on the order of 4×10^5 Hertz (or 400 KHz) and the cost of 1 MB of Dynamic Random Access Memory (DRAM) was approximately $400 (in 2006 dollars). By 2006, commercial processor speeds were on the order of 4×10^9 Hz (or 4 GHz) and the cost of 1 MB of DRAM was $0.0009 [10].

Consistent with each of these features, the following implications were drawn for a theory of cyberpower. With respect to technological advances, it was observed that dependency on cyberspace has given rise to new strategic vulnerabilities. This vulnerability has been dramatized by the specter of a "cyber Pearl Harbor" and the realization that the existing cyberspace is vulnerable to a variety of adversary attacks (e.g., denial of service attacks, exfiltration of sensitive but unclassified information; potential corruption of sensitive data). In addition, due to the diffusion of low cost cyberspace technology, the power of non-states (e.g., individuals, terrorists, transnational criminals, corporations) has been greatly enhanced (see below).

Improvements in cyberspace have also served to enhance the speed and scope of operations. This is manifested in the speed at which global operations can be conducted (e.g., the ability to successfully engage time sensitive targets, any where in the world). In addition, it has led to improvements in the ability to automate command and control, dramatically decreasing the classic Observe-Orient-Decide-Act (OODA) loop process.

In the environmental theories of power, emphasis was placed on controlling key features. For example, in naval theories this entailed the control of key "choke points" (e.g., the Straits of Malacca), while in space power, there was interest in controlling key geosynchronous orbit locations. In the case of cyberspace, the key features of interest are man-made. Thus, for example, there is interest in defending "cyber hotels" where key information and communications technology (ICT) systems are concentrated. In addition, while the choke points in the physical world tend to be immutable, they may change relatively rapidly in cyberspace (e.g., location of extensive server farms).

Finally, national mobilization is a key measure of cyberpower. To ensure that it is available when needed, it is vital to ensure that the US has access to a cadre of cyberspace professionals. This argues for re-examining career progression for cyberspace professionals in the military Services. In addition, it is important to establish links to the private sector where the bulk of cyberspace professionals reside. This suggests that a reservoir of reservists should be established to provide access to this intellectual capital in the event of national need.

It is argued in this white paper that the US Government (USG) has tended to focus on the opportunities offered by changes in cyberspace, rather than the risks that we are assuming. To summarize that dichotomy, Table 1 identifies the opportunities and risks associated with military activities at the strategic, operational, and tactical levels.

As can be seen in Table 1, the risks at the strategic level include loss of technical advantage (due to the diffusion of cyberspace technology), potential rapid change in the operating environment (e.g., possibility that nations such as China could "leap-frog" the US by transitioning rapidly to IPv6), and the vulnerabilities associated with military dependence on key systems (e.g., the GIG). At the operational level, the diffusion of cyberspace technology could result in the US loss of advantage in operational pace. Finally, at the tactical level, advances in cyberspace could generate a new front for adversaries to build resources. These observations suggest that the USG might be assuming significant, unknown risks by failing to take a balanced perspective of key cyberspace trends. It also implies the need to undertake more extensive risk assessments to understand the potential "down-side" of key dependencies.

Table 1. Military Opportunities & Risks in Cyberspace

Level	Opportunities	Risks
Strategic	• NCW-enabled • New "Center of Gravity" opportunities (e.g., deterrence; "virtual conflict")	• Loss of technical advantage • Rapidly changing operating environment • Military dependence on key systems (e.g., GIG)
Operational	• Phasing of operations • Enhanced force structure mix (e.g., cheaper, more precise)	• Loss of advantage in operational pace
Tactical	• Discover and track adversaries using cyberspace	• New front for adversaries to build resources

To begin to deal with these risks, steps should be taken at the strategic, operational, and programmatic levels. At the strategic level, steps should be taken to ensure the resilience of supporting critical infrastructures (e.g., electric power generation and transmission). At the operational level, it is vital to plan to conduct operations against an adversary that is highly cyberwar-capable. This should include the creation of a highly-capable Opposing Force (OPFOR) that would be employed extensively in experiments and exercises. Finally, at the programmatic level, emphasis should be placed on addressing cyberspace implications in the development process. This should include placing higher priority on the challenges of Information Assurance. Overall, an improved analytic capability is required to address each of these issues.

2.2.2. Net-Centric Operations (NCO)

As one aspect of the analytic capability, work is needed to enhance and apply the existing conceptual framework for NCO. As illustrated in Figure 8, the NCO process involves consideration of the interactions among the physical, information, cognitive, and social domains[6]. There is a need to develop better analytic tools for all aspects of this process, particularly in the cognitive and social domains. One potential source of intellectual capital is the ongoing initiative by the Director, Defense Research and Engineering (DDR&E), OSD, to improve human, social, cultural behavior (HSCB) models and simulations.

2.2.3. Mission Oriented Approach to Influence Operations

In the area of influence operations, a strawman framework has been developed to help the community plan for and implement influence operations (Figure 9). This framework represents an extension of the Mission Oriented Approach to Command and Control (C2) that was developed and applied to a variety of C2 issues in the 1980s [16].

This approach begins with the articulation of the nature of the problem of interest. It then poses a sequence of questions. First, what is the operational objective of the

[6] Note that the figure does not explicitly depict the social domain.

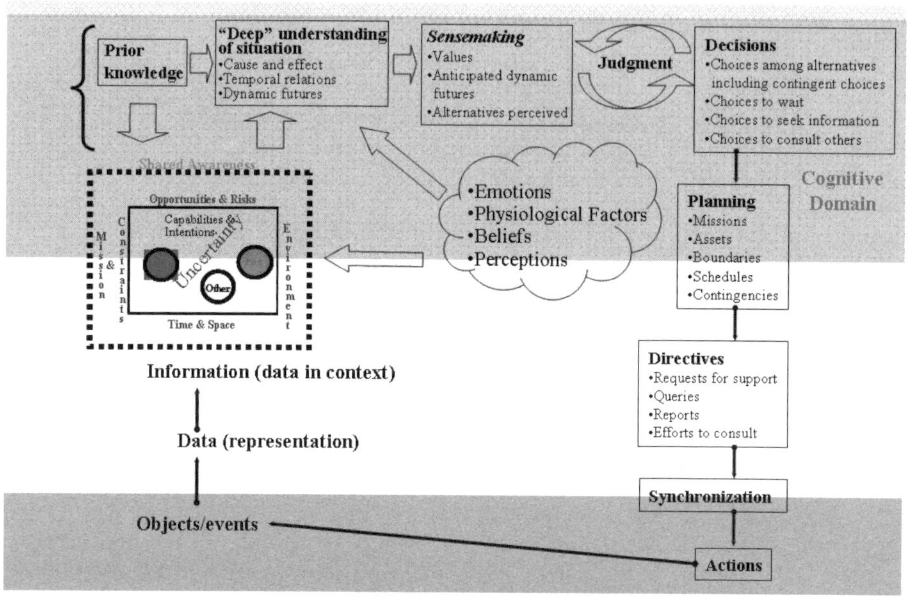

Figure 8. Conceptual Framework for NCO

operation? A reasonable objective may be to establish a trust relationship with the indigenous population (vice "winning their hearts and minds") [17]. Second, how should this operational objective be accomplished? Again, a decision was made to work with surrogate audiences in order to reach the undecided population. These surrogate audiences included the local media, religious leaders, educational leaders, political leaders, and tribal leaders. Consistent with those surrogate audiences, organizations and processes were established to reach out to them effectively. At this point, one can characterize the existing Doctrine, Organization, Training, Materiel, Leadership and Education, Personnel, and Facilities (DOTMLPF) activities and compare them to the operational needs. This will give rise to DOTMLPF shortfalls and the articulation of options to mitigate them. It may also prompt the operator to re-evaluate the operational goals and the operational activities to support them.

This process should be refined and applied to a broader variety of strategic, operation, and tactical influence operations. In particular, it can be used to explore the utility of employing new options in cyberspace to improve future influence operations (e.g., the "New Media", such as the Internet and social networks).

2.2.4. Cyberpower "Rules of Thumb" and Principles

One of the so-called "laws of cyberpower" was formulated by Bob Metcalfe [18]. He postulated that the value of a telecommunications network is proportional to the square of the number of users of the system (n^2). However, there is no empirical data to support this "law". In a recent article [19], it is observed that the value[7] is closer to nlog(n).

[7] To illustrate the differences in these results, assume that one has a network of 100 users. According to "Metcalfe's Law", the "value" of the network is on the order of 10,000. However, the revised "Law" suggests that the value is on the order of 100x2=200.

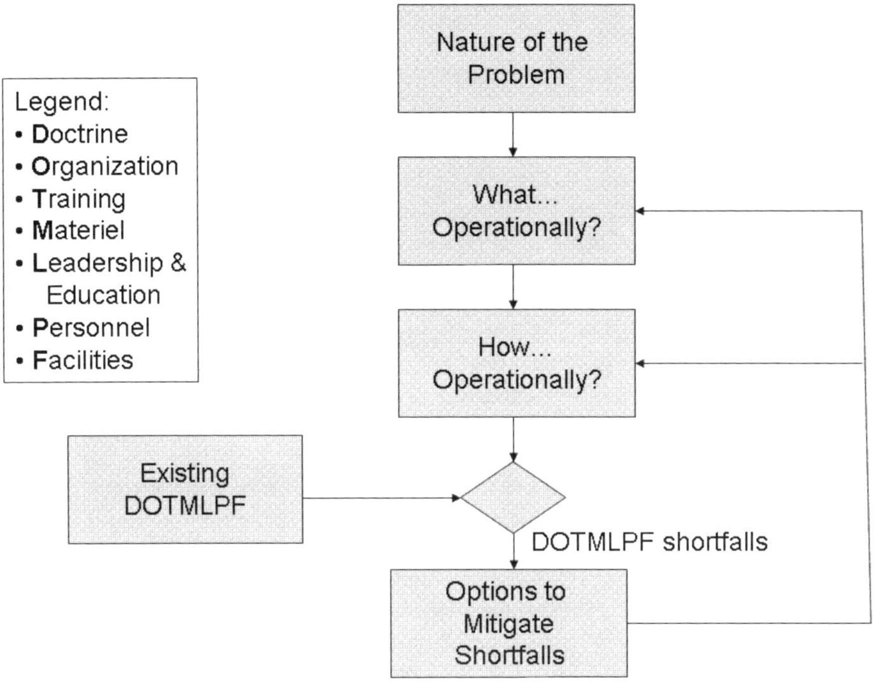

Figure 9. Strawman Framework for Analyzing Influence Operations

From an analytical perspective, the former Office of Force Transformation has supported a number of studies to relate the impact of net-centricity on enhancements in cyberpower (primarily in the military domain). These studies have demonstrated that net-centricity can have a substantial affect on mission effectiveness for selected mission areas. For example, the use of Link 16 by airborne interceptors in M-on-N combat can enhance air-to-air loss exchange ratios by approximately 2.5 [20]. However, the complexity of modern conflict is such that it is difficult to assess the affect of net-centricity on complex missions (e.g., air-land operations; stability and reconstruction operations). This suggests that additional experiments will be needed to assess the quantitative value of net-centricity for complex missions, in which better control is exercised over potentially confounding variables.

2.3. *Theoretical Aspects of Cyberstrategy*

This white paper has identified an extensive list of entities that are being empowered by changes in cyberspace. This list includes individuals, hacktivists[8], non-governmental organizations (e.g., Red Cross), terrorists, trans-national criminals, corporations, nation-states, and international governmental organizations (e.g., the United Nations).

For the purposes of this white paper, attention has been focused on a sub-set of these entities. These include terrorists, trans-national criminals, and a subset of nation

[8] Wikipedia definition: Hacktivism (a portmanteau of hack and activism) is often understood as the writing of code, or otherwise manipulating bits, to promote political ideology…

states (e.g., Estonia, China, Russia). From a USG national security perspective, two key issues stand out. First, is it feasible to achieve "tailored cyber deterrence"? Second, what steps should be taken to deal with cyber espionage?

2.3.1. Terrorist Use of Cyberspace

Terrorists are being empowered substantially by changes in cyberspace. With the loss of physical sanctuary in key areas (e.g., Afghanistan), they have been turning to the sanctuary of cyberspace to perform a variety of key, inter-related functions. These functions include, *inter alia,* recruiting of malleable candidates, raising resources to support their operations, planning their operations (employing such open-source tools as Google Earth), commanding and controlling their operations, conducting influence operations (e.g., disseminating their perspectives of operations in Iraq to sympathetic and uncommitted audiences), and educating and training supporters on a variety of subjects (e.g., interpretations of the Koran; building and deploying Improvised Explosive Devices (IEDs)).

Terrorists have found cyberspace to be an attractive milieu for several reasons. First, the cost of entry is low. One can acquire the latest cyber technology for hundreds-to-thousands of dollars and exploit key open-source software. In addition, terrorists can take full advantage of the extraordinary sums that have been invested by the commercial sector in cyber infrastructure (including communications and navigation systems). Second, cyberpace provides rapid, world-wide reach. Thus, they are able to transcend the limited geographic reach of their prior physical sanctuary and perform the key functions cited above. Third, it has been posited that the next-generation terrorists are being radicalized by on-line interactions[21]. Finally, there is concern that terrorists are developing linkages with trans-national criminals to support their objectives. The trans-national criminals are able to provide terrorists with cyber knowledge while profiting from the relationship.

Recently, a number of reports have been issued that suggest strategies for the USG to pursue to counter the terrorists' use of cyberspace. As an illustration, the Special Report on Internet-Facilitated Radicalization [22] formulated five recommendations to address the cyber threat posed by terrorists[9]. The many actions associated with those recommendations are summarized in Table 2. From the perspective of this white paper on cyberspace theory, some of the more interesting actions involve developing a strategic communication plan based on a compelling narrative, implementing an innovative program on behavior science research, and addressing USG shortfalls in knowledge of culture and language.

---

[9] This report recommended that five steps be taken:
- Craft a compelling counter-narrative for worldwide delivery, in multimedia, at and by the grassroots level.
- Foster intra- and cross-cultural dialogue and understanding to strengthen the ties that bind together communities at the local, national, and international levels.
- Recognize and address the need for additional behavioral science research into the process of radicalization both online and offline.
- Deny or disrupt extremist access to, and extremist efforts through, the Internet via legal and technical means, and covert action, where appropriate.
- Remedy and resource capability gaps in government.

Table 2. Options to Counter Terrorist Use of Cyberspace

Recommendations	Proposed Actions
Craft a compelling, multi-media counter-narrative for world-wide delivery	• Challenge extremist doctrine • Offer a compelling narrative • Use graphic visuals • Deliver the message through authentic sources • Amplify, augment grass-root non-extremist voices
Foster intra- and cross-cultural dialogue at all levels	• Address perceptions, realities of American Muslims alienation, marginalization • Enhance civic engagement • Increase people-to-people exchanges • Deal appropriately with the media
Address need for behavioral science research	• Deepen understanding of the radicalization process • Apply social networking theory
Deny or disrupt extremist use of the Internet	• Employ legal means • Undermine trust that binds adversary networks • Exploit convergence of human intelligence and cyberspace
Address capability gaps in USG	• Address cultural and linguistic deficiencies • Reclaim the high ground • Develop a strategic communication plan • Expand community policing programs

2.3.2. *Criminal Use of Cyberspace*

At a recent workshop on cyber issues at CTNSP, several of the participants focused on the challenges posed by cyber crime.Several of the speakers and panelists emphasized that the threat is real (and expanding). The speakers stated that "We are losing the global cyber war at an accelerated rate." In addition, they stated that "Cybercrime is effective because you can try to commit crimes an infinite number of times, but you need to succeed only a few times." Overall, it was stated that there are three elements of the threat: crime; industrial espionage; and traditional espionage. It was further noted that criminal attack vectors are comparable to those of state attacks.

The speakers also made the following observations. It was recommended that we focus on the "top 25 Common Weaknesses Enumeration (CWEs)". Furthermore, many of the panelists observed that current laws to deal with these issues provide limited value. In addition, it was noted that Web developers and code writers have **no** idea how to write secure code.

This raised the following question: When will the tide turn? It was suggested that we will make useful headway when we implement the following steps. First, it is critical to create safer software. One recommendation was to make business partnerships contractual (e.g., require the company to fix future flaws and security problems in the software). Second, it was observed that we need to stop existing attacks (e.g., implement more effective actions by the US Department of Justice and computer security specialists). However, in order to do so, we need to find the needed talent. As an example, it was observed that China's People Liberation Army periodically runs national talent searches for the best hackers.

2.3.3. Nation State Use of Cyberspace

From a nation-state perspective, different combinations of levers of power are employed to generate desired effects. From a theoretical perspective, these nations formulate their strategy though a mix of P/DIME activities. The effects of these activities are manifested in the areas of PMESII. Tools are being created to explore how alternative P/DIME activities can give rise to differing PMESII effects (see discussion below).

2.3.3.1. United States Use of Cyberspace

Using the P/DIME-PMESII paradigm, one can begin to characterize how cyber changes have empowered the US. In the political dimension, changes in cyberspace have encouraged democratic participation by the population. With respect to the Internet, it has provided a forum for the individual to articulate his views (e.g., proliferation of blogs, contributions to wikis). In addition, political candidates are finding the Internet to be a useful vehicle for raising resources from grass root supporters. Furthermore, Internet sites such as YouTube have enhanced the accountability of candidates.

In the military dimension, the concept of NCO has enhanced effectiveness in selected operational domains (e.g., air-to-air combat). Efforts are still required to quantify the military benefits that are achievable for more complex military operations (e.g., air-land maneuver).

Economically, the commercial sector has seen dramatic improvements in industrial productivity (e.g., Boeing's use of computer aided design tools to support the development of the 777 aircraft and the more recent development of the 787). These cyber-based advancements are giving rise to considerable improvements in responsiveness (e.g., time to market) and cost reductions (e.g., outsourcing "back-room operations" to other nations).

Socially, the development of cyberspace has increased social interactions in several ways. Tens of millions of users participate in social networking sites (e.g., MySpace, FaceBook). In addition, millions of users, world-wide, participate in virtual reality environments (e.g., Second Life). In fact, it has been rumored that terrorist organizations are using virtual reality environments to explore proto-typical operations.

In the information dimension, the Internet has increased dissemination of information, world-wide. Given the US' strong position in entertainment (movies, games) and advertising, it is argued that it provides a strong forum for promoting "soft (or smart) power" [23].

Finally in the infrastructure dimension, many critical infrastructures have been using the Internet to facilitate more efficient and effective operations. However, this constitutes a "double edged sword" because of the potential vulnerability of Supervisory Control and Data Acquisition (SCADA) systems [24].

Overall, it must be stressed that empowerment is more than the sum of the individual PMESII factors.

2.3.3.2. Near-Peer Use of Cyberspace

Various studies of nation-state empowerment provide insights on the projected uses and cyber-strategies of China and Russia. The relevant white papers discuss the recent writings from key conceptual thinkers in those nations and compares and contrasts these strategies. Those nations use a different vocabulary in discussing cyberspace and cyberpower. For example, Chinese writings on the subject focus on stratagems, objective and subjective reality, and the dialectic[10].

Two key aspects of the Chinese view of the Revolution in Military Affairs are particularly germane: "War with the objective of expanding territory has basically withdrawn from the stage of history, and even war with the objective of fighting for natural resources is now giving way to war with the objective of controlling the flow of financial capital."

Furthermore: "If we go our own path to develop military theory, weapons, and equipment, we will develop something never seen before in places that no one has ever thought of before; others will be unable to anticipate or resist our 'self-accommodating systems'."

As an illustration of "self-accommodating systems" against the superior foe, three ways are cited for making a cat eat a hot pepper: "stuff it down his throat, put it in cheese and make him swallow it, or grind it up and spread it on his back. The latter method makes the cat lick itself and receive the satisfaction of cleaning up. The cat is oblivious to the end goal. This is strategy."

2.3.4. Cyber Deterrence

There are three key challenges that must be addressed to deal with cyber deterrence [25]. These include the challenges of attribution, the lack of a cyber-deterrence track record, and the occurrence of unexpected higher-order effects.

The primary challenge is the perceived difficulty of attributing such attacks to a specific attacker (e.g., state, non-state actor). Note, for example, if competitors believe we cannot determine who is attacking us in cyberspace, they may convince themselves that such attacks involve little risk and considerable gain. However, there is a key trade-off that must be weighed: if we demonstrate our ability to detect and attribute cyberspace attacks we may provide intelligence about our capabilities. Thus, we may be posing a greater cyberspace threat to the nation in the future.

The second key challenge is the lack of a known historical track record of US detection, attribution, and response. This poses a series of key issues. They include the credibility of deterrent actions, emboldening potential attackers, and defining publicly what the US considers a cyberspace "attack" and the potential kinds of responses to such attacks.

The third challenge is the potential for producing higher order effects that might result in unintended consequences and possibly undesired consequences. There are three key issues associated with that challenge. First, it is a function of the nature of the attacker's goals and objectives. Second, if the competitor is concerned about unintended consequences, it could enhance the effects of our deterrence activities if it wishes to control escalation or fears "blowback" from its cyberspace operations. Finally, if the competitor's goal is to create chaos, deterrence could be undermined by the potential for unintended consequences.

In addition, there is interest in "tailoring deterrence" [26] to address the variety of adversaries that exist in cyberspace (e.g., non-state actors; state actors). However, there is a debate within the analytic community as to whether tailored deterrence is a viable concept for the full spectrum of adversaries of the US [27]. That issue represents an important element of the research agenda for the community. However, it is

[10] "Reasoning that juxtaposes opposed or contradictory ideas and seeks to resolve conflict".

hypothesized that the full set of P/DIME options should be considered in developing a course of action to respond to a cyber attack. For example, the US might respond to a cyber attack through a variety of levers of power including diplomacy (e.g., a demarche) or economic actions (e.g., restricting the flow of technology).

2.3.5. Cyberstrategy "Rules of Thumb" and Principles

In weighing the cyberstrategy insights, three key insights emerged. First, the "low end" users (e.g., individuals, hacktivists, terrorists, trans-national criminals) have enhanced their power considerably through recent cyberspace trends. A tailored deterrence strategy will be needed to keep these entities in check.

Second, potential near-peer adversaries are aggressively exploring options to exploit attributes of cyberspace. In the near term, this is being manifested through acts of espionage that have resulted in the exfiltration of massive amounts of sensitive governmental and industrial data [28]. In the longer term, the US must be prepared to deal with unique "cyber strategems" that reflect the unique cultural and military history of key nations (e.g., China, Russia).

To deal with the emerging cyber threat, the US must conduct experiments and exercises that feature a creative and aggressive cyber opposing force. It would be naïve and dangerous to assume that future adversaries will not seek to negate the benefits that the US hopes to achieve through net centric warfare.

2.4. Theoretical Aspects of Institutional Factors

This section of the white paper focuses on two critical institutional factors: governance of cyberspace and the legal dimensions of the problem. The section concludes by identifying key institutional issues and principles.

2.4.1. Governance

Table 3 characterizes key governance functions in cyberspace and the organizations that participate in these functions. It can be seen that the mechanisms for governance of the Internet are *exceedingly* complex. Organizational activities often overlap or fit end-to-end, requiring the expenditure of considerable resources in multiple forums to achieve objectives. Consequently, there is a core set of participants (generally in the private sector) that are involved in several of these key organizations.

In an effort to evaluate the performance of Internet governance, we have introduced the following criteria: open, democratic, transparent, dynamic, adaptable, accountable, efficient, and effective. When assessed against these criteria, one can conclude that recent Internet governance has performed remarkably well.

However, as we look to future, the USG will be challenged to alter its position on Internet governance. Preliminary views on this subject are being articulated at the ongoing Internet Governance Forums (IGF). In fact, a recent white paper on the subject [29] made the following observations:

"Internet Governance is an isolating and abstract term that suggests a nexus with an official government entity. The term also implies a role for the US Congress in Internet decision-making. It is a misnomer because there is no true governance of the Internet; only a series of agreements between a distributed and loosely connected group

Table 3. Governance of Cyberspace

Function	ICANN	ISOC*	ITU	OECD	CoE	EU	ISO	IEC	IEEE	W3C	UN
Domain names	●										
International domain names	●		●								
Core Internet functions		●									
Telecommunications standards			●								
World-wide web standards										●	
Product standards			●				●	●	●		
Development			●	●		●					●
Cyber Security**	●	●	●	●	●	●	●	●			

* Internet Society and related organizations (e.g., IETF, IESG, IAB)
** As well as National Governments

of organizations and influencers. A more fitting term may be 'Internet Influence,' or for long-term strategy purposes, 'Internet Evolution'."

2.4.2. Cyber Law

One of the most challenging legal issues confronting the cyber community is as follows: "Is a cyberattack an act of war?" Legalistically, the answer is often presented as one of three possible outcomes: it is not a use of force under UN Article 2(4); it is arguably a use of force or not; it is a use of force under UN Article 2(4).

There are several frameworks that are being considered by the legal community to address this issue. Michael Schmitt has formulated a framework that defines and addresses seven key factors: severity, immediacy, directness, invasiveness, measurability, presumptive legitimacy, and responsibility [30]. Once one has assessed each of those factors, one should employ multi-attribute utility theory to weight each of these factors and come to a determination. An associated challenge is to formulate responses to that attack that are consistent with the legal tenet of proportional response.

Overall, the area of cyber law is in its infancy. Although there have been preliminary rulings on sharing of music (e.g., Napster), there are major issues on the questions of sovereignty, intellectual capital, and civil liberties. These issues will be major areas for research for the foreseeable future.

2.4.3. Institutional Principles

Based on the insights developed during the course of this study, four major strawman principles have emerged in the arena of Institutional Factors.

First, given the complexity of the governance mechanisms, one should seek influence over cyberspace vice governance.

Second, the legal community has barely addressed the key issues that must be resolved in the cyber arena. For example, considerable research is needed to assess the following key questions:

- What is an act of (cyber)war?
- What is the appropriate response to an act of (cyber)war?
- What is the appropriate way to treat intellectual property in the digital age?
- How can nations resolve differences in sovereign laws associated with cyber factors?

Third, there is a need for a framework and enhanced dialogue between champions of civil liberties and proponents of enhanced cyber security to establish an adequate balance. Finally, guidance and procedures are required to address the issue of sharing of cyber information between the USG and industry. This approach should be based on the concept of risk management.

3. Connections

At the beginning of this white paper, it was noted that one of the reasons for a theory was the need to **connect** diverse elements of a body of knowledge. In general, the community is focusing on the issue of connecting the knowledge within a stratum of the pyramid. Even though this is challenging, it generally involves communicating among individuals with a common background and lexicon.

It is far more difficult to have individuals connect **across** the different strata of the pyramid. This requires individuals from different disciplines to work effectively together. In order to do so, it requires a holistic perspective on the Measures of Merit (MoMs) for cyber issues.

Figure 10 suggests a potential decomposition of the MoMs associated with the cyber problem. It identifies four linked sets of measures: Measures of Performance (MoPs), Measures of Functional Performance (MoFPs), Measures of Effectiveness (MoEs), and Measures of Entity Empowerment (MoEEs). Since this field of endeavor is still in its infancy, the material is meant to be illustrative and not exhaustive.

MoPs are needed to characterize the key computer science and electrical engineering dimensions of the problem. A key measure is the amount of bandwidth that is available to representative users of cyberspace. As the bandwidth increases to the megahertz/sec range, the user is able to access advanced features such as imagery and video products. A second key measure is connectivity. For circumstances in which the cyber-infrastructure is fixed, a useful measure is the percent of people in a country that have access to the Internet. However, in many military operations, the cyber-infrastructure and the users are mobile. Under those circumstances, a more useful measure is the performance of Mobile, Ad hoc NETwork (MANET) users (e.g., their ability to stay connected). Third, one can introduce measures of the "noise" that characterizes the cyber-infrastructure. For example, the extent to which the quality of the Internet is degraded can be characterized by the unwanted e-mail that it carries ("spam"), which can subsume a substantial subset of the network's capacity. As an

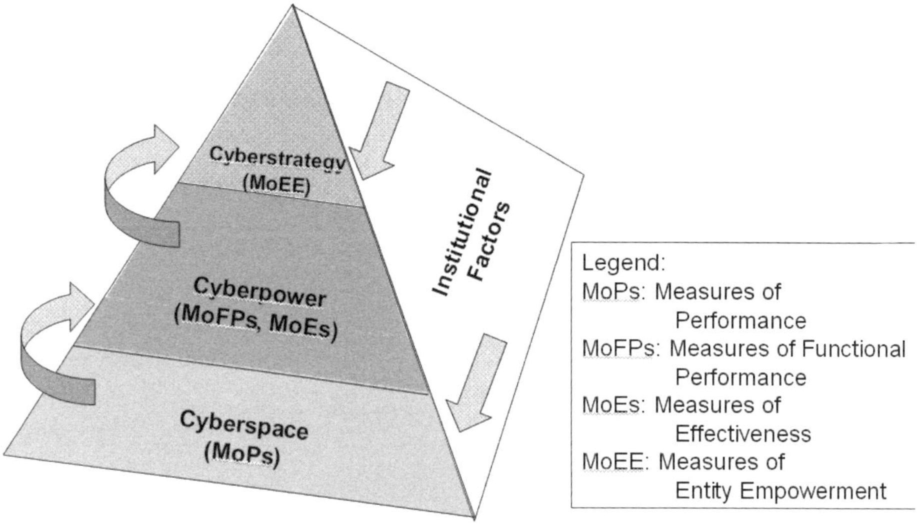

Figure 10. Measures of Merit

example, it has been estimated that in recent months up to 90% of the traffic on the Internet is spam [31]. In addition, the integrity of the information is further compromised by "phishing" exploits in which criminal elements seek to employ the Internet to perpetrate economic scams. Finally, MoPs can be introduced to characterize resistance to adversary actions, including denial of service attacks, propagation of viruses or worms, and illicitly intruding into a system.

It is useful to introduce MoFPs that characterize how successfully selected entities are able to perform key functions, taking advantage of cyberspace. In the case of the US military, the concept of net-centricity is to employ advances in cyberspace to perform essential functions. These include the ability to enhance the performance of increasing levels of information fusion. Similarly, a basic tenet of net-centricity is to propagate commander's intent so that the participants in the operation can synchronize and self-synchronize their actions.

MoEs are needed to characterize how effective entities can be in their key missions, taking advantage of cyberspace. In the context of Major Combat Operations, MoEs are needed to characterize the ability to exploit cyberspace in multiple dimensions. At one extreme, enhancements in cyberspace have the potential to reduce the time to conduct a campaign and the casualties associated with the campaign. At the other extreme, enhancements in cyberspace may substantially enhance Blue loss exchange ratios and the amount of ground gained and controlled.

From the perspective of cyberstrategy, there is interest is characterizing the extent to which enhancements in cyberspace can empower key entities. In the case of nation states, potential MoEEs might include selected PMESII variables. As an example, it might address the ability to leverage cyberspace to influence a population, shape a nation at strategic crossroads, and deter, persuade, and coerce an adversary.

Table 4 depicts candidate MoMs that may be employed in future cyber analyses.

Table 4. Selected Measures of Merit

Measures	Representative Measures
Performance	• System performance (e.g., latency, bandwidth, reliability) • Resistance to adversary attack (e.g., ability to withstand a DDoS attack)
Functional Performance	• Time to create, validate, disseminate messages • Number of meetings held with surrogate groups • Increase/decrease of anti-Coalition graffiti • Who is "waving", where
Effectiveness (against targeted groups)	• Media: Number of positive/negative stories published/aired • Clerics: Tone of mosque sermons • Military: Loss Exchange Ratios
Entity Empowerment	• Improvements in economic reforms (e.g., reconstruction projects completed effectively) • Political reforms (e.g., participation in elections, reconciliation) • Social reforms (e.g., status of critical infrastructures) • Security (e.g., number, severity of insurgent, terrorist attacks)

4. Anticipation

From the perspective of the decision maker, the key challenge is to anticipate what will occur next in the cyber domain and to formulate coherent policy to cope with those issues. To begin to address that challenge, this section deals with four aspects of anticipation. First, it identifies key trends that are expected to characterize cyberspace. Second, it identifies the research activities that should be conducted to address those trends. Third, it briefly identifies the major policy issues that decision makers will need to address. Finally, it discusses the assessment needs that must be addressed to support the formulation and analysis of policy options.

4.1. Cyber Trends

It is extremely difficult to provide quantitative estimates as to how rapidly key trends in cyberspace will be manifested. Thus, the following should be regarded as a partial, qualitative list of some of the most significant potential changes.

First, there is an increased move to adoption of IP-based systems. As a consequence, one can anticipate a convergence of telephone, radio, television, and the Internet. As one example, there is a dramatic use of Voice over IP (VoIP) (with attendant security issues) in the area of telephony. Second, we are seeing the emergence of sensor networks that feature an extremely large number of heterogeneous sensors. As one manifestation, we are seeing the netting of extremely large number of video cameras in urban areas, raising issues in the civil liberties community. Third, we are seeing an inexorable trend towards proliferation of broadband and wireless. An example of this trend was the plan to have city-wide deployment of Worldwide Interoperability for Microwave Access (WiMax). However, this trend suggests the difficulty in predicting when a trend becomes a reality. NEXTEL had made this objective the key to their strategy; however, they have recently observed that the technology has not matured sufficiently to implement it in the near-term [32]. Fourth, we are observing enhanced search capabilities, both for local systems and the entire Internet. One of the keys to this trend has been industrial competition to develop improved search engines (in part, to enhance advertising revenue). Fifth, we are seeing extraordinary efforts to enhance human/machine connectivity. As one example, we are seeing direct nerve and brain connections to computers or prostheses, arising from efforts to treat soldiers injured by IEDs in Iraq [33]. Sixth, we are seeing dramatic increases in user participation in information content. This trend is manifested through the proliferation of video blogs, contributions to wikis, participation in social networks (e.g., MySpace, FaceBook), and involvement in virtual reality environments (e.g., Second Life). Finally, some experts have postulated that we are entering the third phase of the Internet (i.e., phase 1: communicating; phase 2: content; phase 3: collaboration). This third phase is characterized by "cloud computing" where "information is stored and processed on computers somewhere else" [34]. One of the major issues associated with this paradigm is our ability to provide adequate security for the "cloud".

4.2. Opportunities for Cyber Research

As an application of the emerging theory of cyber, Table 5 identifies the major areas where cyber research should be pursued.

Table 5. Areas Where Additional Theoretical Research Is Required

Area	Research Areas
Cyberspace	• Perform technology projections to identify key breakthroughs • Develop techniques to protect essential data from exfiltration, corruption • Formulate an objective network architecture that is more secure, and identify options to transition to it
Cyberpower	• Extend analyses to other levers of power (e.g., diplomatic, economic) • Perform risk assessments to address cyber-dependence • Quantify the Blue-Red information duel
Cyberstrategy	• Conduct research on "tailored deterrence" • Explore options to address cyber espionage
Institutional Factors	• Perform research on cyber influence; legal frameworks; balance between security and civil liberties
Cyber Assessment	• Develop analytical methods, tools, data, and intellectual capital to assess cyber issues

4.2.1. Cyberspace Research

In the area of cyberspace, improved technology projections are needed to identify key breakthroughs that may substantially affect MoPs for cyberspace. Second, it is inevitable that malevolent actors (e.g., insiders, adaptive adversaries) will gain access to the USG and defense industrial base cyberspace. This suggests that research is needed to protect the essential data in cyberspace from exfiltration or corruption.

Finally, additional research is needed to formulate an objective architecture for cyberspace that is inherently more secure than the existing architecture. Consistent with that effort, there is a need to address the challenging issue of transitioning from the existing to the objective architecture.

4.2.2. Cyberpower Research

Due to resource constraints, this evolving assessment of cyber theory has not adequately addressed all the levers of power (e.g., political, diplomatic, economic). As an initial step, assessments should be completed for these other levers of power. Second, existing assessments of the military lever of power have focused almost exclusively on the potential benefits that can accrue by creatively employing cyberspace. It is equally important to perform risk assessments to understand the potential downside of relying extensively on cyberspace. This includes conducting experiments and developing the methodology, tools, data, and intellectual capital required to perform military risk assessments. Similarly, it is important to conduct research into the potential benefits and risks associated with leveraging cyberspace developments for non-US military capability (e.g., NATO allies that are pursuing Network Enabled Capabilities (NEC)). Finally, in the area of information, additional research is needed to quantify the information duels that are likely to occur with potential adversaries.

4.2.3. Cyberstrategy Research

To deal with the challenges posed by the full array of entities empowered by enhancements in cyberspace, it is vital that the information-enabled societies conduct research on "tailored deterrence". This concept suggests that key alliances, such as NATO, must develop a holistic philosophy that understands each of the potential adversaries (e.g., its goals, culture, risk calculus), develops and plans for capabilities to deter these adversaries, and develops a strategy to communicate these concepts to the potential adversaries.

4.2.4. Institutional Factors Research

Theoretical research is needed to address key gaps in institutional knowledge in the areas of governance, legal issues, sharing of information, Internet regulation, and civil liberties.

First, in the area of governance, the USG must reassess the role of the Internet Corporation for Assigned Names and Numbers (ICANN) in the governance of the Internet. It is clear that, in the future, the USG must be more adroit in the area of "cyber influence" vice governance. This will require a thorough re-examination of all the institutional bodies that affect cyber governance and the development of a USG strategy to interact with them.

Second, "cyber legal" issues are in their infancy. The current situation is non-homogeneous with inconsistent laws in various sovereign nations (e.g., German hate-crime laws; limited signatories to the Council of Europe Convention on Cybercrime [35]. In particular, there is a need to clarify the issue of espionage in cyberspace (e.g., What is it? What rights of response are left to the victims?). In addition, there is a need to adopt a consistent model that can be applied to determine whether a cyber attack is an act of war.

Third, there is continued controversy about the sharing of information between the USG and the private sector. Research is needed to determine what information should be shared, under what circumstances.

Fourth, it has been observed that regulatory agencies, such as the Federal Communications Commission, have the authority to regulate Internet Service Providers (ISPs) to redress selected cyber security issues. However, to date, regulatory agencies have been reluctant to address these issues.

Fifth, the recent debate about the Foreign Intelligence Surveillance Act (FISA) court has mobilized the civil liberties community to raise the specter of "Big Brother". As a consequence of the actions of civil liberties organizations, key USG programs have been terminated or modified (e.g., DARPA's Total Information Awareness (TIA), DHS's Multi-state Anti-Terrorism Information Exchange (MATRIX)). Research is needed to clarify the appropriate balance among actions to deal with adversaries while still protecting civil liberties.

4.2.5. Cyber Assessment Research

As discussed below, our ability to perform cyber assessments is extremely uneven. As a consequence, research efforts are required to develop analytical methods, tools, data, services, and intellectual capital to address key cyber issues in the areas of cyberpower, cyberstrategy, and infrastructure issues.

Table 6. Selected Policy Recommendations

Area	Issue/Recommendations
Cyberspace	• **Security:** USG should adopt a "differentiated security" approach • **Human capital, R&D:** Enhance cyber education & training; establish cyber labs; increase resources
Cyberpower	• **Net Centric Operations risks:** Employ an OPFOR that is highly cyberwar-capable • **CNA:** Reduce classification, enhance Integration • **Influence Operations:** Adopt a holistic, multi-disciplinary, Interagency approach • **SSTR:** Adopt I-Power approach
Cyberstrategy	• **Organization:** Create a new organization to formulate/oversee policy on cyber issues • **Deterrence:** Formulate, implement "tailored deterrence" • **Espionage:** Conduct policy/legal review
Institutional Issues	• **Governance:** USG should develop, implement Internet *influence*

4.3. Cyber Policy Issues

Several major policy issues have been singled out that require further attention. For the purposes of this preliminary cyber theory, these issues have served to focus the boundaries of this study, although we have also addressed a number of lower priority policy issues. Consequently, emphasis has been placed on assembling the intellectual capital required to illuminate those issues.

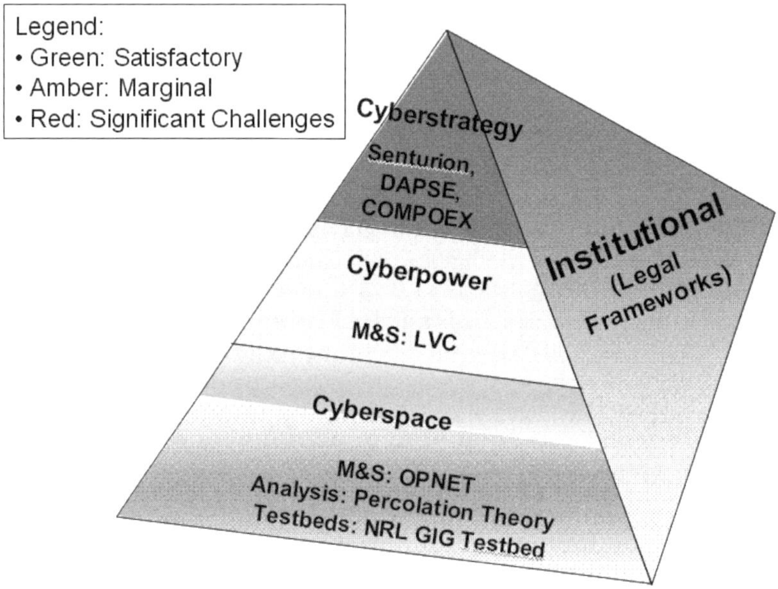

Figure 11. Subjective Assessment of MS&A for Cyber Policy Analyses

In Table 6, these issues have been aggregated into the categories of cyberspace, cyberpower, cyberstrategy, and institutional factors. Most of these issues are extremely broad and contentious; consequently, additional analyses will be required to address them adequately.

4.4. Cyber Assessment

One of the major challenges confronting the analysis community is to develop the methods, tools, and data needed to support cyber policy decision makers. Figure 11 suggests the relative maturity of key tools in the areas of cyberspace, cyberpower, cyberstrategy, and institutional factors.

In the areas of cyberspace, there are several tools that the community is employing to address computer science and communications issues. Perhaps the best know is the OPNET simulation [36] that is widely employed to address network architectural issues. From an analytic perspective, techniques such as percolation theory [37] enable one to evaluate the robustness of a network. Looking to the future, the National Research Laboratory (NRL) has developed a GIG Testbed to explore the myriad issues associated with linking new systems and networks.

In the area of cyberpower, the community has had some success in employing live, virtual, and constructive simulations. For example, in assessments of air-to-air combat, insights have been derived from the live AIMVAL-ACEVAL experiments, virtual experiments in the former McDonnell Air Combat Simulator (MACS), and constructive experiments using tools such as TAC BRAWLER. However, the community still requires better tools to assess the impact of advances in cyberspace on broader military and informational effectiveness (e.g., land combat in complex terrain).

In the area of cyberstrategy, a number of promising initiatives are underway. In response to recent tasking by STRATCOM, a new methodology and associated tools are emerging (i.e., Deterrence Analysis & Planning Support Environment (DAPSE) [38]. However, these results have not yet been applied to major cyberstrategy issues. In addition, promising tools are emerging from academia (e.g., Senturion; GMU's Pythia) and DARPA (e.g., Conflict Modeling, Planning & Outcomes Experimentation (COMPOEX)). However, these are still in early stages of development and application.

Finally, as noted above, there are only primitive tools available to address issues of governance, legal issues, and civil liberties. Some tools are being developed to explore the cascading effects among critical infrastructures (e.g., National Infrastructure Simulation and Analysis Center (NISAC) system dynamics models [39]); however, they have not yet undergone rigorous validation.

5. Summary

Consistent with the macro-framework that has been adopted to characterize the cyber problem, this section summarizes the key insights in the areas of cyberspace, cyberpower, cyberstrategy, and institutional factors. The section concludes by identifying the next steps that should be taken to refine the theory of cyberpower.

5.1. Key Insights

5.1.1. Cyberspace

Cyberspace is an environment that is experiencing exponential growth in key MoPs. There is an extraordinary diffusion of knowledge among all the stakeholders of cyberspace, including malevolent users. As a consequence of this diffusion of knowledge, cyberspace is being degraded by "noise" (e.g., proliferation of spam) and a broad variety of cyber attacks. The most troubling of these attacks includes Distributed Denial of Service, exfiltration of data, and the potential for corruption of data. In each instance, recent experience has demonstrated that these attacks are relatively easy to implement (e.g., technically, financially) and extremely difficult to attribute.

These vulnerabilities arise from the basic architecture that has evolved from the original ARPAnet. A new cyberspace architecture may be required to halt the perceived erosion of security. However, there will be substantial difficulties in transitioning from the current architecture to one that is more robust against adversary action.

5.1.2. Cyberpower

As cyberspace evolves, it has the potential to enhance each of the levers of national power. This white paper has focused on two of these levers: military and information.

In the area of military power, it was observed that studies are underway to characterize the extent to which enhancements in cyberspace can enhance key MoEs. These studies tend to be unambiguous in the area of air-to-air combat where experiments suggest that enhanced digital communications can enhance loss-exchange ratios by a factor of approximately 2.5. Although studies of other military operations have also been undertaken, the results are generally confounded by other factors (e.g., mobility, protection).

To complement these experiments, an assessment of theories of environmental warfare was undertaken that critically reassessed the theories of land, sea, air, and space theory. Based on that assessment, it was concluded that a theory of cyberpower should focus on four key factors: technological advances, speed and scope of operations, control of key features, and national mobilization.

From the perspective of "information", this white paper has addressed influence operations from a strategic and tactical perspective. Based on prior experiences and an adaptation of earlier analytical frameworks, an approach was developed for linking operational objectives and processes to DOTMLPF requirements. These assessments suggest that developments in cyberspace can substantially affect future efforts to enhance influence operations (e.g., implement precision guided *messages*).

5.1.3. Cyberstrategy

The evolving theory of cyber has identified a range of entities that will be empowered by enhancements in cyberspace. These include: terrorist groups, who are employing cyberspace to, inter alia, recruit, raise money, propagandize, educate and train, plan operations, command and control operations; hacktivists, who are employing cyberspace to conduct "cyber riots" (e.g., Estonia) and implement exploits in cyberspace; transnational criminals, who pursue a variety of techniques (e.g., phishing, denial of service attacks) to raise substantial funds (reputed to be more than the money derived from drug trafficking); and nation states, the most advanced of whom are employing cyberspace to enhance all dimensions of PMESII activities.

However, changes in cyberspace have given rise to unintended consequences. Many of the entities at the "low end" of the entity spectrum (e.g., terrorists, hacktivists, transnational criminals) are making life more dangerous for information-enabled societies. In particular, these entities tend to be much more adaptable than nation states, causing the latter to respond, belatedly, to the initiatives of the former. In addition, research about selected near-peers (e.g., China, Russia) suggests that they have new perspectives on cyberstrategy that will present information-enabled societies with new challenges in cyberspace.

5.1.4. Institutional Factors

From an institutional perspective, issues are emerging that will affect all aspect of cyber theory. This white paper has highlighted the challenges that exist in cyber governance, legal issues, exchange of cyber information between governments and industry, and the balance between national security and civil liberties.

From a theoretical perspective, one of the major challenges emerges from the difficulty in characterizing and responding to an attack in cyberspace. As demonstrated by recent events, it is extremely difficult to attribute an attack to an adversary that chooses to act anonymously. In light of that ambiguity, it is difficult to formulate a coherent response to such an attack. For example, it is still unclear how an alliance, such as NATO, might respond in the future to a cyber attack against one or more of its members.

5.2. Next Steps

This effort constitutes an evolving theory of cyberpower. To refine this product, it is recommended that the following steps should be pursued.

5.2.1. Define

There is still confusion about the definitions for the key terms in a theory of cyberpower. However, the community should find it relatively straightforward to go from the current base to agreement on key terms (e.g., "cyberspace"). However, additional work is still required to establish the linkage between cyber terms and the terms associated with information operations.

5.2.2. Categorize

The "cyber pyramid" has proven to be a useful taxonomy in "binning" key concepts. However, there is still a need to develop specific cyber frameworks and models to explore key policy issues that confront senior decision makers.

5.2.3. Explain

It is anticipated that this evolving theory of cyberpower will be incomplete. Additional efforts are needed to address key issues that are beyond the scope of this white paper. In the area of cyberpower, there is a need to assess how potential changes in cyberspace will affect political, diplomatic, and economic functionality and effectiveness. In the area of cyberstrategy, there is a need to assess the extent to which key entities are empowered by advances in cyberspace and cyberpower. These include individuals, NGOs, transnational corporations, selected nation states, alliances (e.g., NATO), and international organizations (e.g., UN). Finally, in the area of institutional factors, there is a pressing need to assess the effect of changes in cyberspace on the balance between civil liberties and national security. In assessing these issues it would be useful to employ a risk management approach.

5.2.4. Connect

Currently, we have relatively little understanding about the appropriate Measures of Merit to employ in cyber assessments nor the relationships among those measures. For example, we do not have a clear understanding about how changes in cyberspace (e.g., MoPs such as bandwidth or resistance to enemy countermeasures) impacts the US's levers of power (i.e., P/DIME) or empowerment (i.e., PMESII). At a minimum, it is important to develop preliminary relationships so that a decision maker can understand the implications of how potential changes in cyberspace or institutional factors will affect cyberpower and cyberstrategy.

5.2.5. Anticipate

As noted in this white paper, cyberspace is in the midst of explosive, non-linear change. It is vital that more detailed technology assessments be undertaken to anticipate and understand potential break-throughs in cyberspace (e.g., the analogue of discovering giant magnetoresistance or fundamental changes in the architecture of the

Internet). Furthermore, efforts should be made in the development and application of models, simulations, and analyses to assess the impact of these changes on cyberpower and cyberstrategy. These developments in methodologies, tools, and data should provide decision makers with the analytic support needed to explore the long-range affect of alternative cyber options.

References

[1] Dr. Harold R. Winton, Air War College, Maxwell AFB, "An Imperfect Jewel", presented at Institute of National Strategic Studies (INSS) workshop on theory of warfare, NDU, Washington, DC, September 2006.

[2] Jim Holt, "Unstrung", The New Yorker, October 2, 2006.

[3] William Gibson, "Neuromancer", Ace Science Fiction, 1984.

[4] Deputy Secretary of Defense Memorandum, "The Definition of Cyberspace", May 12, 2008.

[5] "The National Military Strategy of the United States of America – A Strategy for Today, A Vision for Tomorrow," Joint Chiefs of Staff, 2004.

[6] Joint Publication 3-0. "Joint Operations", Joint Staff, 17 September (incorporating change 1 13 February 2008).

[7] G.E.P. Box, "Robustness in the Strategy of Scientific Model Building, in "Robustness in Statistics", R. L. Launer and G. N. Wilkinson, editors, 1979, Academic Press: New York.

[8] John Markoff, "Military Supercomputer Sets Record", New York Times, June 9, 2008).

[9] Jeremy M. Kaplan, "A New Conceptual Framework for Net-Centric, Enterprise Wide, System-of-Systems Engineering," CTNSP/NDU, Number 29, July 2006.

[10] Sally Adee, "37 Years of Moore's Law", IEEE Spectrum, May 2008.

[11] H.J. Mackinder,"The Geographical Pivot of History", 1904.

[12] Alfred T. Mahan, "The Influence of Sea Power Upon History (1660 – 1783)", Little, Brown and Company, Boston, 1890.

[13] Giulio Douhet, "The Command of the Air", translated by Dino Ferrari, Coward-McCann, 1942.

[14] Colin S. Gray and Geoffrey Sloan, "Geopolitics, Geography, and Strategy", Routledge, 30 November 1999.

[15] David S. Alberts and Richard E. Hayes, "Power to the Edge", Command and Control Research Program, June 2003.

[16] David T. Signori and Stuart H. Starr, "The Mission Oriented Approach to NATO C2 Planning", Signal Magazine, PP 119 – 127, September 1987.

[17] Colonel Ralph Baker, "The Decisive Weapon: A Brigade Combat Team Commander's Perspective on Information Operations", Military Review, May-June 2006.

[18] George Gilder, "Metcalfe's Law and Legacy", Forbes ASAP, 13 September 1993.

[19] Bob Briscoe, Andrew Odlyzko, Benjamin Tilly, "Metcalfe's Law is Wrong", IEEE Spectrum, July 2006.

[20] Daniel Gonzales, et al, "Network-Centric Operations Case Study: Air-to-Air Combat With and Without Link 16", RAND, National Defense Research Institute, 2005.

[21] Marc Sageman, "The Homegrown Young Radicals of Next-Gen Jihad", Washington Post, page B-1, June 8, 2008.

[22] Homeland Security Policy Institute, "NETworked Radicalization: A Counter-Strategy", GWU, Washington, DC, May 2007.

[23] Joseph S. Nye, Jr., "Understanding International Conflicts: An Introduction to Theory and History", New York: Pearson-Longman, 2005.

[24] Siobhan Gorman, "Electricity Grid in U.S. Penetrated by Spies", Wall Street Journal, April 8, 2009, page 1.

[25] Kevin Chilton and Greg Weaver, "Waging Deterrence in the Twenty-First Century", Strategic Studies Quarterly, Spring 2009.

[26] N. Elaine Bunn, "Can Deterrence Be Tailored?", Strategic Forum, Institute of National Strategic Studies, NDU, No. 225, January 2007.

[27] AFEI Conference on Cyber Deterrence, Tysons Corner, VA,Nov 1-2, 2007.

[28] John Markoff, "Vast Spy System Loots Computers in 103 Countries," New York Times, March 29, 2009.

[29] OASD(NII)/DOD CIO Globalization Task Force, "Development of an Internet Influence/Evolution Strategy for the Department of Defense",October 19, 2007.
[30] M. N. Schmitt, *Bellum Americanum:* "The US view of Twenty-first Century war and its possible implications for the law of Armed Conflict", Mich. J. Int. Law 19, 4(1998), pp. 1051-1090.
[31] John Soat, "IT Confidential: Is There Anything That Can Be Done About E-mail?", Information Week, February 17, 2007.
[32] Audi Lagorce, "Clearwire, Sprint Nextel Scrap WiMax Network Agreement", Market Watch, November 9, 2007.
[33] Michael J. Riezenman, "Melding Mind and Machine", the Institute, IEEE, June 2008.
[34] Geoffrey A. Fowler and Ben Worthen, "The Internet Industry is on a Cloud – Whatever That May Mean", Wall Street Journal, March 26, 2009, page 1.
[35] Convention on Cybercrime, Budapest, Hungary, November 23, 2001 (//conventions.coe.int/Treaty/EN/Treaties/Htm/185.htm).
[36] Emad Aboelela, "Network Simulation Experiments Manual", Morgan Kaufmann, publisher; 3rd edition, June 2003.
[37] Ira Kohlberg, "Percolation Theory of Coupled Infrastructures", 2007 Homeland Security Symposium, "Cascading Infrastructure Failures: Avoidance and Response", National Academies of Sciences, Washington, DC, May 2007.
[38] Strategic Multi-Layer Analysis Team (Nancy Chesser, Editor), "Deterrence in the 21st Century: An Effects-Based Approach in An Interconnected World, Volume I", sponsored by USSTRATCOM Global Innovation and Strategy Center, 1 October 2007.
[39] COL William Wimbish and MAJ Jeffrey Sterling, "A National Infrastructure Simulation and Analysis Center (NISAC): Strategic Leader Education and Formulation of Critical Infrastructure Policies", Centre for Strategic Leadership, US Army War College, August 2003.

Glossary

Abbreviation	Definition
C2	Command and control
COMPOEX	Conflict Modeling, Planning & Outcomes Experimentation
CNA	Computer Network Attack
CNO	Computer Network Operations
CTNSP	Center for Technology and National Security Policy
DAPSE	Deterrence Analysis & Planning Support Environment
DARPA	Defense Advanced Research Projects Agency
DIME	Diplomatic, Information, Military, Economic
DoD	Department of Defense
DOTMLPF	Doctrine, Organization, Training, Materiel, Leadership & Education, Personnel, Facilities
GIG	Global Information Grid
HSCB	Human, Social, Cultural Behavior
IAB	Internet Architecture Board
IED	Improvised Explosive Device
IETF	Internet Engineering Task Force
IGF	Internet Governance Forum
INSS	Institute for National Strategic Studies
IO	Information Operations
IP	Internet Protocol
IRTF	Internet Research Task Force
ISOC	Internet Society
JMEM	Joint Munitions Effectiveness Manual

JTRS	Joint Tactical Radios System
JWICS	Joint Worldwide Intelligence Communications System
MACS	McDonnell Air Combat Simulator
MANET	Mobile Ad Hoc Network
MoE	Measure of Effectiveness
MoEE	Measure of Entity Empowerment
MoFP	Measure of Functional Performance
MoM	Measure of Merit
MoP	Measure of Performance
NATO	North Atlantic Treaty Organization
NCO	Net Centric Operations
NCW	Net Centric Warfare
NDU	National Defense University
NEC	Net Enabled Capability
NMS-CO	National Military Strategy for Cyber Operations
NRL	Naval Research Laboratory
OLPC	One Laptop Per Child
OODA	Observe-Orient- Decide-Act
OS	Operating System
OSD	Office of the Secretary of Defense
P/DIME	Political/ Diplomatic, Information, Military, Economic
PMESII	Political, Military, Economic, Social, Information, Infrastructure
QDR	Quadrennial Defense Review
R&D	Research & Development
SME	Subject Matter Expert
SOA	Service Oriented Architecture
SSG	Strategic Studies Group
SSTR	Stability, Security, Transition, Reconstruction
STRATCOM	Strategic Command
TIA	Total Information Awareness
TCP	Transmission Control Protocol
UDP	User Datagram Protocol
USG	United States Government
VOIP	Voice over Internet Protocol
WIMAX	Worldwide Interoperability for Microwave Access

The Virtual Battlefield: Perspectives on Cyber Warfare
C. Czosseck and K. Geers (Eds.)
IOS Press, 2009
© 2009 The authors and IOS Press. All rights reserved.
doi:10.3233/978-1-60750-060-5-53

Sub Rosa Cyber War

Martin C. Libicki[a,1]
[a]*RAND Corporation*

Abstract. Cyberspace offers the prospect of *sub rosa* warfare, in which neither side acknowledges that they are in conflict with one another or even that one side has been attacked at all. This is possible for two reasons: first, because the battle damage from some types of cyber attack may not be globally visible, and second because attribution can be very difficult. The reason that both sides may keep matters *sub rosa* is to maintain freedom of actions, on the theory that public visibility may complicate negotiations and lead to escalation. Nevertheless, *sub rosa* warfare has it dangers, notably a lack of the kind of scrutiny that may promote actions which cannot bear the light of day, and the overconfident assumption that no third party is aware of what is going on between the hackers of both sides.

Keywords. Cyber space, cyber war, *sub rosa*

Introduction

The last twenty years have seen a burgeoning knowledge base on cyber this and cyber that. We know a good deal more about how to get into other people's systems – and we know a good deal more about how to keep others out. Computer users are far more conscious of security considerations – they have had little choice in the matter. Computer security has risen in the ranks of government – and alliance – issues.

Nevertheless, it is fair to say that this growth has been concentrated at the tactical, which is largely to say technological but also the management end. Shelves are filled with books on how-to, but far fewer tomes explain why to. There is very little open material on how to integrate cyber operations with kinetic operations, which is to say how to use cyber operations to advance the ends for which kinetic operations used to be the exclusive means. Quite possibly, there may not be much behind the green door either, because there is very little intelligent discussion within the literature of professional military integration of *hypothetical* capabilities. As for strategic discussion, there is some, but a great deal is built on the premise that cyber warfare is kinetic warfare (or nihilistic terrorism) by other means. Well, it's not; the two are quite different. One might say they are as checkers and chess – which only look the same because their terrain is the same and some of the pieces have the same name.

This essay expounds on one of the more interesting differences. To wit, cyber war offers the prospect of *sub rosa* warfare, which is a form of combat in which the participation of both sides, or at least one side, is obscured to third parties. *Sub rosa* warfare

---

[1] Senior Management Scientist, RAND Corporation. Note, the following represents the author's view, not those of RAND or its sponsors or clients.

has some aspects of intelligence operations, and some aspects of special operations – although it is neither. Of note, *sub rosa* warfare is almost impossible to conduct with tanks, much less nuclear weapons.

1. Embracing Ambiguity

Ambiguity, one can argue, is the essence of cyber combat and the exploitation of ambiguity may be central to *some* strategies of cyber warfare. There is a natural human tendency to assume that ambiguity is an epiphenomenon, something that obscures reality, a measurement error, as it were, and thus a temporary irritant to true understanding. Dust the surface, and the essence of the activity shines through and we see things for what they are. Some of this flavor comes from Clausewitz: fog and friction are what differentiate war on paper from war in the field. Those of Platonic bent may see the former as reality and the latter as shadows on the cave's wall. Clausewitz warned that one could not assume fog and friction away; they were embedded. What he did not argue was that fog and friction were central, with all that violence being peripheral – much less that the manipulation of fog and friction played much of a role in operational planning.

With cyber, the opposite can be true. In some cases, ambiguity is central and damage to systems is an artifact. If so, one should embrace ambiguity and not treat it as something of an embarrassment.

1.1. Definitions

Before going further, a few boundary markers may be useful.

First, cyber war will be defined as consisting of computer network (more broadly, systems) attack and defense. An attack succeeds when the target's use of its own systems is hampered – either because such systems fail to work or work very efficiently (disruption) or because systems work but produce errors or artifacts (corruption).

This definition specifically excludes computer network exploitation, which meets neither of these criteria. It is fair to say that CNE accounts for the great preponderance of computer network operations carried out among states and similarly serious non-criminal organizations. Yet it is a different phenomenon. Spying is not an act of war. It never has been, and there's little reason to change that. Furthermore, spying is inherently *sub rosa* and its motives are almost self-explanatory. This is not to say that CNE does not matter – it does – or that it is not interesting – it can be. But, it's not the subject of this essay.

Second, retaliation will be assumed to be an option in wake of a cyber attack, but that such retaliation will be limited to the cyber realm. This is not to say that retaliation must always be in kind, or that both attacker and retaliator can carry on mischief in cyber space without concerning themselves about escalation into the physical realm. Indeed, escalation is a major motive for keeping things *sub rosa*. However, it is difficult to keep physical retaliation *sub rosa*, and considering as much takes us into a different conversation.

Third, the essay limits itself to state-on-state cyber war, mostly because it best frames consideration of attack and retaliation. Non-state actors generally cannot be deterred, or even mildly dissuaded, by putting their systems at risk – because they do not have systems. Furthermore, although *sub rosa* warfare has a legitimate rationale of sorts, the usual approach to non-state actors generally involves the application of justice, and there are serious problems with *sub rosa* applications of justice that transcend cyber war issues.

1.2. Roshomon

What makes it possible to speak of *sub rosa* attacks is that information systems are generally invisible. The artifacts of a system – such as a personal computer – may be seen, but while some of what systems do is reflected back to the user, a great deal of what they do takes place inside and is not reported. Within an organization, all of the artifacts of computation and even most of the direct results may be hidden from the public, and what they do is visible only to the extent that the owners wish it so. Thus, damage to such a system is often invisible, even if some of the second-order effects may be quite visible. The contrast with physical warfare needs no further elaboration.

Of note, therefore, is the possibility that the target, the attacker, and third parties hold completely different perspectives on the nature of any one cyber attack.

The target includes the system operators, those they may call on for support, and those they report to. All three of *these*, incidentally, may hold views of what happened that diverge from one another's perspectives. In general, the target ought to have some idea that something went wrong, perhaps why, and what the consequences were. As a general rule disruption attacks are easier to see than corruption attacks. However, the target may not know (at least not immediately) what the attack vector did, how it managed to work past defenses, who controlled (or at least launched) the vector, and what the purpose was.

This plausible membership of the set of people who know they have been attacked merits further consideration. One can imagine an attack that only systems administrators notice – one that, for instance, requires them to put in overtime for the purposes of restoring a system's prior functionality and integrity. That being so, how damaging, which is to say, how consequential and thus how strategic can such an attack be? Thus, one must either posit a greater affected population that, nevertheless, keeps silent (such as the intelligence community) or a user base that senses damage but is misinformed about why. Some potential contexts include (1) systems that were planned to go on-stream or start new services but were prevented from doing so (a common enough phenomenon without hackers), (2) systems that went down because of what was believed to be human error, accident, or Mother Nature, or (3) systems that appear to function normally but produce bad information that the public at large is unaware of (e.g., scrambled payments for medical reimbursement checks). For systems operators to keep silent about the last one is quite risky unless they have confidence that they can correct things later without others being the wiser.

The fourth, but partial, possibility is that the problems are blamed on hackers but the hackers are identified as non-state actors (and thus subject to prosecution rather than retaliation). Such an attack is only barely *sub rosa*, in part because many third parties may believe otherwise (it was a state attack) and some of the implications of *sub rosa* attacks, discussed below, do not apply.

The attacker includes the hackers and those they report to (assuming their reports are honest and complete). The attacker will know the attack vector, how it evaded security (or at least the security features they saw), and what the purpose of the attack was (or at least one of the attacker's bosses will know). Depending on what kind of sensors it has emplaced in or near the attacked system, it may know something about the direct damage, but there may be a great deal it does not know, especially with respect to programs that the target system may have to route around or make up for damage.

Third parties include the public of the target country, the public of the attacker country, third-party states, and third-party publics. If the attack (and perhaps retaliation) were really – which is to say, successfully – *sub rosa* they will not know much about what, if anything is going on.

As we demonstrate, the difference in what each of these parties (broadly defined) knows, coupled with (presumably) their reluctance to share such information, makes the rest of the story possible.

Note that *sub rosa* attacks are not defined simply as those where the attacker's identity is unknown or uncertain – although, if the attack itself is unheard of, the identity of the attacker is moot.

Indeed, it should take little imagination to understand how much of cyber war is subject to ambiguity: not only *did something happen* – but *was it an accident (bad software, human error, Mother Nature), what was the damage, what was left behind, who did it, how they did it (including when they did it, and where was the point of access),* and, most importantly in the long run, *can they do it again?*

2. Basic Principles

Next, we turn to some basic principles of computer network attacks, not necessarily to say anything original, but to emphasize a few things by way of foundation for what comes later.

With one type of exception, the DDoS attack (more on this a little later), attacks are enabled by vulnerabilities on the part of the target.

One can assert, for starters, that there is no forced entry in cyber space. If someone has gotten into a system – or more particularly into the no-go area of a system – from the outside, it is because that someone has persuaded the system to do what its operators did not really want done and what its designers believed they had built the system to prevent. Nevertheless, in any contest between a computer's design and use-model (such as a user's intuition that email is information not instructions) on the one hand and its software code, on the other, the code always wins. Whoever gets into a system gets into a system through paths that the software (to include protocols and firmware) permits. The software may have flaws or may have been misconfigured (for instance, the permissions the administrator established differ from the permissions that the administrator thought had been established). Yet, a system is what it is, not necessarily what it should be. Such a divergence, when it has security implications, is a vulnerability. Whatever the methods, manual or automated, hackers' use, an attempt to take advantage of a vulnerability to gain access to a system or to get it to accept rogue instructions is called an exploit.

A system's integrity dictates how badly a system can be hurt by attacks in cyber space. One might even argue that a system's integrity is a more important determinant of success than the quality of the adversary's exploits—after all, no vulnerabilities, no exploits; no exploits, no cyber attacks.

Thus, in theory, all computer mischief is ultimately the fault of the system's owner – if not because of misuse or misconfiguration, then because of using a system with security bugs in the first place. In practice, all computer systems are susceptible to errors. In that sense all systems are somewhat opaque, unpredictable, and thus, ambiguous. The divergence between design and code is a consequence of the complexity of software systems and the potential for human error. The more complex the system – and they do get continually more complex – the more places there are in which errors can hide. Every information system has vulnerabilities—some more serious than others. The software suppliers themselves find a large share of these vulnerabilities and issue periodic patches, which users are then supposed to install – some more expeditiously and correctly than others (notwithstanding those hackers who observe patch releases, reverse engineer them quickly, determine the vulnerabilities the patches were supposed to fix, develop an appropriate exploit, and use it against those slow to patch their systems). Hackers find some vulnerabilities and then spring corresponding exploits on unsuspecting users who have otherwise done everything correctly. Literally thousands of exploits are sitting around. Many of the more devious ones require physical access to the target system. Most of the ones that reach the news do not work on well-patched systems.

In a sense, cyber attacks rely on deception – persuading systems to do what their designers do not want them to do. Fortunately, deception can be its own undoing. An exploit, if discovered, signals to sysadmins that something is not right. If good logs are kept, sysadmins may be able to determine where something unusual took place in the interaction between the hacker and the system. Changes in files (data or instructions), or the presence of unexpected files can also be telling. The process is hardly perfect; it is possible to determine a specific vulnerability and miss the broader design flaw of which the specific vulnerability is just an instance. Nevertheless, any one sysadmin can take advantage of an international community of system defenders with a common interest in minimizing outstanding vulnerabilities.

In contemplating cyber space, it may help to differentiate system peripheries from the system core. Peripheries may be said to contain user equipment; that is, equipment whose function and parameters are established by users. Peripheries, if not air gapped or protected via consistent encryption, tend to be repeatedly vulnerable largely because users are rarely trained in or focused on information security. User systems and privileges can be taken over through password cracking, phishing, social engineering, downloads from bad Web sites, use of corrupted media such as zip drives), etc. Sadly, the security of the periphery as a whole is often no better than the security of the most feckless user. The core, by contrast, is what sysadmins control—monitors, routers, management devices, machinery (such as weapons), and databases. Sysadmins are (or should be) trained and sensitive to security issues; they also set the terms by which users (and their systems) interact with the core. Although it is good personnel practice to sensitize users to security issues, it is good engineering practice to assume that users will not always be sensitive. While it is possible to protect the core from insecure users, it is less clear whether networks can function when enough user systems are compromised badly enough, even though network administration is a function of sysadmins. In general, it is hard to compromise the core in the same precise way twice, but the periphery is always at risk.

DDoS attacks are, as noted, a partial exception to the rule that a system can be attacked only if it has vulnerabilities (the Mafia-Boy attack of February 2000 apparently did take advantage of a certain class of vulnerabilities, since largely cleaned up). However, it is hard to conceive of a *sub rosa* DDoS attack in the sense that the public does not notice. So, we can disregard the exception for our purposes.

sense that the public does not notice. So, we can disregard the exception for our purposes.

2.1. The Attacker's Motive for Going Sub Rosa

An attack can be *sub rosa* only if the effects are limited to entities (such as state entities whose outputs are opaque and who believe in keeping secrets) or if the attacks could conceivably be ascribed to something other than hacking. The target has a good deal to say about whether an attack is *sub rosa;* yet, if attackers want to leave open the possibility of a *sub rosa* attack they have to avoid having such attacks affect the broad public but in ways that cannot be credibly ascribed to accident. They cannot take credit for an attack, which means that it cannot be used for certain forms of coercion.

The overall motive – for both sides – for keeping matters out of the press is that cyber warfare is a negative-sum game. Although this may be said generally true for warfare, it may be doubly true for cyber war (CNE, importantly, aside). Simply put, there is very little to be directly gained, which is to say, seized, in cyber war, unlike kinetic warfare where at least one side can entertain the possibility of a smash and grab (e.g., Kuwait's oil fields). Cyber war cannot even disarm the other side's cyber warfare capabilities, and while it can disarm kinetic warfare capabilities, it can only do so for a limited amount of time. Thus were there to be an extended cyber war, it would inevitably be a contest of attrition, a test of who can, in Wellington's terms, pound longest before someone's spirit gives out.

To go into particulars; an attacker may wish to limit its attacks to those that offer the target the opportunity to keep quiet in part to forestall retaliation. The attacker believes that while the state's elites may be able to handle things rationally – for instance, understand when they have been back-footed and thus retreat from some position – the same cannot be said for the target's publics. Thus, informing such publics will put pressure on the state to retaliate publicly when state elites may think other courses are less costly to the state. Worse is the possibility of escalation; elites may have a consensus among themselves to keep things in the cyber realm, but the public may not favor such limits. More generally, getting one or both publics involved introduces the possibility that events may spin out beyond the elites' ability to keep things under some sort of control. Many observers of war – for instance, of Gelb's book, *The Irony of Vietnam, the System Worked* – have concluded that state decision makers often prefer to risk losing a war than to risk losing *control* over a war. Finally, if the war is controlled, it is possible for elites of both sides to engineer a de-escalation of hostilities. All this manages the risk that the attacker faces in a cyber confrontation – for both sides.

One should also note the possibility that the effects of the cyber attack can be fit into the attacker's narrative *but only if* the results of the attack can be blamed on something other than the attack. Of course, if no one notices the effects of the attack, there's nothing to narrate about. A *sub rosa* attack whose effects are felt but not explained tends to shed focus not on the attacker but on the incompetence of the target – one unable, for instance, to protect sensitive health records from being scrambled.

As noted, a high form of *sub rosa* warfare is to make the attack look like an accident. One should not count too highly on anything more than momentary success; investigations tend to be pretty good at getting at root phenomena.

Incidentally, for some purposes the attacker may want its identity known to its opposite number. A few tricks such as mailing a letter before the attack that is received afterwards, leaving a "Kilroy was here" in the target machine, or revealing knowledge that only a penetration could provide should suffice.

Here are a few scenarios for a *sub rosa* attack:

One, they can be used to put others on notice that their systems are not so reliable that they can afford to engage in such a fight. Consider this. In step one, an attacking state creates anomalous behavior in a key system, be it government or a government-linked entity. The act (rather than the attacker, which is kept as ambiguous as possible) gets the attention of the leaders of the target state, which perceives its infrastructure at risk.

Subsequently, the system owners and their engineers claim that it was an accident and vow that such an act will never happen again. They get large sums of money to work hard on the problem. After this team starts to claim success, the attacker again creates anomalous behavior, preferably to the first victim, but perhaps to another comparably important system. This signals that problems persist (admittedly, step two is hard, precisely because the target state is working diligently against the possibility – certainly on the attacked system and quite likely on similar others). This not only reduces the credibility of the target's information system security, it also, and more importantly, reduces the credibility of those who promised to achieve that security.

Yet the attacker does not reveal itself or what it has done. This is unnecessary and even gets in the way. Doing so would make getting back at the attacker a more visible centerpiece of the target's strategy than simply misleadingly reassuring those who know they rely on the attacked system. Indeed, the attack itself is not so much the issue as it is to foster a general sense that the other side's information systems are fragile and unreliable. The attacker's message then becomes not "Cower before us!"—which requires identifying "us"—but the more impersonal, "You live in glass houses; are you sure you want to invest so much in stones?"

Perhaps the whole point of the attack is to make the target extra wary of expanding or opening up its networks, especially to outsiders, such as allied militaries, other government agencies, or support contractors. Further wariness may result from making the attack appear to come from a trusted source. Such a strategy presumes a skewed response from the target: not that networking should not be done naively but that networking is bad. It is easy to see why such a strategy can backfire and thus why cyber strategists, thinking over an extended period, must keep second and nth-order effects in mind.

Scenario two, cripple, test, or exercise someone else's military. Cyber attacks on the target's military may be used to impede the target's ability to respond to crises. A large, successful attack may retard the target's ability to wage war; if the target's military deployment can be delayed long enough (e.g., after everything has been decided and after the aggressor's forces have dug in for defense), the target's military intervention in a crisis started by the cyber attacker may be deemed pointless.

Such an attack can be a prelude to aggressive military action, or it can be in response to fears, however ill-founded, that the target is about to start something. In the former case, if the attack disarmed the target's military enough to allow successful kinetic combat, the *sub rosa* nature of the cyber attack may be temporary, and basically irrelevant if the cyber attack is quickly followed by violence of an obvious sort ("quickly" because the effects of *any* cyber attack are temporary and measured in hours or days). However, if the cyber attack fails to dent the target's military capability the attacker may call off its dogs and has no reason to publicize what it has done. In the latter case, cyber attack as pre-emption, the result of a successful cyber attack may be exactly nothing – in contrast to the violence that might have happened if the target's systems were intact. Since the effects of the cyber attack are temporary, war may take place anyway later – or not, if the cyber attacker (who is the presumed impending victim of the target's military) has used the time gained to rush to the front, so to speak, and discourage the war's outbreak.

Complicating this logic are attacks that look like they are meant to cripple another's military but are not. For instance, what if the cyber attacks were meant to persuade the target military that war was imminent, draw it to the ramparts for no reason, and repeat the cycle often enough to exhaust or spoof the target (as Egypt did when it carried out exercises in early-1973 but not attack until October of that year)? In contrast to physical feints, however, cyber feints may be poor strategy. By hardening the target's systems, every attack makes a subsequent attack more difficult. The choice of targets, if not masked by noise, may also suggest what the attacker finds important to disrupt and thus hints at how the cyber attacker would fight if war turned physical.

Attacks may be launched on military systems to see how well their operators react, in preparation for some later, larger attack. Can enemy sysadmins determine what happened and why? What workarounds do they use? Will corruption be detected? If the target knows it has been so tested, should it retaliate? Conversely, attacks may well reveal a great deal about the attacker and what it knows about the target's vulnerabilities.

There are solid grounds for believing that attacks on military can retain their *sub rosa* character. The attacker has the usual motives for keeping quiet, with the possible exception that it may wish to whisper about the attack to the target's allies so as to reduce their faith in the target's military. For the target, on its part, to reveal that it was attacked – and successfully so – is apt to reduce rather than increase confidence in its military capabilities. The latter may not have much of a choice if the damage is so widespread that a universe of witnesses defeats all thoughts of keeping them silent. The target may also broadcast the attack for purposes of supporting a "hate the enemy" campaign, regardless.

Cyber attacks that cripple intelligence assets do not have to lead to war. They may be justified if they blind the target's systems long enough for the attacker to carry out operations (e.g., moving missile parts) safe from prying eyes. Perhaps needless to add, intelligence assets are extremely hard targets for cyber war.

Coercion – especially against democratic states – normally requires the damage to be publicly visible and clearly associated with the coercer and its cause. Adversary actions need not affect the public, though, if there are other ways to compel governments to accede to demands. Indeed, the opposite may be true: the less the public knows, the easier it may be to garner concessions, especially invisible ones.

The case for *sub rosa* cyber war for the purposes of coercion rests on the belief that publicly visible attacks could lead to more popular pressure on the state to stand firm than to concede. The attacker counts on the possibility that the target's leaders are less afraid to make concessions whose true rationale can be hidden than to be blamed when, say, the economy hits an air pocket. As long as the new policy (which contains concessions) does not appear unwise *per se* or does not contradict earlier policies too much, the target's leadership need merely hide the fact that their policy choices were driven by fear. Keeping mum has other advantages for the target. Reducing the public itch for revenge (or their desire to demonstrate resolve) may facilitate negotiations or mutual de-escalation. Obscuring the fact or at least the damage from the attack may also mask the state's vulnerabilities from the eyes of third parties (presumably, the attacker will have a better sense of which vulnerabilities it had, in fact, exploited).

One ought not forget in all this that the *sub rosa* strategy has a serious Achilles heel from the cyber attacker's point of view. It assumes or, more to the point, *requires* that the target reacts as expected and maintains its silence. This requires that the cyber attacker have sufficient insight into the target to operate below the threshold past which it decides to mobilize against the cyber attacker – an act that generally requires the target being open about the attack and its consequences. The larger the cyber attacker's gain vis-à-vis the target, the less likely the target is to restrict its own activities. In effect, the attacker's strategy is hostage to the target's behavior, the basis for which we now turn.

2.2. Should the Target Reveal the Cyber Attack?

The likelihood that any attack is visible is the likelihood that the effects of an attack are visible multiplied by the likelihood that these effects will be publicly ascribed to a cyber attack (rather than to error, accident, or bad design). Both parts of the equation are anything but given. CNE is rarely apparent until an investigation reveals it. Corruption may go unnoticed until it reveals itself as a discrepancy between what a system is doing and what it should be doing. Sometimes even disruption may go unnoticed; for example, if a sensor is silent, is it silent because it has nothing to report, or has someone tampered with its reporting channel? If it is not people but machines or other processes which consume certain services, their loss may be noticed only when the processes they feed behave incorrectly.

Normally, full disclosure is the best policy. It is too easy for governments to believe they can control information much better than they actually succeed in doing – witness Chernobyl. Post hoc revelation eats at government credibility—not to mention competence, if playing catch-up with events makes the government look bad. Screaming helps mobilize the citizenry to support the government and (less cynically) pay attention to information security. It raises the seriousness level of the whole cyber war contest and thus gives the government more scope for implementing domestic security measures that the citizenry would otherwise object to. If the fact of the damage is evident, but not the cause, revealing the cause may enhance the credibility of infrastructure owners by switching attention from their own fecklessness as sysadmins to factors (portrayed as) outside their control. Revelation is necessary if the target state is going to respond visibly, either with retaliation or without (using legal, diplomatic, or economic measures, for example). Going public provides an opportunity to be clear about the aims of the response; it also subjects them to the test of knowing whether it can bear scrutiny. Incidentally, revelation may also be necessary for *sub rosa* retaliation: just because the retaliator did not want to make a fuss about how it hit back does not mean that the attacker (as target of retaliation) will do likewise.

Yet silence may still be golden. Revelation may expose the fecklessness of the target's system security, reducing the public confidence in it and making it a target for repeat attacks (a case for discretion comes from the public's tendency to overestimate the risks of cyber insecurity; there is considerable agreement that the public is wildly inconsistent in how it reacts to low-probability, high-impact risks). Evidence to support the attack claim may reveal sensitive information about system security.

2.3. Should Cyber Retaliation Be Obvious?

In cyber space, the target can hit back against the attacker, and no one (aside from the security establishments on either side) need be the wiser. This sort of *sub rosa* retaliation tends to make more sense if the attack is not public or if public attribution is not viable. In the latter case, the evidence behind attribution may be of the sort that is not easily released or not easily argued if released. *Sub rosa* retaliation avoids having to make the choice of what to reveal. This is no small matter. Reveal one's forensics and one has given all attackers a clue about what to avoid leaving behind the next time. Information about sources and methods is among the most closely-guarded secrets of the intelligence community. Furthermore, the attacker, as the victim of retaliation, could be under subsequent public pressure to counter-retaliate. If the effects of retaliation were not obvious, the attacker could therefore conclude that letting things drop after the retaliation is wiser than carrying on.

States that would employ *sub rosa* retaliation have to manage the expectations of those in the know who are looking for revenge. Retaliation could still convey the target's displeasure over the attacker's leadership and could change the latter's calculus to discourage further attacks.

Sub rosa retaliation, however, may be too seductive, particularly if the retaliator feels no need to convince the attacker of its guilt—after all, the attacker knows that it struck first, right? One danger is that, if the intelligence or law enforcement agency does not need to worry about defending its attribution to others, its case to national command authorities (that is, those who control the retaliation capability) may go unchallenged. The agency may thus claim its attribution is correct when the evidence suggests a higher degree of skepticism is warranted. Furthermore, a decision to retaliate *sub rosa* – like the decision to attack *sub rosa* – takes certain targets off the list (e.g., power plants) or at least demands they be hit in ways that do not look like a hit (which then fails to communicate displeasure reliably). The remaining targets may be those thought to be important to the other side's intelligence and law enforcement communities but do not directly affect the public at large. Finally, the entire strategy rests on the attacker's willingness not to make a fuss. Again, but in reverse this time: the wisdom of the strategy is hostage to the discretion of the state that (supposedly) engineered the attack in the first place.

2.4. Sub Rosa Retaliation against a Sub Rosa Attack Has One Big Advantage

To wit, the requirements for attribution are not nearly so great. One does not even need that much confidence in the quality of attribution. So, in hitting back, one may consider two possibilities. One; it was the attacker that suffered retaliation. Two; it was an innocent third party.

Take the first case. The attacker, knowing that it started things, will have a fairly good idea of why it was hit and take the message (subject to all the other caveats). If retaliation is to be reliably read as retaliation by the attacker, the "accident" would have to occur rather quickly after the original attack. Thus, the capacity to retaliate has to be maintained at a fairly high degree of readiness (that is, one must ascertain that the vulnerabilities still exist and that the victim's reaction will be roughly as predicted). Furthermore, the normal deliberation that might take place after an attack to increase the odds that the retaliation was well-directed would have to be short-circuited. All the previous caveats about the difference between what you think others do not know and what they actually do not know also apply.

In the second case, the innocent third party, unaware of what may have motivated an unprovoked attack (which the retaliation may look like to the victim) can only trot out its usual suspects and look for forensic evidence. As noted, this requires the original attack be unknown to any but the attacker and the target.

Managing the consequences of any venture that assumes ignorance among others is always contingent on third parties not spilling the beans. For instance, if retaliation against a third-party state is discovered by the attacking state, the attacker now has a very valuable piece of information – who attacked the third-party state. If the attacking state can figure out how to profit from implicating the retaliating state, it may well do so. Telling the third-party state that it started things may not be the smartest move, but it may be able to downplay its own role to suggest the retaliator over-reacted and was stupid about things to boot. It may maintain its innocence but circulate hints that make it easier for the innocent victim to identify the attacker (finding something is a lot easier when you know exactly what you are looking for). Or, the attacking state may blackmail the retaliator lest its actions be revealed to the innocent victim. The assumption that no one in the third-party state knows about the original attack may be in error; it is not unknown for two states with little in common but their dislike of the United States to swap intelligence (Iraq and Serbia, for instance, traded information on how to defeat U.S. aircraft and avoid anti-radiation missiles). More generally, the original attack may not be so secret prior to the attack or its existence may be revealed after the fact. Such revelation may be deliberate (perhaps someone here in the know is bothered by the retaliation or the possibility that it was misdirected), or simply reflect the universal difficulty of hiding secrets. Finally, the retaliator may have overstated its ability to keep itself anonymous. The third party does not have to know who did it, but it may have serious enough suspicions to affects its relationship with the retaliator – and if it did not know why the retaliator acted as it did, it may be angrier than if it understood that retaliator's motivation.

Again, perhaps *sub rosa* cyber war may be too clever by half – and one does not gain points for upholding the rule of law in cyber space by being sneaky.

What about being even cleverer and making retaliation look like an accident? The last technique is a variant of the first. Not only is the retaliation anonymous but it appears to be an accident. It is two steps rather than one step removed from something that the innocent third-party victim may find actionable. Again, the true attacker will presumably suspect that the accident was too closely timed to the original attack to be an accident, while the innocent victim of misguided regulation will have even less indication of what happened much less why. Indeed it would be most cool if the reprisal could be made to look like something caused by the original attack going haywire – all the dissuasive impact, and none of the risk.

It is unclear how to make an attack look like an accident in the first place. True, many attacks are initially hard to distinguish from accidents – which argues against hasty reactions all around. But there are techniques that can distinguish the two. If the problem is faulty software (such as the DSC bug that crippled phone service in the 1990s) then the fault can often be replicated by simulating the conditions at the time of failure. Human error can often be detected in various process logs. The greater the pain, the greater are the resources likely to be devoted to its elucidation. Thus, safeguards against the victim's (whether the original attacker or the unfortunate third party) detecting that it has been attacked may be temporary.

Finally, a state that wishes to establish principles – such as, do not hack – and then enforces them surreptitiously communicates either that it lacks sufficient faith in such principles or the strength to maintain and defend them openly.

Subtlety, nay sneakiness, in retaliating against a cyber attack absent strong attribution is normally difficult, but the exigencies of cyber space – the high level of ambiguity everywhere in the medium – only make things harder. Thus, while there are some notional ways to ways to work around the attribution problem they require a great deal of certainty about matters (the effect of cyber attacks, or the perception of attackers and third parties) that stand in stark contrast to the *uncertainty* about who did it. This leaves us with approaches that our British friends might call frightfully clever, with the emphasis on "frightful."

2.5. Ending Sub Rosa Warfare

How does one end a war that one does not admit one is fighting? In general, wars can end in one of four ways: through the destruction of one or both parties, through a formal peace agreement, through an informal tacit peace agreement, or as a series of bilateral decisions not to attack.

Cyber war generally lacks the power to destroy one or more parties to a conflict, and all the more so when the warfare is *sub rosa* – which not only takes certain types of attacks off the table (notably those that put pressure on populations), but also lies below the level where either side has reason to escalate into at least explicit warfare.

A formal peace agreement that pledges each side to halt an activity appears inconceivable if neither party admits to being the victim, much less the perpetrator of the acts in question. Yet, the transition from *sub rosa* to explicit cyber war is easy to make. Third parties may discover as much and make their findings public. Each side may also discover reasons for changing its mind and announcing as much.

An informal peace agreement requires that each side of a fight that is not public is nevertheless willing to discuss such secret maneuverings with its counterpart. At a minimum this requires some confidence on each party's part that it is not kidding the other.

Both formal and informal peace agreements in cyber space, however, can be problematic to enforce, or even state the terms of. Monitoring peace pacts in cyber space poses challenges not found in physical space. If either side still believes it can, if unpunished, reap unilateral advantages from new attacks, then attribution and damage assessment will likely remain as difficult afterward as they were beforehand (if attacks are extended to include CNE, the odds that one or another side finds attacks useful only go up). Each could cheat by shifting from visible disruption attacks to more-subtle corruption attacks.

Unfortunately, tacit de-escalation presents many of the same validation problems as negotiations – only made worse by the fact that there would only be a rough consensus rather than an explicit statement of what actions were and were not considered a violation. How could one tell that the other side is even cooperating, without clarity on what constituted cooperation?

In physical wars, peace pacts are often followed by unilateral disarmament (after World War I, for instance, Germany's army was limited to 100,000) or multilateral disarmament (for example, the Washington Naval Treaty). But disarmament in cyber space is virtually meaningless because cyber war is less about arms (exploits) than about vulnerabilities. So, disarmament cannot bulwark a peace agreement that applies to cyber space.

Mutual transparency may help keep the peace (in much the same way that formerly warring sides exchanged hostages), but no state (not even a friendly one) exposes the secrets of its security architecture to another. Besides, if the war is still *sub rosa* both sides have amply demonstrated the virtue of transparency. If it did, the transparency would have to be bilateral rather than public, lest mischievous third parties profit from the new-found knowledge. Even then, each side could attack the other from third parties outside the transparency agreement.

Thus, the least problematic outcome is for neither side to find any especial reason to commit serious resources to breaking the systems of the other. This may ensue because the broader ends that led at least one of them into cyber war in the first place have been met or because further cyber war will get no party closer to meeting them than the last spate of cyber war did.

The part of the equation in which one side decides that the effort no longer pays is not strategically problematic because it does not require the other side to recognize that anything has changed. But it is hard to believe that the party that quit making the effort would not hope to see some rewards for its restraint. As long as the one side had not made either explicit (that is, negotiated) or implicit commitments to restraint, the other side would not be able to hold up some future system malfunction as evidence that it had been lied to or cheated. Furthermore, if the other side still found advantage in computer attacks – or if it was engaged in other forms of hostilities – it may have no motive to acknowledge such restraint. But if the other side also finds that the advantages of hacking have waned or that they are trumped by the rewards of friendly engagement, it too might work itself into a *modus vivendi*.

3. Conclusions

Cyber space is a medium in which the absence (or, more specifically, unimportance) of physical artifacts permits a form of warfare that is generally unavailable in other media. The closest analog to *sub rosa* warfare would be a campaign of espionage, but even there, the potential exposure of the saboteurs (whilst hackers can be sheltered by the attacking country or in the anonymity of the Internet) makes it hard to keep matters quiet for terribly long. That such *sub rosa* warfare is possible, however, makes it neither probable nor particularly wise. Paradoxically, maintaining *sub rosa* warfare requires the tacit assent of the other side, and is therefore quite fragile. More practically, the very shadowy nature of the whole enterprise (coupled with the difficulty of getting policymakers to understand the requisite ins and outs of cyber war in general) creates enormous temptation to take risks without adequate political consideration of their cost.

The Virtual Battlefield: Perspectives on Cyber Warfare
C. Czosseck and K. Geers (Eds.)
IOS Press, 2009
© 2009 The authors and IOS Press. All rights reserved.
doi:10.3233/978-1-60750-060-5-66

Warfare and the Continuum of Cyber Risks: A Policy Perspective

Andrew CUTTS
U.S. Department of Homeland Security

Abstract. At the highest levels of national government, two of the most important decisions to get right are properly prioritizing among competing missions, and balancing between short-term and long-term objectives. The most consequential and highest risk threat is attack by one or more nation-states intent on projecting power, and who are willing to damage or destroy critical information infrastructure by cyber means in order to achieve this objective. Threat actors falling into this category have the necessary time, resources, sophistication, and access to do so. This category certainly includes cyber warfare. Today, nation-states are beginning to understand in concrete terms the potential benefits and costs of cyber attacks used as a means of projecting national power. It may not take a great deal of a nation's cyber resources, planning time, or technical access to achieve limited national objectives.

In the U.S., cyber defense of critical infrastructures is largely a homeland security mission. It may be that defense always lags the most potent offense. But the goal is an *effective* defense, not a perfect one. To get ahead of the most serious national cybersecurity risks, including that of cyber warfare, a country's cybersecurity leadership must seek an appropriate balance of resources, energy, and focus between those threats that are most frequent and those that are most consequential. The historical bias in dealing with cyber risk has been to look at it through the lens of commerce, not national security – and to reinforce the emphasis on short-term thinking rather than long-term strategy. One way to overcome this bias is simply to emphasize efforts that mitigate the most consequential risks. A nation's cyber leadership could decide, for example, that it should apply significant early resources to mitigating the national security risk associated with defending critical infrastructure against nation-state threats.

Keywords. Cyber warfare; critical information infrastructure; cyber risk; cybersecurity policy; cybersecurity; homeland security; cyber attacks

Author's note

This paper offers a framework for thinking about and debating vital national cybersecurity policy issues. It makes no attempt to settle those issues. Its primary purpose is diagnostic. To the extent it offers prescriptions, it does so only to forward thoughtful debate and discussion. It does not reflect a settled position of the Department of Homeland Security or of the U.S. Government.

Homeland Security

Introduction

The cyber attacks against Estonian networks in 2007 were a wake-up call for information-based societies in general, and for the North Atlantic Treaty Organization in particular. Those attacks demonstrated that protecting classified networks and defense-related communications, while very important, is insufficient for an information based nation-state.[1] They forecast the risk that critical information infrastructure owned and operated by the private sector, including that which supports energy, transportation, banking, communications and the media, could in the future be the target of cyber attacks by a strategic opponent. The defense and security of these networks is in the national and public interest. A country's national security and economic well-being are at stake.

NATO as an organization can do much to establish and enhance a common cyber defense among its members. Yet it is largely up to each nation to protect its own respective networks and infrastructure from cyber attack. In this respect, each of our countries is very much in the same boat. To a greater or lesser extent we each face similar challenges in protecting nationally vital information infrastructure.

1. A Simple Conceptual Framework

At the highest levels of national government, two of the most important decisions to get right are properly prioritizing among competing missions, and balancing between short-term and long-term objectives. This is true in monetary policy. It is true in energy and environmental policy. It is vital to military success. And it is no less vital to national cybersecurity efforts.

Dr. Paul Bracken, a professor at Yale University, once wrote of the need to "model simple, and think complex."[2] Simple models help us to think about the complexity of what they reveal. A simple model[1] the author has found useful for thinking about and communicating the need to balance between competing national cybersecurity priorities is the cyber risk continuum in Figure 1 below[2].

The graphic depicts a range of cyber threats, increasing in both potential consequence and risk from right to left[3]. In this case, consequence can be thought of as the potential for harm to a nation's security or economic well-being. These threats are not unique to the U.S. but rather are faced to some extent by any information-based society. The graphic conveys that the threats of highest consequence, and their associated risks, are also likely to occur with the least frequency.

On the far right of the continuum are nuisance threats including "script-kiddies" and hackers. Next in increasing consequence is cyber crime. Criminals, of course, are

---

[1] The term "model" is used to convey that this is not a quantitatively-driven plot of data.

[2] The model is not backed by a data-driven assessment of risk. It is based on anecdotal evidence.

[3] This paper frequently uses the term "cyber risk", which differs from a "cyber threat". Risks are combinations of threats, vulnerabilities and consequences. A discussion of specific vulnerabilities is outside the scope of this paper, which focuses on two of the three factors in determining overall risk – namely "threats" and levels of "consequence". The paper assumes vulnerabilities exist, and that threat actors differ in their capability and motivation to exploit these vulnerabilities in ways that might harm a country's economic or national security.

Homeland Security

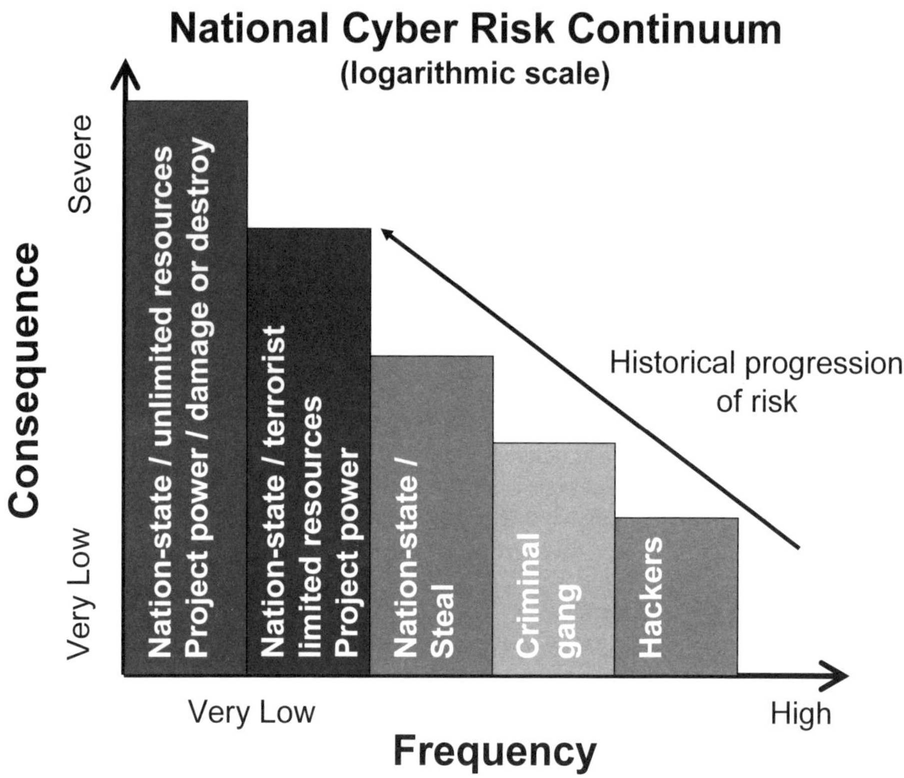

Figure 1.

motivated to make money and are willing to break national and international law to do so. Cyber criminal gangs are increasingly sophisticated and the economic cost of their illegal activity – most notably in the banking and finance sector – has greatly increased over the last five years.

Nation-states that are capable of and motivated to steal intellectual property or state secrets (espionage) pose the next threat on the chart. This is due largely to the potential resources, sophistication, and motivation they apply to achieve their objectives.

The next most consequential threat on this continuum is that posed by either nation-states or non-state actors, including terrorist organizations, which might be motivated to project power through cyber attacks on critical infrastructure. This category includes threat actors who have limited resources, time, and access to accomplish their objectives.

The most consequential threat is attack by one or more nation-states intent on projecting power, and who are willing to damage or destroy critical information infrastructure by cyber means in order to achieve this objective. Threat actors falling into this category have the necessary time, resources, sophistication, and access to do so. This category certainly includes cyber warfare.

Homeland Security

It is important for national decision-makers to understand that cyber warfare exists on a continuum of risk. It is not going too far out on a limb to say that risks have tended to increase over time. Effective mitigation of lesser threats does not necessarily mean that more consequential threats, including cyber war, are also mitigated. It is obvious, but should be stated nonetheless, that a nation's tolerance for cyber risk relates directly to its economic and security dependence on information infrastructure.

2. Limitations of the Model

It has been wisely said that "all models are wrong, but some are useful."[3] In that spirit, the author acknowledges that the continuum is overly simplified. It shows only a few threat categories from among many combinations of threat actors and capabilities. It treats these as distinct from one another, whereas in reality nation-states, terrorist organizations, hackers, and cyber criminals might use one another, perhaps unwittingly, to achieve their respective objectives. And of course it is often impossible to tell one threat actor from another in cyberspace, given the inherent difficulty of attributing attacks to their ultimate source. Still it is useful as a construct for thinking about the problem, and for conveying some of its strategic implications to senior decision-makers.

First, it conveys the simple progression of the cyber threat – a key component of risk. Early on in the development of cyber infrastructure, we were only exposed to risk from script kiddies and hackers. As more businesses connected to the Internet, sophisticated criminal gangs emerged. Within the last few years nation states are reported to have stolen massive amounts of data from defended networks. In other words, the progression of risk has steadily moved to the left. The author believes the underlying reason for this is economic. At each step the benefits of cyber attacks to those who conduct them have outweighed the costs they incur.

This was acceptable when the potential consequences were low. But they are increasing. The current state of affairs is clearly unacceptable. In the U.S. it has been called a national security crisis.[4] Today, nation-states are beginning to understand in concrete terms the potential benefits and costs of cyber attacks used as a means of projecting national power. It may not take a great deal of a nation's cyber resources, planning time, or technical access to achieve limited national objectives.

As every businessman knows, past performance is not necessarily an indication of future activity. One cannot predict that just because the progression of risk has moved steadily and swiftly up the continuum, it will continue to do so. But an obvious question for national policy makers is this: "What set of factors might stop this progression?" A corollary question: "Should we assume the progression will continue unless the economics of the problem changes – unless costs to potential attackers can be introduced to clearly change their cost/benefit calculation?"

One way to change, or at least slow, the vector of risk is to raise the cost of attack by enabling a better defense. In the U.S., cyber defense of critical infrastructures is largely a homeland security mission. It may be that defense always lags the most potent offense. But the goal is an *effective* defense, not a perfect one. The exact nature of that defense will vary from country to country, sector to sector, network to network, and threat to threat. It will depend on strategic objectives, an assessment of strengths and

Homeland Security

weaknesses, available resources, national capacity for research and development, and tolerance for risk, among other factors. To be effective, a national cyber defensive capability, commensurate to the level of risk, should exist before a country experiences the high consequence threat on the left hand side of this continuum.

3. An Apt Metaphor

Several years ago a United States Senator, The Honorable Robert F. Bennett, gave a keynote speech at a conference on cyber conflict. He used a sports metaphor to convey a key point. Referring to Wayne Gretzky the famous hockey player, Senator Bennett alluded to the fact that Gretzky had such a sense for the flow of the game, he could anticipate ahead of other players where the puck would be – and he skated to that position on the ice. He arrived before the puck, and was then in position to help the team score.

The author believes the Senator's point was this: an information-based society like the U.S. cannot protect its information infrastructure from the worst cyber risks unless it makes a concentrated effort to get ahead of the threat. This point is vitally important. Yes, our countries must address the threats we face today; we cannot neglect them. But for some countries it could take years, perhaps decades, to build an effective defense against the most consequential risks on the continuum. It is one thing to protect government networks. It is entirely another to protect non-government networks against nation-state cyber threats. Building a national capacity to do so will not happen overnight. And that raises another vital question for any national policy maker: "How long do we have before the most consequential threats might materialize?" Whatever a country believes that timeframe to be, if it has no effective defense in place before then, it assumes a very great risk indeed.

4. A Policy Imperative

The cyber threat never stops. Our respective operational echelons and cyber defenders have no time to come down from the ramparts. Their typical day is filled with efforts to mitigate current and near-horizon threats. But over-the-horizon risks will not disappear. Operational cadres may not have the time or present capability to deal with them, but these risks deserve more than a fleeting glance when operations allow. A country whose tolerance for cyber risk is low should devote the resources necessary to understand the most consequential threats and address the risks they pose.

Nobel Prize-winning economist Herbert Simon once said, "Short term thinking drives out long term strategy, every time."[2] This insight is the economic corollary to the Gretzky metaphor. It certainly rings true in the field of cybersecurity.

The national cyber risks faced by an information-based society are great. They may seem far in the future; but the most consequential risks must be mitigated today with action that is direct and decisive, not oblique or incremental, regardless of their frequency. Proactive steps to mitigate over-the-horizon risks will be much less costly to commerce and national security than allowing these threats to materialize. Recent

Homeland Security

experience in other policy domains, including finance and hurricane preparedness, has proven this point.

To get ahead of the most serious national cybersecurity risks, including that of cyber warfare, a country's cybersecurity leadership must seek an appropriate balance of resources, energy, and focus between those threats that are most frequent and those that are most consequential. Creating the conditions in government where infrequent threats can be understood and addressed is easier said than done.

In each of our countries, the organizations that have defensive cybersecurity responsibilities perform one or more of three different missions. They fight cyber crime. They defend government networks, including those that are used by civilian agencies, the military, and the intelligence community. And in some cases they must help protect non-government networks that qualify as critical national infrastructure. As we all know, these are mostly owned and operated by the private sector. They include the data, hardware, software, and control systems that undergird our financial markets, the generation and distribution of electricity, modes of mass transportation, and our vital telecommunications. They support thousands or millions of competitive business models - each one unique; and they are operated mostly with economics in mind.

This last mission raises two additional policy questions for many of us. The first is this: "Against which cyber threats on the continuum should our governments be held responsible for protecting the private sector?" At every point on the continuum, commerce is vital. So are civil liberties. Clearly, the bias at the lower end of the risk spectrum should be weighted toward private enterprises taking the lead for managing these risks as they relate to individual business models. Equally clearly, no private enterprise – no matter how well capitalized – can bear the cost of defending itself against destructive nation-state attacks. In this case the opposite ends of this continuum are a bit like the opposite poles of a political spectrum; it is fairly easy to see what exists at either end, but it is much harder to characterize the middle.

5. Drawing the Line between Security and Defense

This leads to the second question: Where on this continuum should a country's leadership draw the line between security and defense? Where does one stop and the other start? A country cannot debate this question forever. Leaving it unanswered leads to a situation in which no one – not the defense community, not the security community, nor the private sector, is clear about responsibilities. Lack of mission clarity leads to lack of authority, resources, and capabilities. And that comes at the expense of neglecting the high end of the consequence spectrum.

Figure 2 depicts one way of thinking about the difference between cyber "security" and cyber "defense" – at least for a western democracy such as the U.S. It shows two parallel lines - both of which are drawn rather subjectively[4]. The area below the grey

[4] Exactly where the security and defense boundaries should be drawn in relation to the continuum of threats is worth careful policy debate and consideration. In this case they are drawn for illustrative and discussion purposes only; in reality they could be higher or lower. Moreover, it may well be that the higher end of the "security" boundary is not static across all threats, but rather it increases in stair-step fashion as the threat increases. By this is meant that the private sector might be enabled to participate at higher and higher levels of capability in their own (and the national) defense as threats and risks escalate.

Homeland Security

line is labeled "security". In this case, the term "security" indicates that the clear bias should be toward expecting private enterprises to bear primary responsibility for managing risks in this range. Naturally they would do this primarily out of fiduciary responsibility to their stockholders – but also in some cases as part of a regulatory framework.

This does not mean that the private sector should be *solely* responsible for mitigating risks associated with threats at the lower end of the spectrum. Cyber crime, for example, is an area where the private sector must work with law enforcement to address the threat adequately. Many national Computer Emergency Readiness Teams (CERTS) also provide incident services to the private sector that help them mitigate risks even from the lower tier of cybersecurity threats.

It does mean, however, that both government and industry should agree that the primary metric through which the risks associated with these threats are cooperatively managed is that of maintaining competitive business models. Mitigation efforts must sustain profitability for individual businesses, value chains, and complete sectors of business activity.

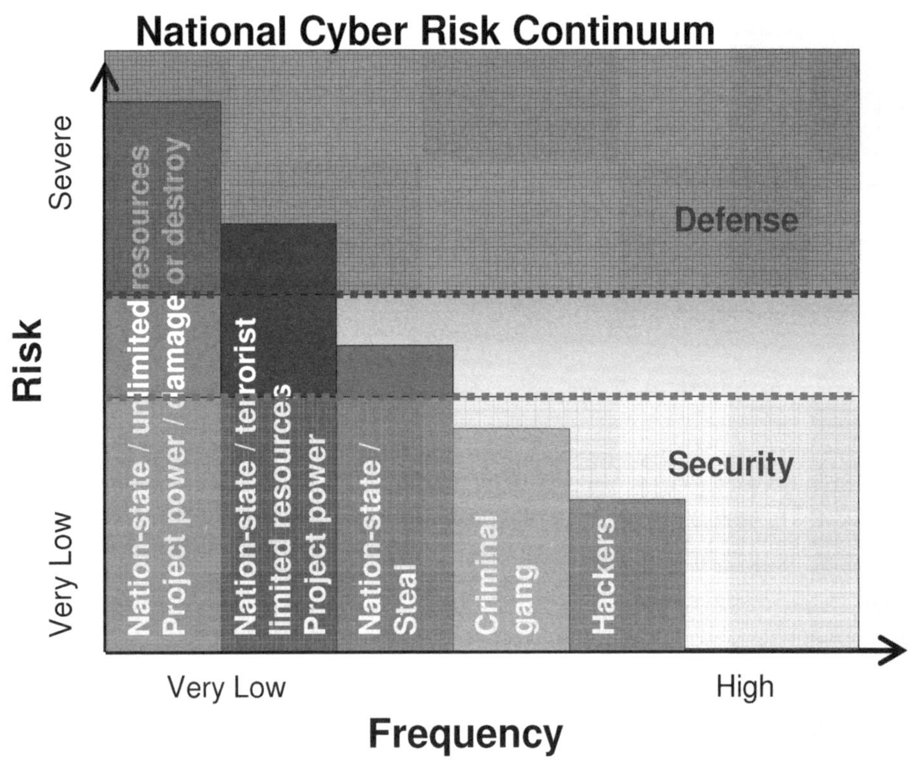

Figure 2.

Homeland Security

Note that on this graphic the "security" domain extends across the entire spectrum from left to right. One important conclusion is that private enterprises – at least those that operate critical infrastructure - must have some responsibility, presumably backed by demonstrated capability – to assist in mitigating all categories of cyber threat - even those that contribute to the most consequential national cyber risks.

The extent of that responsibility is a question of strategy. At its most basic, the central question is this: should privately-owned critical information infrastructure be defended against the most consequential threats from "within the enterprise" – meaning by experts who operate and know the importance of each byte on those networks on a daily basis? Or should they be defended by government cybersecurity experts, using nationally developed capabilities from outside the normal perimeters of most business networks? Both have advantages. Both also present very difficult challenges. A balanced strategy might include both, but the question then becomes: what is the appropriate balance between the two? Answering this question is fundamental to national cybersecurity efforts.

The area above the blue line is labeled "defense". In this case the term conveys that risks resulting from these threats are systemic and could be nationally debilitating. Government should be primarily accountable for addressing this tier of threat, even though operational responsibilities must inevitably be shared between government and the private sector. Indeed, the threats at this end of the spectrum are so potentially severe that their mitigation should be thought of as a *mission to be performed* rather than as *something to be managed.*

These threats are simply beyond the scope of private sector capability to adequately address. Both government and industry should agree that the primary lens through which these high-tier threats are addressed is one of national security. If national security is at risk, commerce is also at risk. There should be no question about priority in this domain. National security must take priority over commercial interests for those who are assigned responsibility to manage the highest consequence threats, whether that assignment is given to the defense community or the homeland security enterprise.

6. The Sticky Middle Ground

In between is a shaded area in which the delineation between security and defense is blurred. It is in this area where roles and responsibilities between government and industry are most difficult to define. Addressing threats that fall into this domain offer the most difficult decisions. This is true for two reasons.

First is the strong potential for conflicts of perspective and for competing interests. The private sector must compete; fiduciary responsibilities require businesses to maintain their focus on the bottom line. In most cases, their risk horizons are invariably short. On the other hand government must support and enable commerce, but not at the expense of providing for the common defense – a constitutional requirement in the U.S. It must take necessary steps to manage long-term risks. Inevitably, tension exists between these differing perspectives and responsibilities.

Second is the extent to which mitigation decisions for risks in this middle tier involve unknowns. The complexity of cyberspace; its tendency to create unforeseen

Homeland Security

interdependencies; the way it immediately transmits and links impacts of decisions made remotely across great distances and geographic, political, and organizational boundaries; the potential for small, hidden vulnerabilities to result in highly leveraged consequences; and its newness as a domain for conflict, all greatly increase the fog of risk management and crisis decision-making. This is especially true in the domain of risks that occupy the middle of the consequence spectrum.

This leads to another important insight. It is incredibly easy to get bogged down in mitigating mid-tier risks. Recall that the progression of threat – and by extension the progression of risk- reached this middle tier by starting at the bottom end of the consequence spectrum. The historical bias in dealing with cyber risk has been to look at it through the lens of commerce, not national security – and to reinforce the emphasis on short-term thinking rather than long-term strategy.

These factors, together with the tendency for conflicting perspectives and difficult decision-making against this middle tier of threats, *create an operational environment in which the struggle to devote any meaningful time and effort toward getting ahead of the most consequential threats is a real challenge.*

One way to meet this challenge is simply to emphasize efforts that mitigate the most consequential risks. A nation's cyber leadership could decide, for example, that it should apply significant early resources to the left end of the continuum – to mitigating the national security risk associated with defending critical infrastructure against nation-state threats. Over time it could capitalize on these resources by applying them against lesser risks. In this way it could ensure that it does not grind to an operational halt short of accomplishing its highest strategic priorities. It also could gain the most value from its resources.

Other ways to emphasize the most consequential threats and risks: (1) develop a long-term war-gaming practice that continually refines the policy, legal, economic, operational, and technical issues associated with the high end of the continuum; (2) ensure planning scenarios for exercises and war-games focus on these threats; and (3) require that national and sector risk assessments cover the entire risk continuum.

A logical step for any country's cyber leadership is to undertake a continual effort to assess risk across this spectrum. Part of this effort should include identifying the subset of discrete vulnerabilities in critical information infrastructure, which if exploited would have the most debilitating consequences to national or economic security. Developing an appropriate strategy for mitigating each of these discrete risks should be a joint effort between government and the private sector. But government owns the responsibility, and should have the authority, to say when or whether the most severe risks have been acceptably mitigated.

7. Capability v. Mission

Of course, delineating risk responsibilities between government and the private sector is only part of the solution. In large bureaucracies such as in the U.S. the imperative to clearly define defense from security exists for another reason, and that is the need to clarify missions between government agencies, and assure each mission is supported by adequate capability.

If one's mission responsibilities are unclear, it is impossible for one to know if his capabilities are sufficient.

Homeland Security

Figure 3 shows the threat continuum overlaid against an assessment of capability vs. mission. It supposes the existence of an organization with a given risk mitigation capability. If this organization is assigned a mission of defending critical information infrastructure against criminal threats and theft of intellectual property, Mission (A), then its capability is sufficient. If however, its mission includes defending the same infrastructure against destructive cyber attacks at the high end of the consequence spectrum, Mission (B), then its capability is woefully inadequate and leaves unmitigated risks.

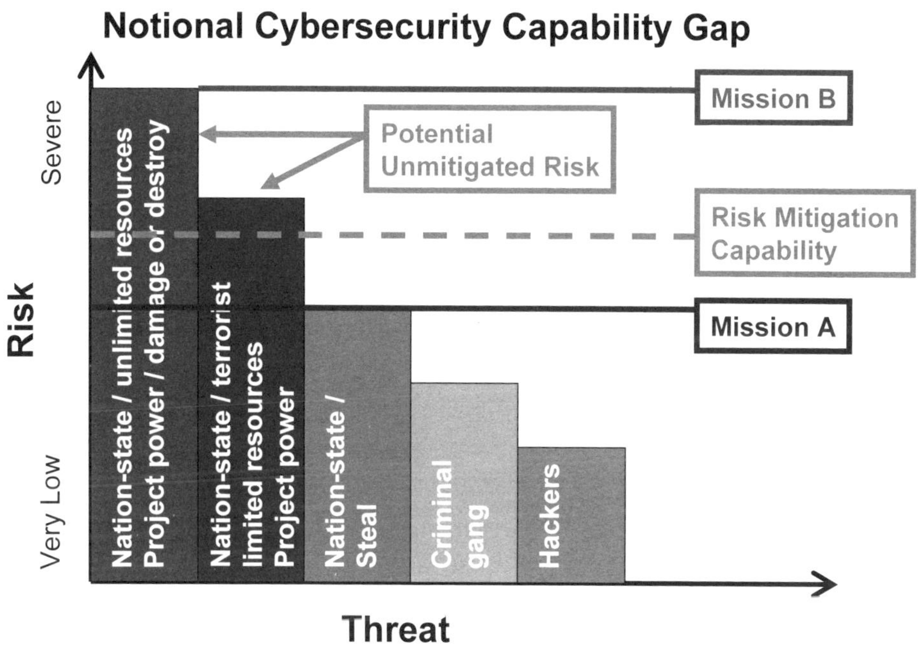

Figure 3.

By inference, it is vitally important for national leadership to understand which threats on the continuum it is most interested in addressing, and receive a straightforward assessment of national capabilities mapped against those threats. This is true for individual government organizations as well. Any delta between mission and capability deserves the focus of decision-makers.

8. Summary

For which discrete categories of threat on the continuum are the defense community, the security community, and the private sector ultimately responsible? Policy leadership in any country must make that clear. Only then can operational leadership ensure that short-term actions and operational objectives measure up against appropriate long-term strategy.

Homeland Security

References

[1] "Defending Against Cyber Attacks". 2009. North Atlantic Treaty Organization. 9 July 2009 http://www.nato.int/cps/en/SID-D30474D0-A97D6B01/natolive/topics_49193.htm

[2] Bracken, Paul. "Net Assessment: A Practical Guide," *Parameters* (Spring 2006), p. 100. http://www.carlisle.army.mil/usawc/parameters/06spring/bracken.pdf

[3] Box, George E. P.; Norman R. Draper (1987). "Empirical Model-Building and Response Surfaces". Wiley, p. 424. ISBN 0471810339. Wikiquote. 9 July 2009 http://en.wikiquote.org/wiki/George_E._P._Box

[4] Gross, Grant. "U.S.Government Focuses on Securing Backdoors in Tech Product". *Security Central* (September 2008). Infoworld. 9 July 2009 http://www.infoworld.com/d/security-central/us-government-focuses-securing-backdoors-in-tech-products-853

THE INTERNATIONAL INSTITUTE
FOR STRATEGIC STUDIES
ARUNDEL HOUSE
13-15 ARUNDEL STREET
LONDON WC2R 3DX

Homeland
Security

The Virtual Battlefield: Perspectives on Cyber Warfare
C. Czosseck and K. Geers (Eds.)
IOS Press, 2009

© 2009 The authors and IOS Press. All rights reserved.
doi:10.3233/978-1-60750-060-5-77

Cyber Terrorism: A New Dimension in Battlespace

Major J P I A G CHARVAT
SO2 Course Director
Centre of Excellence Defence Against Terrorism

Abstract. This paper discusses the concept of terrorism, who the terrorists are and develops an understanding of why they conduct the activities they do. Understanding the *mens rea* of the attacker will allow consideration of the type of attack they may plan and the effect they are likely to try and achieve. It looks at the main motivations of terrorist groups and discusses their use of the Internet for various aspects of a terrorist campaign such as propaganda and recruitment. It will consider the various tactics that have been used and how the Internet has provided a new opportunity for terrorists to conduct their campaigns and how it has been adapted by them for their purposes. It examines the potential threat of a cyber attack by terrorist organizations and how they can use the Internet and Cyber Space to attack a target with similar results to a conventional physical attack. The paper will briefly discuss some of the possible defences against this form of terrorism.

Keywords. Terrorism, Terrorist motivation, Cyber attack, Terrorist use of the Internet

Introduction

Since the 2001 attacks on the United States there has been a significant effort from NATO and its partners to address the issue of International Terrorism. Terrorism has many different types and there is no profile that consistently fits either a terrorist organization or an individual terrorist. Terrorism is an adaptive threat and will constantly look for weak points in the state or organization it is attacking. Cyber Space and the Internet are providing an emerging battleground that many terrorist organizations are trying to exploit as a means of furthering their campaigns or actual attacks using an electronic medium. As developed societies become increasingly reliant on electronic communications, control systems and commerce the potential for a terrorist to hit the target becomes a more realistic possibility. As with a conventional attack or command and control medium, the Internet now offers genuine targets that will become attractive to certain terrorist organizations for certain acts. Just as when defending against conventional terrorism, cyber systems can be secured and any potential threat investigated using electronic evidence to catch the instigators.

1. Defining Terrorism

Most people would say they know what terrorism is, but surprisingly there is no internationally agreed definition. There are literally hundreds of different definitions in current use with the use of violence or threat of violence being the only general common theme [1]. The only other elements to appear in more than 50% of definitions are "Political" and "fear, terror emphasized" [2]. This is a hamper on international cooperation as some terrorist organizations are seen as legitimate fighters by some countries. Terrorism does differ from other crimes in its *mens rea*; it is done with a purpose and a specific strategic outcome in mind.

If there has been significant international and intellectual disagreement about a definition for terrorism itself, the disagreement as to what, if anything, constitutes Cyber Terrorism is even more diverse. This paper will consider the battle against terrorism in Cyberspace, Cyber Terrorism as a form of attack and the terrorist use of the Internet as a tool for physical action. It will also consider areas of both anti and counter terrorism within a cyber environment.

It is important that we consider who the terrorists are. There are literally hundreds of groups of varying size and ability, which, to some extent, warrant the label of terrorists. Terrorism has 4 classic motivations [3]. Firstly there are single-issue terrorists, those who believe in a particular cause and are prepared to use violence to protest their message in the hope of ending the issue, which sparks their grievance. Animal Rights and Anti-Choice over abortion are the two most prevalent of such issues. Vivisection researchers or family planning workers have been the targets of sustained campaigns and assassinations in protest over these issues. Although generally small and with a low lethality rate, these groups could find the cyber world particularly to their liking as in the cyber environment they can effectively punch above their weight. Ideological Terrorists are those who use violence to promote their political ideology, usually from the far left or right. While these groups were most active in the Cold War, there are still several of such groups still active[1] and other active groups that have evolved from their beginnings as an Ideological Terrorist group.[2]

Nationalist terrorists have been the most lethal of all terrorist groups over the last 40 years[3] [4] and are still active in several major campaigns worldwide. These are terrorist groups who seek independence from a state or to cede from one state to another because of ethnic or geographic grievances. Very few modern nations are made up of just one ethnic group and in many areas of the world this has led to ethnic tensions that have spilled over into terrorism. The LTTE in Sri Lanka and the PKK Kongra/Gel[4] terrorist organization in Turkey are the most active of such groups.

Religio-Political terrorist groups tend to be more lethal as they believe they are acting for God or on a divine order and that those not of their belief are against God [3]. There are extremist groups spanning all major religions and some minor cults who have resorted to terrorism. These terrorists have abused their religion and act outside it, they must not be confused with the religion they misrepresent in their claims. Although many religions do accept that there are circumstances for justifiable violence or warfare, none, with the exception of a doomsday cult such as Aum Shinrikyo, would

---

[1] FARC (Revolutionary Armed Forces of Columbia) are probably the most active today.

[2] The PKK Kongra/Gel terrorist organisation, active in Turkey, began seeking a socialist revolution in that country.

[3] Religio-Political groups have been more lethal over the last 5 years.

[4] The EU lists LTTE and PKK as terrorist organizations.

apply this to the indiscriminate targeting of civilians or security forces outside the legal conventions of legitimate warfare.

As with any definition labelling model, there can be hybrid terrorist groups that either evolve their motivations or have multiple aims. The Provisional IRA are an example, they were a Nationalist group as they wanted Northern Ireland to cede from the United Kingdom to the Irish Republic but were also an Ideological group as they wanted Ireland to become a Socialist state.

The terrorists themselves must be considered, understanding their psychology is important in understanding how to defeat them. There is no clear profile of a terrorist, they come from all walks of life and have varying levels of education, employment and wealth. One common factor is that they are not mentally unstable, terrorist organizations want activists with the ability to think and be reliable. The level of intelligence may decide the role of the terrorist, as will any specialist skills such as chemistry or IT, and the organization will require College level members as well as those with more basic standards of education. We must accept that most terrorist groups are made of skilled and intelligent people who are acting out of genuine belief (self-formed or indoctrinated) and not a group of clueless idiots. This must be considered in the cyber defence plan against terrorism, they will study, take time, plan and employ experts of the highest calibre to achieve their aim.

2. Terrorist use of the Internet

Terrorists use the Internet for a variety of reasons that although nothing new *per* se make their business far simpler and have a far wider reach than through non-electronic means. Cyber Space offers many areas of potential exploitation to terrorist organizations. It is a tool for recruitment, radicalization, propaganda and fund raising as well as offering quick and simple command and control. Undoubtedly the Internet and electronic mediums offer many advantages to terrorists, given the ease of use and the increased reliance of developed societies on the Internet the potential for terrorist exploitation is expanding daily.

The first area of concern is the terrorist use of the Internet. By this I mean a medium for many aspects of terrorism other than as an attack tool, which will be discussed later. There are many features of terrorism that can be conducted through the Internet, although this is primarily an information and communication medium it does allow a greatly extended reach and far quicker communication. Terrorism is a politically motivated act and therefore requires publicity and a forum of communication. The key areas of exploitation in modern terrorist use are propaganda, recruitment, radicalization, communication and research. The Internet allows small groups or individual terrorists the opportunity to reach literally millions of people very easily. "Perhaps one of the most promising features of the Internet is that it gives voice to many who have been unable to buy or generate media attention" [5].

Propaganda is essential to terrorism, these are rational people who carry out their campaigns for a political or social end state, they must have a medium to explain their message and 'justify' their actions. Before the Internet this was relatively difficult to achieve for a mass audience. Television and print media would run stories on terrorists but these would be subject to editorial control and sometimes to legal restrictions. Books, magazines and pamphlets would only hit a small audience and would only really be read by those with an interest in the terrorist's cause. The World Wide Web is unregulated and accessible to almost everyone. Internet Cafes are increasingly popular and allow access to the public without the need for the ownership of a computer or subscription to an Internet Service Provider. Simply by adding the correct key words or links via a seemingly mainstream site, the terrorist organization can easily display its message without regulation. It can allow the key grievances, which motivate the terrorists, to be publicly aired, and as they control the content, these can be backed up with 'proof' manufactured and edited by the organization. These sites look official and will use multimedia to attract attention and create an air of legitimacy for the site visitor. They also have the advantage that their Information Operations are not bound by truth or conventions, which allows, with appropriate editing, a seemingly convincing piece of propaganda without any real basis. By May 2005, using only the US State Department's list of terrorist organizations, there were over 4500 terrorist supporting websites [5], rising to 5500 in 2007 [6].

An example of how easily websites can be used for misinformation is the German *Bund Deutscher Juristen*[7]. This website for the German Lawyers' Association ran an article about their Chairman, Dr Claus Grötz, quoting that he said the possible use of testimony gained after light torture could be used in German courts. There was public outcry and headline news calling for his resignation. The site had only been set up 2 days before the 'story' broke and neither the Association nor Dr Grötz actually existed. Although this is not a terrorist example, it highlights the possibilities that a terrorist organization could exploit. Lazy journalism meant that the story was not corroborated and given to the German public as authentic by the mainstream media. To have this about a State's action relating to a terrorist situation could be used by the terrorist organization to gain public sympathy and international condemnation of the victim state.

Figure 1. Screenshot of BDJ website and translation

Recruitment and Radicalization are an essential element for a terrorist organization. The Internet provides a greatly enhanced forum for this. The ability for terrorists to find and groom young people is demonstrated in Forum and Chat Room websites. This provides a largely unregulated medium for terrorists to meet and groom potential recruits. Often they will monitor those Forums and Chat Rooms that may have a relevance to their motivation, grievance or cause. This could be an animal rights Chat Room where extremists Single Issue terrorists may use the opportunity to engage anyone who shows thoughts or emotions along the same lines. This form of contact can be well orchestrated and involve several people, effectively keeping the potential recruit in an air lock away from the terrorist proper until they are deemed ready. Initially a pro-terrorist 'chatter' will engage the potential recruit in fairly mainstream conversation and ask a few probing but seemingly innocent questions over a period of time. They will use this time to pass on pro-terrorist messages and try to affirm the potential recruit's feelings to that particular cause. The will also post messages against the terrorist's targets in an attempt to convince the potential recruit that the terrorist message is accurate.

Once regular contact is made, the initial contact will assess the potential recruit and pass them on to a groomer. Those selected for such grooming have already displayed some form of agreement with the terrorist cause and also the personality traits that suggest they may be willing to take an active part in any struggle. Still at this point the potential recruit is unlikely to know she or he is in contact with anyone other than a fellow chatter of a like mind and engaging in serious conversation. The groomer is likely to be very knowledgeable about the cause and will start to feed strong propaganda about the terrorist's motivation. This is again a filtering process to find out those who would continue towards direct action from those who have strong views but would never cross from political protest. Those who are regarded as strong believers in the terrorist cause, and displayed the correct personality traits to suggest they would join are then passed on to their first proper contact with the terrorist organization itself. This process can take a long time, as the groomer has to be certain they have the right people and the potential recruit is not going to be made aware that he or she is in contact with a terrorist until they are ready.

The groomer will pass the potential recruit onto a recruiter who, at that stage will, for the first time, make indications that they are from a terrorist organization. From this point the skill set of the potential recruit will be examined and their commitment finally checked before they are in a position to ever actually meet or know the identity of anyone they have been engaged with.

An example of this, in its early stage was monitored on a mainstream Muslim Youth website in the United Kingdom.[5] Hussain was a 15-year-old schoolboy who posted a school project for comment from other members of the forum site. In his post he expressed mixed feelings and uncertainty about how the West saw Islam and the true nature of the Jihad. Hussain stated that he believed Jihad was a personal struggle and it was against Islam to kill. He also expressed that he felt as though the West regarded him as a terrorist because he was Muslim and that there was significant anti-Muslim sentiment in Western Society. The first two replies agreed with Hussain that in Islam it is forbidden to kill innocents. OBL4Caliph entered the debate and began saying that he was an authority on Islam and that Jihad was a duty for all Muslims and that it was a requirement to kill those who opposed the religion. During the ensuing posts it was clear that most forum members said OBL4Caliph was wrong. However his language and argument were more structured to a youth's mind and he began to try and convince Hussain. While clearly Hussain had used the Internet for a sensible and

[5] This case was monitored by the author, the website name is withheld as it has no association with terrorism and quickly banned OBL4caliph from the site. There is no evidence OBL4Caliph directly represented or was a member of any terrorist organization.

reasonable purpose, canvassing views of like-minded people about his thoughts as a confused teenager, he had inadvertently shown a little potential in his thought, which led to a potential terrorist grooming.

Terrorism has evolved and in the beginning of the 21st Century we are dealing with a new type of terrorist organization as well as the classical groups. Traditionally terrorist organizations have been just that, an organization with a leadership and strict control. Attacks and campaigns would be planned and authorized by the leadership as part of a coordinated approach to their policy. The emergence of more 'networked' organizations with a horizontal leadership has made the Internet a breakthrough in Command and Communication. Obviously e-mail is an instant form of communication and can easily be encoded. Seemingly innocent messages can be sent that only the recipient would understand, although a coded message is as old as terrorism itself, email allows much greater speed in delivery and an almost guaranteed receipt. An example was a mail sent by the 9/11 hijacker Mohammed Atta:

> "*The semester begins in only three more weeks. We've obtained 19 confirmations for students in the faculty of law, the faculty of urban planning, the faculty of fine arts and the faculty of engineering.*
>
> *Best wishes from the Professor to all of you!*
>
> *Mohammad*" [7]

Clearly this message seems innocent, but knowing the events of 9/11 and who wrote it, we know this set the attack dates (it was written 3 weeks before the attacks), the 19 confirmations were the 19 terrorists who were carrying out the atrocities, the 'faculties' were codes for the targets and this confirmed which ones, and of course the 'professor' is Osama Bin Laden.

The Internet also offers a unique opportunity for people to meet likeminded others despite geographical separation. Via forums, chat or other emerging methods such as Twitter or Facebook, unconnected and uncommon 'radicals' could easily contact other uncommon radicals and find a common virtual space online.

Messages can be hidden in pictures or a made to look like Spam mail so not to attract attention. Terrorists have also been known to set up an email account and change the password daily, the cells and terrorists will know the username and daily password. Messages can be written, saved as a draft and then accessed by the whole network without being sent. This greatly reduces the possibility of interception or an evidence trail before or after an attack.

Simple symbolism is also a known method of terrorist communication. Many terrorist organizations have websites or will publish video clips on video sharing sites. These can contain a hidden code; simple graphics such as a terrorist holding his or her AK47 in the left hand will be displayed. Having the same graphic with the rifle in the right hand can be a signal for a terrorist cell and be almost undetectable to the intelligence services monitoring it.

Training and Research & Development are essential elements for a terrorist attack. The Internet provides easy and immediate information sharing and research capabilities to a Terrorist organization. There have been numerous examples of terrorists posting training manuals on the Internet, which explain how to conduct attacks and make explosives using readily available high-street ingredients. This has a greater appeal to network organizations with horizontal hierarchies, such as al-Qaeda. These organizations would be content for cells, even with no formal contact with the organization proper, to conduct attacks in their name. Their philosophy is espoused on-line and those who are radicalized to support it could become competent terrorist without physically going to a training camp. Publications such as the Terrorist Cookbook provide the know-how that a budding terrorist would need.

Fund raising for or by a terrorist organization is another area where the Internet provides a quick and simple medium for the organizations. In the modern world of electronic banking this can be achieved directly or indirectly, both via legitimate transitions and illegal means. The Internet offers many opportunities for front business and pseudo-charities to raise monies. It also allows easy transfer, internationally, to make the tracking and freezing of suspect terror funds very difficult. In some examples, charities have been set up for disaster relief, such as the devastating earthquake in Pakistan in 2005. While these charities have done some relief projects, some of the funds were siphoned off to terrorist organizations. If $100,000 is set aside to build a school, the actual relief could be $800,000 with 20% going to terrorism. These 'charities' can be easily set up by terrorist supporters and be indistinguishable from genuine relief work. The Internet allows a wide target audience for donation requests and easy transfer to the 'relief' fund. At a time of such disaster considerable amounts of money can be stolen in this way.

3. Terrorist Cyber Attack

There are many who argue that there is no such thing as Cyber Terrorism proper. Terrorists can use the Internet as discussed above, but Terrorism needs to be a tangible physical attack. I would disagree, there are many features of modern life that are reliant on Cyber Space and present a new opportunity for terrorist direct action, and these opportunities are increasing as we become reliant on computers. There are the possibilities of attacking electronic means such as web-defacement, malware, data mining, training and Denial of Service. As SCADA becomes more widely used and controls important key infrastructure, an attack on these could be as physical as a bomb.

Web-defacement is an easy way to annoy a target or gain propaganda. Unlike a spurious site, such as the BDJ, these attacks alter the data and information on an official site. The ISP and web domain name would prove to be official if they were checked.

While simple altering of an official website gains little more than small scale propaganda, there is a potential to cause real panic. The often given scenario is a terrorist defacing the Homeland Security website and advising people to leave a major city due to a chemical leak. If picked up by the press or enough people, this false message could gain legitimacy and cause panic, with untold casualties in the ensuing rush to leave and the obvious financial implications that would cause. This sort of attack is relatively unlikely to succeed as other forms of warning would not back it up, but if a TV network accidentally picked it up it could be damaging. The increase in popularity of 'New Media' does allow a greater potential for this type of scenario as it as unregulated and can lack the responsibility of confirming a source that traditional media would have. Social contact media such as Twitter could inadvertently spread false information and be taken as genuine. Although most users are responsible and corrective information will quickly appear a botnet giving mass 'tweets' may give legitimacy through weight of numbers [8]. This is a very simple action for a skilled hacker Cyber Terrorist with very little risk or cost.

Malware is an obvious weapon that a cyber expert terrorist could unleash. Viruses and works can bring down systems and networks and cause great disruption to the target. These could render an important operating system temporarily useless or make it malfunction. The potential loss of data through such a virus or work could have a huge implication if targeted correctly.

Data mining is an appealing prospect for a terrorist organization and an area where INFOSEC becomes a priority for governments and security services. Increasingly personal and financial details are held on record in electronic files. Often, for ease of use and legitimate information sharing, these are held on networked systems. Terrorist organizations could attempt to hack in to these systems to gather information about potential targets, financial details or indeed information altering to damage the victim organization. This could be used to identify key individuals to target for assassination or kidnap. It could also find details, which it used to discredit or blackmail key personnel to help with a terrorist activity. Identity theft could allow a terrorist access to bank, identity documents or access control passes which could be severely damaging and greatly assist in an attack. Given the ease and size of modern data transportation mediums, such as flash USB sticks, the loss of this information must be guarded against in all electronic forms. In the United Kingdom the membership list for the right wing British Nationalist Party was made available on the Internet, much to the embarrassment of several high profile members whose membership of this party was proscribed by their employers[6]. There are many such types of information, which could be of use to a terrorist organization. They can also learn about the schedules and locations of targets. According to an al-Qaeda training manual captured in Afghanistan "Using public sources openly and without resorting to illegal means, it is possible to gather at least 80% of all information required about the enemy"[6].

Seemingly innocent programmes and tools on the Internet can provide valuable information to a terrorist planning a physical attack. These are in everyday use by millions of Internet users for wholly legitimate purposes. Programmes are easily available that offer satellite imagery of the world. These offer exact details of often-sensitive areas and can allow accurate target selection and area knowledge without the risk of reconnaissance. Indeed the al-Qaeda computer expert Muhammed Naeem Noor Kahn was captured in Pakistan in July 2004 with a computer filled with photographs and floor diagrams of buildings in the US [8].

If the Internet is brought down in a country or region it will do immense damage to the population, infrastructure and financial sectors of that society. The Internet relies on bandwidth and if targeted with a Distributed Denial of Service, DDoS, this can block the selected servers and effectively jam the Internet. This is relatively simple to do. If enough emails are sent in a short enough time, or enough hits are made simultaneously on a selected website, the bandwidth will fail to cope with the amount of data it is being requested to move and simply clog up, rather like a main street in rush-hour. One method of doing this is through a Botnet. A Botnet is a network of computers that have been taken over as slaves by one master computer. From the master computer a terrorist could direct all the Bots in the Botnet to email or log onto a website simultaneously. A Botnet can be millions of computers all over the world with the

[6] Clearly this is a non-terrorist related example and no inference relating the BNP to terrorism is drawn in any way.

slaves showing no apparent symptoms. Botnets are relatively easy to set up for an experienced hacker and are available for purchase. Although not a terrorist attack, Estonia was the victim of a sustained DDoS attack in 2007. As Estonia prefers to operate as paperless as possible, effectively closing the Internet had a huge effect in that country.[7] This was combined with some civil disturbance relating to the moving of a Soviet war memorial in Tallinn city centre. This could be used by a terrorist organization to bring down a major banking system, Bishop's Gate and the Baltic Exchange bombs in the early 1990's in London highlight the attraction of a purely economic target to a terrorist organization. What the Provisional IRA achieved with a bomb could now be done to a much larger target area with only a laptop.

Supervisory Control And Data Acquisition (SCADA) systems control many major infrastructure systems and are increasingly being relied upon. These provide one of the greatest vulnerabilities to a purely Cyber Terrorism attack with potentially massive physical effect. SCADA systems are used to run power plants, control dams and even city traffic flow by controlling the traffic light system. If these can be accessed by a terrorist organization they can effectively take control of that facility. One of the main vulnerabilities for this is through a mole employee or a disgruntled worker. If a terrorist organization placed members as regular employees of a facility it is possible that they could, given the patient nature of sleeper cells, gain a position of trust in the facility and gain access to the SCADA computer system. From there they could initiate an attack to break a key facility or cause other forms of damage. An example of the potential this has happened in Queensland in Australia. Although not a terrorist attack a hacker got into the Sewerage SCADA in Maroochy Shire Council on Australia's Sunshine Coast. Vitek Boden put in glitches and deliberately released millions of liters of raw sewerage into the water system and sparked an investigation. He was a former employee who had access to the required passwords and knew the system. After his dismissal the council had failed to change the passwords and effectively allowed Boden access to the system [8]. His motives were personal but this shows what a terrorist could do.

4. Conclusions

The face of modern terrorism is ever changing and it seeks new methods of carrying out attacks, propaganda campaigns and recruitment. Cyber Space in certainly a new area of a battlefield and one that terrorist organizations are striving to exploit. There are many advantages to the terrorist to use the Internet for a myriad of essential threads to maintain a terrorist campaign. This threat must not be overlooked, as many societies move more areas of life and infrastructure to computer control and networked systems, the reliance on then is ever increasing. Unfortunately this reliance creates a fairly soft target. The terrorist no longer needs to physically be in the same country, let alone the site of an attack if it is conducted through cyber space. Information is far more accessible and available instantly, something a terrorist could exploit. Taking a photograph of a military instillation will raise suspicion and risk alerting the authorities to the terrorist or that an attack is being planned. Looking at the same site on a computer would leave the terrorist completely anonymous and undetected.

Terrorists such as Younis Tsouli, the infamous Terrorist 007, have become vital to terrorist organizations. Tsouli was not a fighter that offered much ability to conduct a physical attack, but he was committed to the al-Qaeda cause. He set up countless websites and video sharing forums to promote a pro-al-Qaeda message and demonize the US and UK. He operated from a London flat and was financed by another terrorist, interestingly who he never actually met. Tsouli was arrested with millions of files and videos with terrorist propaganda and training aids, which he was the central hub in distributing and putting on line. It was assessed that his arrest as a major blow for al-Qaeda, as much as any active field commander.

It is not all bad news though, by using electronic means, the terrorists can leave a signature and be monitored or arrested based on electronic evidence. It can also be used to monitor terrorist 'noise' [9] as an intelligence-gathering tool. Like any form of attack, a cyber attack is likely to leave some form of signature or evidence that if properly monitored or collected can be used as a counter terrorist tool. There is also some potential to use cyber means to attack the terrorists back, however this has some ethical dilemmas and is not within the scope of this paper.

Defence Against Cyber Terrorism will differ little form solid cyber defence and security. It is however important to understand the terrorist mindset and how they are likely to use the medium for their purpose. It must be conducted as part of the wider force protection and be conscious of the massive potential for compromise. It is highly unlikely anyone would take a large paper file, with highly sensitive information useful to a terrorist, outside a secure office, but this seems to be ignored when that information and a thousand times more is held on a 5cm USB stick. It is not only the system that requires improved security if the defence against Cyber Terrorism is to be successful. The often used comment, TPIBKAC, The Problem Is Between Keyboard and Chair and that there is no patch for stupidity hold true.

Having attended several workshops on Cyber Defence, one area of concern is that many 'experts' believe Cyber Terrorism simply won't happen. It is foolish to write it off, at best it may currently be more 'potential than problem', but it is short sighted to exclude it from risk assessment. Firstly, it may have already, we don't know who may be in place ready to attack a SCADA system, secondly we are increasing our reliance on these systems and their attractiveness as a target grows. Without doubt, the modern terrorist needs the Internet as much as the AK47 and it is a factor we would ignore at our peril.

All views expressed in this article are the authors and do not necessarily reflect or represent the views of COE DAT, CCD COE, NATO or the UK MOD or Government.

References

[1] Record, Jeffery: Bounding the Global War on Terrorism, Strategic Studies Institute, US Army War College, Leavenworth, 2003

[2] Schmid, Alex and Jongmans, Albert et al: Political Terrorism: A new guide to Action, Authors, Concepts, Data Bases, Theories and Literature, Transaction Books, New Brunswick, 1988

[3] CSTPV St Andrew's University Certificate in Terrorism Studies

[4] COE DAT Information Collation Managemant Cell database

[5] Weimann, Gabriel: Terror on the Internet, USIP, Washington DC, 2006

[6] Weimann, Gabriel: WWW.AL-QAEDA: The reliance of Al-Qaeda on the Internet[7]

[7] COE DAT Cyber Terrorism Couse IV Mar 09

[8] COE DAT Strategic Communications Workshop May 09

[9] Huizing, Harry: Cyber Terrorism Briefing Note, COE DAT, Ankara, 2008

[10] Krone, Troy: Gaps in cyberspace can leave us vulnerable, Platypus Magazine (edition 90, Mar 2006)

[7] Thanks to Prof Weimann for his kind permission to use this article.

[11] COE DAT Cyber Terrorism Workshop Oct 07
[12] Bunker, Robert J: Networks, Terrorism and Global Insurgency, Routledge, Abingdon, 2005
[13] Hennessy, Joh L and others: Information Technology for Counterterrorism, National Academies Press, Washington DC, 2003
[14] Hoffman, Bruce: Inside Terrorism, Columbia University Press, New York, 2006
[15] Huntington, Samuel: The Clash of Civilizations, Free Press, London, 2002
[16] Laqueur, Walter: The New Terrorism: Fanaticism and the Arms of Mass Destruction, Oxford University Press, New York, 1999
[17] Sageman, Marc: Understanding Terror Networks, Penn, Philadelphia, 2004
[18] Stern, Jessica: The Ultimate Terrorist, Harvard University Press, Cambridge MA, 1999
[19] Tuman, Joseph S: Communicating Terror, Sage, Thousand Oaks, 2003
[20] Whittaker, David (ed): The Terrorism Reader 3rd Ed, Routledge, London, 2007
[21] Wilkinson, Paul: Terrorism Versus Democracy, Routledge, London, 2006

The Virtual Battlefield: Perspectives on Cyber Warfare
C. Czosseck and K. Geers (Eds.)
IOS Press, 2009
© 2009 The authors and IOS Press. All rights reserved.
doi:10.3233/978-1-60750-060-5-88

Borders in Cyberspace: Can Sovereignty Adapt to the Challenges of Cyber Security?

Forrest HARE[a,1]
[a] *School of Public Policy, George Mason University*

Abstract. The new US administration has begun efforts to securitize the substantial problems the United States is currently facing in cyberspace. Recently, President Obama ordered his National Security Council to conduct a rapid review of existing measures being undertaken by the federal government, and provide recommendations for additional ones. Many stakeholders in the US government and private industry are watching these actions closely as there seems to be broad acceptance that the issues call for more extensive security measures. However, many issues will complicate effective securitization of threats in cyberspace. For example, not all stakeholders agree on the priorities or where the focus of security measures should be yet cyber security is a "trans-sovereign" issue affecting both developed and developing countries in an interdependent manner.

Because actors in cyberspace enjoy relative anonymity and can threaten inter-connected targets around the globe, there is considerable debate as to whether the concept of borders is relevant to the challenges of cyber security. Regardless the focus of the debate, the concept of borders is important because they define the territory in which national governments can employ sovereign measures. To analyze borders in the context of cyber security, this paper asks the question, "Is there an important role for the concept of borders, if not physical lines, in improving national security in cyberspace?" To explore the question, the paper takes two approaches. The first is a comparison of the cyber security issues to international drug trafficking in an effort to explore how sovereign measures used to combat drug trafficking may be applicable to improving cyber security. The second approach is an examination of the issue from the perspective of the Heal and Kunreuther Inter-Dependent Security Model with an attempt to inform the cyber security decision process of national governments as they consider options to invest in a higher level of security.

The paper will argue that, whether the problem is addressed from the standpoint of criminal behavior like drug trafficking, or cyber attacks in an interdependent, global domain, borders can be a potentially useful construct to address cyber security issues and inform national policy decisions, regardless of the physical location of relevant nodes. However, sovereign powers must be careful not to use the concepts of borders to curtail the progress our nations have made to connect and better the world via this evolving and expanding environment.

Keywords. Cyber Security, borders, agent-based modeling, interdependent security model

[1] Corresponding Author: Forrest Hare; E-mail: fhare@gmu.edu

Introduction

In the United States, there have been multiple initiatives to raise awareness and securitize the nation's vulnerabilities in the medium of cyberspace[2]. Many stakeholders in the US government and private industry are watching these actions closely as there seems to be broad acceptance that the issues call for more extensive security measures. Two initiatives are noteworthy- the Congressional Commission on Cybersecurity for the 44th Presidency, and the Obama administration's 60-day cyber review. The commission's report [2] identified cyber security as "one of the major national security problems facing the United States (pg 1)." In addition, President Obama ordered his National Security Council to conduct a 60-day review of existing measures "to ensure that U.S. Government cyber security initiatives are appropriately integrated, resourced, and coordinated with Congress and the private sector [3]."

In many cases, for example in response to the recent violence on the U.S./Mexico border, an important component of any security response is often a call to "defend the borders". However, such measures are problematic when attempting to respond to threats in cyberspace. One of the biggest issues with applying border control measures in cyberspace is the amount of inter-connectivity with other nations in the medium. Because cyber security measures must be internationally coordinated, the question often arises as to whether the concept of borders is relevant in the domain [see 4,5]. This article explores the question, "can borders, as components of sovereignty, be relevant to addressing cyber security?" To explore this question, I will use two different analytical constructs. First, I will compare the problems with securing a nation in cyberspace to the problems of combating drug trafficking. Researchers have raised similar concerns about stemming the tide of illegal drugs crossing national borders. If these two problems appear similar in their challenges, then perhaps we can draw lessons for cyber security from border-related measures to combat drug trafficking. Second, I will apply the Interdependent Security model, on which I will elaborate in a later section, to the problem of national cyber security. If this model can be considered a valid construct, perhaps it can also point to a role for borders as the nation-state actors in the model choose cyber security investment strategies. At a minimum, it would highlight in which countries sufficient measures are being taken and where they aren't, thereby highlighting the boundaries between them. Based on these dissimilar analyses, I will argue that, regardless the challenges of applying security measures "at the borders," as concepts of sovereignty, national borders remain relevant components of state-level responses to security threats in cyberspace. This analysis will not be an attempt to find answers to empirical questions, but rather provide new frameworks beyond the purely technical/legal aspects to address cyber security and borders.

The next section will provide some background and frame the discussion more precisely. There are several challenges to even effectively securitize threats in cyberspace. For example, the nature of "cyber security" as a national security issue is ambiguous and there is a heightened potential for a security dilemma in the domain. The goal of this section will be to frame the problem in such a way that it effectively bounds the ensuing discussion and informs the questions of borders as they relate to the

---

[2] For the purpose of this article, "securitization" is understood as the process outlined by Buzan et al. [1]. Namely, an issue is presented as posing an existential threat to a designated referent object (in this case, a nation-state) requiring emergency measures and justifying actions outside the normal political bounds (pgs 23-24).

domain. After addressing these issues, I will present the two analytical frameworks and highlight their relevant findings. The article will conclude with a synthesis of the issues and a discussion of policy implications. Regardless the findings from this work, sovereign powers must be careful not to use the concepts of borders to curtail the progress we have made to connect and better the world via this evolving and expanding environment.

1. Framing the Issue

Effective securitization of threats in cyberspace can be complicated by many issues. First, there is little agreement as to what the security issue in cyberspace actually is. This is a common problem with issues of security that must compete to be on the public agenda. Arnold Wolfers [6] called national security an "ambiguous concept," and because of the unknown nature of actors and their motives in cyberspace, the ambiguity is only heightened in this domain. Different actors will securitize the problems according to their perceptions or agendas. For example, while one nation may assert that an existential threat is posed by a denial of service attack against their fragile banking infrastructures, another may highlight fundamentally different issues. Some policy advocates would include threats from websites critical of government regimes to be a component of cyberspace security. A. Strelstov [7], a member of a Russian delegation to the UN, identified that, "undermining a state's economic and social systems and psychological manipulation of a population for the purpose of destabilizing society," is also a component of what the Russians call international information security (pg 8).

An additional concern is the heightened potential for a security dilemma in cyberspace. As characterized by Herz [8], a security dilemma may arise as one nation's efforts to arm themselves in defense may provoke another nation to do likewise, thereby creating a greater threat. Buzan [9] goes further to identify that some ambiguous measures may actually be attempts to gain more power vis à vis potential adversaries. This challenge of what Buzan terms the "power-security dilemma" is most difficult to counter in cyberspace. When fielding tanks and anti-aircraft missiles, their presences can be declared as defensive measures and made visible to the public. However, it is much more difficult to make public or confirm the defensive nature of measures a country may employ to improve security in cyberspace. Assertions that the actions are also offensive will be difficult to counter because any offensive potential would be difficult to disprove and offensive use would be difficult to identify. Complicating the issue further, attacks within in the domain can easily be masked and attributed to a nation-state, when, in fact, they may be the actions of non-state actors (or vice versa).

Overcoming both these factors requires a common understanding of the issues. Any effort to securitize a situation requires a threat agent, a victim, and an understanding of how the threat agent causes an existential threat to the victim. In cyberspace, the threat agents can be criminals, hackers, terrorists, and nation-states[3]. The greatest challenge is determining who is conducting the attack before extensive forensics have been accomplished. The targets of these actors are also diverse. They

[3] For a good survey of the myriad of malicious actors in cyberspace, see Denning [10], Gorman [11], and Kramer et al. [12].

may be in the business of stealing personal identities to commit fraud, conducting industrial espionage, engaging in cyber extortion of critical infrastructure owners, or preparing for and conducting a deliberate conflict accompanied by actions in cyberspace. Any analysis and body of policy recommendations that attempt to incorporate every possible combination of these malicious actors and their attack methods would be hard pressed to escape the trap of ambiguity. Therefore, to narrow the scope to a level appropriate for this analysis, national-level cyber security will entail the following:

> Attacks and infiltrations by either state or organized non-state actors against government and critical infrastructure systems (privately and publicly owned) to gain knowledge of a national security value and/or attempt to degrade/disrupt such systems.

National security is about existential threats to the state. Obtaining knowledge of a national security value can create an existential threat by allowing potential adversaries to gain the knowledge to develop effective counter-measures to a nation's advanced military and other defenses. In addition, cyber attacks that degrade the ability to command and control national security assets and attacks that disrupt critical infrastructure have direct implications to national security. This infrastructure may be civilian, military, or both. In the United States, for example, the Department of Defense relies heavily on the nation's public and private cyber infrastructure backbone for communications purposes [13].[4]

Some security measures are currently in place to protect against the threats articulated above. Such measures are employed by both government agencies and the private sector owners of much of a nation's critical infrastructure [see 14]. An obvious measure to defend against the theft of sensitive information would be to place all critical information and correspondence on closed systems that are not connected to the publicly accessible Internet. In the United States, for example, this would entail containing the information within the national security system architecture managed by the National Security Agency and Defense Information Systems Agency[5]. Certainly, governments secure much of their critical information in this manner. However, it is also the case that, as we become more reliant on the Internet for collaboration on all activities, especially between the public and private sector, it is becoming increasingly difficult to keep critical information controlled in this manner. A recent incident regarding a potential loss of design information for the F-35 Joint Strike Fighter highlights this problem. The information was stolen from private, proprietary industry networks (meaning no government access or frequent auditing), and it apparently contained several terabytes of design data on the future air defense capability for several nations [15]. Remaining disconnected from the greater cyberspace could be a measure employed by critical infrastructure owners and operators also. The control networks could be closed, proprietary systems with no remote access. In fact, older generation control systems employed tailored protocols and were only managed through proprietary, closed systems because there was no Internet available at the time.

---

[4] Note that the focus for this article does not include industrial espionage unrelated to national security, hacking for pleasure, identity theft, and the use of the Internet for training, messaging, and internal transactions of bad actors. Though these can all be considered criminal acts in their own right, they are outside the scope of this discussion.

[5] For an overview of the U.S. National Security System, refer to the CNSS website at www.cnss.gov

However, the trend has been to install remotely maintained systems employing common OS architectures to leverage the connectivity benefits of the Internet [16]. Therefore, these critical infrastructure systems have assumed a risk common to all those dependent on the effective functioning of the Internet.

The United States, as a sovereign country, certainly has the inherent right to control all of its borders in any domain [17]. With the above considerations, it is clear the public sector cannot manage all necessary security actions alone. Private companies are an important part of the dynamic that is absent in other areas of national security where the actions of the military, or law enforcement, dominate the response options. We have no early warning radar system or Coast Guard to patrol the borders in cyberspace. Unlike in other domains, information of an attack will come first from those being attacked. Therefore it is highly unlikely that a government organization, unless it is actually the target of a cyber attack, will have greater situational awareness. An effort must be made to incentivize the private sector to invest in cyber security as well. In many cases, national security depends on it. But if none of the measures being employed have a border patrol component, does that necessarily mean that borders are not significant in cyberspace? The next two sections will introduce two different frameworks to address this question. In the first of the two analytic frameworks, I will compare the problems of securing a nation against cyber threats to the challenges of securing a nation against international drug trafficking.

2. Comparison with Drug Trafficking

In a way, the world has become a victim of its own developmental successes. Over the last two decades, we have seen an incredible amount of openness in commerce and the exchange of ideas. However, with openness comes much vulnerability. Several authors have highlighted the fact that increased globalization and economic interdependence have been accompanied by greater economic disparities. Globalization has also created an environment where it is much easier for clandestine transnational actors (CTAs) to operate [see, for example 18][6]. As such, it has been increasingly difficult to secure nations against the growing, non-traditional threats from these CTAs. However, it is possible that efforts to do so can inform the challenges of cyber security. But how similar are the issues?

From my perspective, there are at least six factors in combating drug trafficking that compare to the challenges of cyber security. First, CTAs engaging in the drug trade are based in countries with few legitimate economic opportunities [18]. Legal activities, such as growing subsistence crops have little chance of competing with the lure of income from growing poppy seeds in countries like Afghanistan and Thailand. A similar trend has been developing in cyberspace. Several developing countries have become sanctuaries for cybercriminals, or transit points for malicious actors located in other regions. A recent arrest in Romania highlights the growing hacking community in Eastern Europe. A hacker nicknamed "Wolfenstein" is suspected of breaking into US Department of Defense computer systems and planting malware [20]. Brazil has also been identified as a growing contender for the cyber crime capitol of the world. In 2004, the Brazilian federal police claimed that it was home to 80% of the world's hackers [21]. In addition to these countries, researchers consider China to be a growing threat

[6] The term "clandestine transnational actors" is adopted from Andreas [19].

[22]. But due to problems of attribution, it is difficult to tell if the actions are state-sponsored, or private actions [23]. Regardless the location, or identity of a specific attacker, a nation-state may be involved as the sponsor. Many countries would be interested in information about rivals thereby acting as customers to those providing national security-related data obtained through cyber espionage.

Second, but closely related to the previous point, illegal drug control regimes rely almost exclusively on the coercive actions of national governments, but the trade is conducted in areas where actions at the state level are often ineffective [24,18]. Because each country is sovereign and reserves the right to draft and enforce it's own laws, international drug control actions must contend with widely differing legal regimes. Compounding the problem, developing countries have much less effective law enforcement. The same barriers confront cybercrime responders. For example, as recently as 2004, hacking was not considered a criminal offense in Brazil [21]. Only recently have such vehicles as the European Cyber Crime convention and other bi-lateral agreements led to improvements in synchronizing legal regimes to combat cyber offenses.

Third, the Internet itself is an excellent source of knowledge on how to engage in the drug trade [24]. One can easily find instructions for how to make such drugs as LSD or methamphetamine (see, for example, www.homemadedrugs.net) by browsing websites or conversing with others in forums. Just as easily, one can find the tools required to break into computer systems, as well as instructions for their use in news groups (see, for example, www.sectools.org). During the conflict between Russia and Georgia, for example, there was substantial evidence that the attacks on Georgian governmental websites were directed via web forums [25]. In neither the case of drug trafficking nor hacking is formal training required or even available. The only actors who may have received formal training would have done so as former security officials.

Fourth, customs agents have to sift through ever increasing amounts of legitimate goods and people to find illegal drugs. According to Stephen Flynn [26], a border security expert and former US Coast Guard officer, customs agents must patrol a continuous stream of peoples and goods at more than 3,700 terminals at over 300 points of entry. As he states, it "[i]ntercepting the ripples of danger in this tidal wave of commerce is about as likely as winning the lottery (pg 57)." Similar challenges exist in cyberspace. With well over 3 Tbps traversing international routes between the US and the rest of the world, it is virtually impossible to differentiate legitimate Internet traffic from traffic with a malicious purpose [7]. Information that has been stolen from somewhere, or that contains commands that will "flip a switch" in such a way as to cause severe damage to a critical infrastructure system, is extremely difficult to identify. Intercepting it requires previous knowledge that the information should not be traveling across the Internet in the first place. In other words, you can train a dog to identify marijuana, but it is unlikely it can be trained to identify the difference between Bayer aspirin and a generic or that the prescription has expired and belongs to someone else.

Fifth, efforts to combat retail drug transactions are constrained by civil liberty concerns. Victims, who could be considered accomplices in an illicit transaction, can hide behind privacy rights [24]. Oftentimes the victims of cyber espionage may choose to cover the event as well, but for slightly different reasons. Since cyber crimes can be

---

[7] 3 Tbps is an educated guess based on analyzing the Telegeography data source used for the second analysis of this paper. An exact figure would be extremely difficult to obtain and would only be valid for a few seconds.

hidden from the public by both the victim and the perpetrator, a company that has been infiltrated may chose not to report an event for fear of assuming liability for the actions [27]. They may also be concerned about a reduction in customers' trust in their ability to safeguard sensitive information. In fact, it may not be until the recipient nation of stolen data has built an exact replica of a system, for which they have obtained the design secrets, that there is any indication that a theft has occurred.

Lastly, there is little agreement on what exactly constitutes the "evil to be eradicated" when assessing and implementing counter-drug trafficking measures [24]. For example, the current debate surrounding the drug cartel violence in Mexico centers on the role of the United States in creating the problem [28]. What is worse, trafficking drugs or supplying criminal gangs with automatic weapons? In the current conflict across the border, the drug cartels are armed with a much more powerful arsenal than the local police. Law enforcement officials confront similar challenges when combating the growth in cyber crime. Many criminals use the defense of, "I just did it to see if I could get into the system and didn't know what I was getting." Being a red, white, grey, or black hat may depend on a person's perspective. Many of these actors see themselves as beneficial to the network security industry and downplay their influence on cyber criminals (which they often once were). Movies, such as "Hackers" continue to glorify the actions of teen-agers who break and enter systems in cyberspace, when such actions against a physical facility would be clearly viewed as trespassing. As stated above, as recently as 2004, hacking was not even considered a criminal offense in Brazil.

There may be several more similarities, but is should be evident from this presentation that there are many conceptual similarities between these two types of non-traditional threats to national security. With this in mind, I will discuss how borders have played a role in the international war on drugs to determine if such measures can illuminate the complexities of countering attacks through cyberspace. As stated, in this age of globalization, it is virtually impossible to detect contraband crossing national borders. The US has, for example, 106,000 miles of physical borders and shorelines and over 400 million people transit those borders yearly [26]. Though we cannot completely secure the borders against drug smuggling, they still seem to play an important role in efforts to combat the trade. The recent measures by both the US and Mexican governments along their shared borders highlight the political importance of actions taken to secure borders against the movement of drugs. Arguable the measures were enacted due to a perceived loss of control of the borders to the drug cartels. Peter Andreas, Harvard Professor and the author of *Border Games* [29], asserts that border control measures are an important symbolic and perceptual response that the state is defending its sovereignty and its citizens from an existential threat. By "sending in the troops," the state can demonstrate its moral resolve and commitment to maintaining its territorial integrity. Even if there is little empirical evidence that any measures enacted to defend the borders against the flow of drugs has an effect at reducing the inflow to the US, there is tremendous pressure to take action. Besides demonstrating resolve, the visible actions remove pressure to confront the more difficult but root causes of the drug trade-the insatiable demand.

In addition to the largely symbolic nature of recent actions on the US-Mexican border, law enforcement officials do attempt to achieve a deterrent effect with their actions. In this and other border regions, measures have gone beyond the dedication of more personnel. Enforcement measures have relied on improved surveillance technology but also such measures as "pushing out" borders [19]. For example, in both the case of European Union countries and the United States, the immediate neighbor countries are enlisted to "thicken" the border defenses. In the case of the United States, the problem of drug trafficking is not limited to the immediate border. The drugs originate in several source regions and many are funneled through Mexico. According to Andreas [19], by supporting Mexican efforts deep in Mexican territory, a larger buffer zone is created while supporting the smooth flow of legal cross-border commerce. In the European Union, member countries encourage neighboring countries to improve coordination with their law enforcement efforts by making tighter law enforcement actions pre-conditions for admission to the EU. Lastly, efforts can also focus on commercial trans-shippers of legitimate goods who depend on speedy transit of international customs facilities. Flynn [26] suggests that it is in the interest of transnational shipping companies to tighten their own logistics and transportation procedures. As the logistics infrastructure continues to improve and widen the markets for perishable goods and "just-in-time" deliveries, shippers are under increasing pressure to maintain delivery schedules. Therefore, they have a tremendous incentive to avoid any potential delays that could be created if they are found to be lacking controls on their cargo. Customs officials could use this incentive structure to their advantage and encourage commercial trans-shippers to help reduce the potential smuggling of illicit drugs.

These three example measures could have implications for patrolling the borders in cyberspace. First of all, as threats in cyberspace become increasingly securitized, we can expect the same pressure for national governments to take action as they have done in the wars on drugs. This will undoubtedly entail largely symbolic actions to attempt to secure national borders in the domain. In the case of the US, such efforts may be cast as an attempt to "regain the control of cyberspace" it ostensibly maintained during the early years of Internet development. At the time, it was managed by the US Department of Defense and then the National Science Foundation. The symbolic gestures to "regain control" can be reified by technological border control points, attempting to thicken the cyber borders, or both.

For example, a border control point could be established at the terminus between undersea cables and fiber optic lines. At these points, customs, law enforcement, or other agents of the federal government could employ any of several technical solutions such as deep packet inspection devices or Anagran flow management devices [17,30]. Other solutions suggest labeling traffic to identify countries of origin and destination [31]. The intent here is not to debate the technical or practical feasibilities of such measures[8]. Without employing any such measures, there is no empirical evidence available to determine their efficacy, or if they will slow Internet traffic appreciably. The point here is to show that several measures have been researched and, enacting any or all would, at a minimum, be symbolic statements to assert sovereignty over national territory in cyberspace.

More practical measures would mirror the defense-in-depth approach taken by Europe and the United States to combat drug trafficking and other CTA activities. For example, nations with more developed legal and law enforcement regimes could encourage neighboring nations to improve their legal regimes. Developed nations could also provide technical support to others' national cyber security centers. One unique characterization of cyberspace is that neighboring nations in the domain are not always

---

[8] For an in-depth discussion of a multi-agency Internet Border Inspection Station concept, see Upton 2003. For details on the Anagran technology, see http://www.anagran.com/products_fr_1000_intelligent.php.

physically contiguous. However, that should not limit the possibilities for cooperation. As with drug trafficking, the focus must be both on nations where attacks have historically transited, and those where the attacks are perceived to be originating. A recent effort in Europe to "thicken the cyber borders" has been the broad adoption of the Council of Europe Convention on Cyber Crime. Six countries signed, ratified and entered it into force by 2004. However, since that time, an additional nineteen countries, to include the United States have adopted the convention[9].

Additional defense-in-depth measures could focus on the cyberspace common carriers- the Internet Service Providers and Backbone companies. They are the carriers of the legitimate traffic in which the contraband is hidden. Like the international trans-shippers of physical goods, these are the commercial interests that would be adversely impacted by tightened border controls that may result from the emplacement of government-monitored border inspection devices. Employing such a suite of inspection tools, which would adequately provide for the protection of civil liberties, would invariably slow Internet traffic. Therefore, ISPs would be expected to have a great incentive to support the improvement of self-regulated inspection regimes. If they can be motivated to improve their internal procedures to help law enforcement combat cyber attacks, then there will be less pressure for more restrictive national-level. Perhaps, the absence of a real threat of employing federal border security measures has contributed to neglect on the part of ISPs to better control the activities of their customers. Regardless the exact point of entry of goods and people in any domain, states have sovereign rights over all their territory and can also pursue legal recourse against cyber crimes committed anywhere within their borders. Though effectiveness has been limited, we must continue to rely on state responses for the foreseeable future.

This section of the paper used a comparative case study approach to identify lessons that could be taken from the fight against international drug trafficking and applied to cyber security. The next section of the paper will take a fundamentally different approach and explore the use of a game-theoretic construct and novel quantitative methodology to address the issue. The analysis expands on a theory that has been previous employed to research situations in which the security of one actor depends substantially on the actions of other actors in their system.

3. Interdependent Security Theory

In his book, Micromotives and Macrobehavior, Nobel laureate Thomas Schelling [32] described the concept of binary, "either-or," choices that create externalities on the decisions of others. To explain the concept, he described several different situations where the question was not about how much anyone does, but how many make one or the other choice. For example, the decision to follow daylight savings time or joining a boycott would be considered binary. The interesting implication of Schelling's model is the potential to "tip' the collection of decision makers from one decision to the other. This tipping effect could reduce the potential social costs when not enough actors initially make the socially beneficial choice. The model can also be applied in situations where actors must coordinate security decisions. Economists Kunreuther and Heal [33] built on this concept of interdependent decision-making after the events of

---

[9] Assessment based on a table from the Council of Europe website at: http://conventions.coe.int/Treaty/Commun/ChercheSig.asp?NT=185&CM=&DF=&CL=ENG

9/11. They introduced a game-theoretic approach to explore more deterministic outcomes of a class of binary choices, the interdependent security (IDS) investment decision. Kunreuther and Heal's focus has been on the existence of Nash-equilibria in the decision whether or not to invest in security measures. Of particular note, in an IDS model, a fundamental assumption is that an indirect attack, or an attack that originates from within the system due to a failure by a partner actor, cannot be defended against. For example, no matter how much an airline invests in security at their terminal location, if a partner airline allows a bag with a bomb to be transferred to their airplane bypassing the terminal inspection, the investment is for naught.

Arguably, this analytic framework can be applied to the cyber security problem at the nation-state level. In this case, the actors making a security investment decision would be national level governments. The investment could be undertaken by private or public agents, but the security action would be those measures required to secure the critical infrastructure cyber systems and the information systems containing data of a national security value, in other words, the targets of the attacks specified early in the paper. Attackers could be any actor that has the ability to hold the above targets at risk either through a direct attack on a country from a location therein, or indirectly through other countries' national IT systems.[10] In other words, attribution of an attack is not necessary to use this framework.

Understanding that inter-connectivity and inter-dependence is a complex issue at the national level, this analysis must generalize these aspects of the model. Specifically, I make several important assumptions. First, I must assume the state, or an agent acting on its behalf (such as the Department of Homeland Security in the US), can maintain some level of control over the actions of the owners and operators of the national critical infrastructure. In other words, the national government must have the ability to ensure a cyber security investment is executed. Secondly, the analysis assumes that it is possible to invest at a level that significantly protects these systems from failure or information theft that would create an existential risk to the nation. In other words, applying the theory at a national level means that not every single attack must be thwarted as national systems do have some resilience. Lastly, it assumes that actions taken to secure a nation's critical assets in cyberspace can be made visible to others[11].

Even with the above assumptions, there are two additional challenges to applying the IDS theory in this situation. First, it is difficult, if not impossible, to obtain the empirical data regarding the security investment decisions of these agents (especially if many nations do not oversee cyber security at the national level), and the interdependent nature of the decisions by so many actors can be difficult to calculate as their decisions are updated. In such situations, an agent-based model can be useful. Agent-based models employ object-oriented software programs to model the behaviors of inter-acting agents endowed with specific parameters that govern their behavior. Such tools can also model the dynamics created by changes in the behavior of the agents. They are ideal for game theoretic decision problems amongst many actors. In order to conduct this analysis using an agent-based modeling technique, the first decision that must be made is the size and composition of the sample. Who are the representative agents making the investment decisions? Conceivably, I could have

---

[10] In other words, the exact identity or sponsor of the hostile act is not required for this model to function.

[11] For such an assumption to be valid, it would imply close coordination of efforts amongst those who have chosen to make the necessary investments. For example, partner nations could have liaisons working in each other's national response centers.

constructed a model of all the nations of the world. It would not have created a computational problem for an agent-based, but it would probably not closely reflect the "real world" in cyberspace[12]. Instead, I identified a more practical sample based on the most highly connected nations in the world. The sample network used for this analysis is comprised of the 21 largest countries, in measure of 2008 bandwidth connectivity, as reported by Telegeography[13]. While this sample does not necessarily represent existing cyber security cooperation, it is intended to capture two features. First, nations with the most extensive international Internet connections can be assumed to be those for which cyber security is most critical. Secondly, it identifies which countries have the greatest imperative to work together due to their existing high level of inter-connectivity. A graphic depiction of the sample network is presented in Figure 1. Again, this is only a representative sample to illustrate the potential for this type of analysis. The nodes are representatives of the 21 countries with the highest amount of Internet traffic, and they are connected by non-valued links symbolizing the overall traffic route (i.e., there may be several physical connections between the nodes).

The particular structure of this network has a potential benefit for most of the involved nations. For example, assuming that the vast majority of international traffic

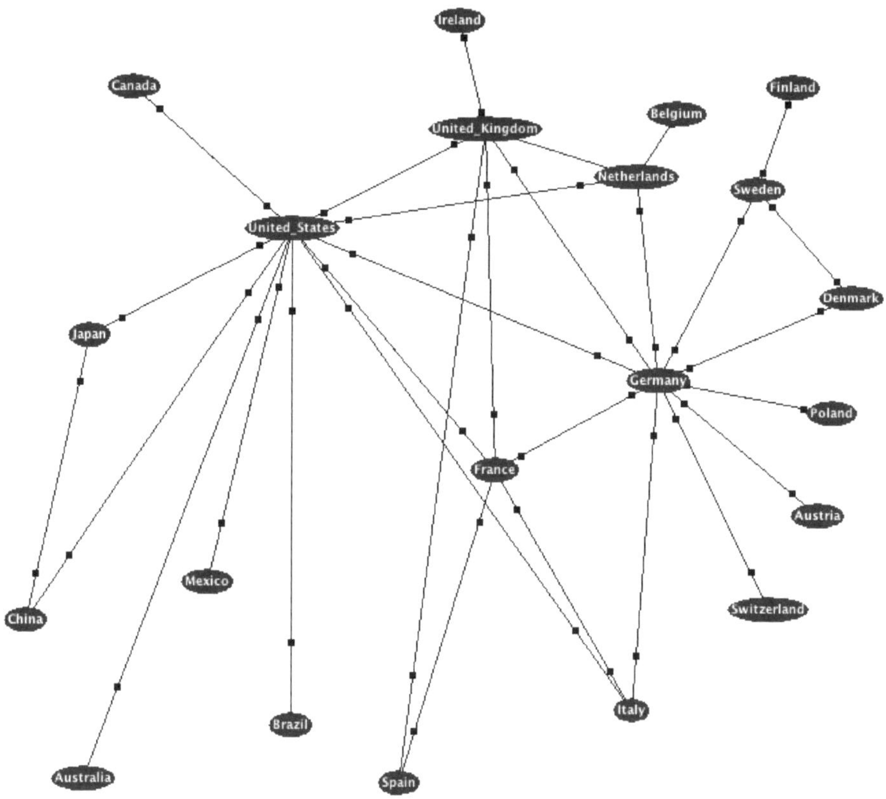

Figure 1. Sample Network of Nations

[12] For example, the investment decisions of the Maldives may have little effect on those of the United States.
[13] http://www.telegeography.com/products/map_internet/index.php

must flow through one of the few central actors in this network, a cyber attack that successfully breaches the critical systems in any nation must, in many cases, flow through (or originate) in one of the central countries in order to reach one of the other nations connected to the world only through that central nation. Though this does not reduce the external risk for the few central actors, the structure greatly reduces the externality problem encountered by the rest of the countries in this network. Therefore in practice, though the sample contains 21 agents (nation-states), the interdependent nature of the security investment decision for most "non-central" nations is reduced immediately to a two-player game minimizing risk estimation[14].

Constructing an agent-based program for this model is fairly straightforward. However, an accurate calibration can be challenging. A complete discussion of the construction and calibration is contained in Appendix 1. Based on the data collected from several cyber security sources, the model was run at varying levels of externally generated risks by altering the probability of an indirect attack. The output of the model is most easily interpreted by comparing graphic depictions of the investment decisions over time. Encouragingly, the series of graphics in Figures 2-4 suggest that the agents in this sample network, the highly connected nations, could behave in a manner consistent with the IDS theory.

For the model run depicted in Figure 2, the probability of an indirect attack is set at 0.2. In this instance, the probability is sufficiently low that all the agents in the model find it feasible to invest in security to thwart direct attacks from "outside the system." Skipping Figure 3 for the moment and moving to Figure 4, we find the opposite results. In this case, only a handful of minimally connected nations choose to invest. Since no other nations choose to join these actors once they have decided to invest, the system contains both investors and non-investors at equilibrium. The more interesting result is obtained when the risk of indirect attack is at a point in between. In Figure 3, the probability of indirect attack is set at 0.3. In this scenario, all but the central actors find

Figure 2.

Full investing on first round

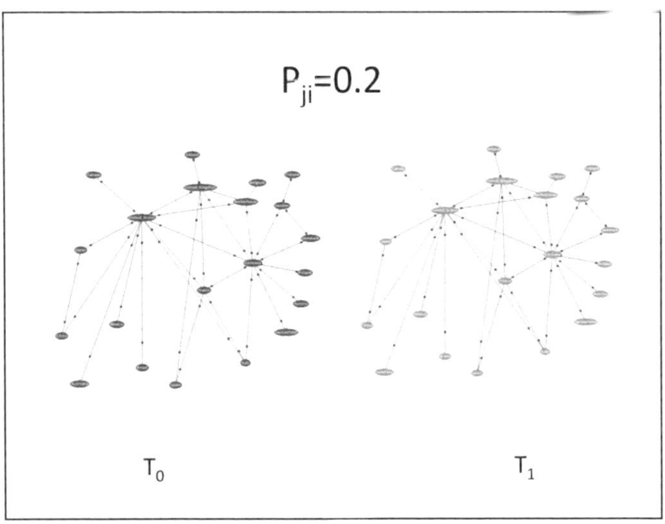

[14] Interestingly, the sample network does exhibit features that make it theoretically feasible to extend these findings to a much larger set of nations. The network appears to be a scale-free form. For a good discussion of the implications of this characteristic, see Barabasi [34].

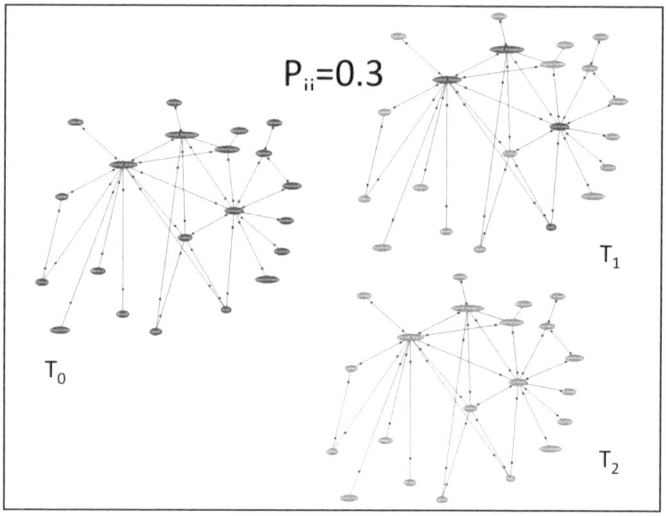

Figure 3. Full investing after several rounds

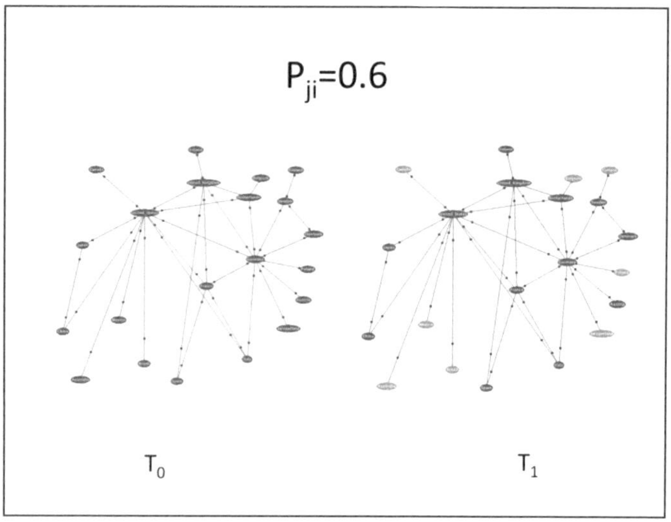

Figure 4. Mixed investing at equilibrium

it advantageous to invest regardless the decision made by other agents. After T_2, the risk of indirect attack to the central actors has now been reduced to the same level faced by the periphery in T_1. As a result, the central actors now find it in their interest to invest. In other words, the conditions are such that the system will cascade to a state of full investing as predicted by the IDS game-theoretic model.

These results demonstrate how the actions of one country to improve their national cyber security, whether enacted publicly, privately, or in concert, could positively impact the decisions of other countries confronted with similar challenges. Though we are far from such a reality, the results suggest that when it is possible for a nation to secure its territory within cyberspace, then borders become increasingly important components of cyber security. Namely, they become important as concepts that delineate which portions of cyberspace may be more secure than others. There may not be a cyber security measure enacted directly at physical border post, but when state actors can determine at which point their activities stop and the actions, or inactions, of a neighbor occur, than they can better determine the relative values of enacting their own security measures.

Hopefully, this model demonstrates the importance of coalition building. The border that may be important may not be national borders, but the borders between the investing coalition and others. Until the entire system becomes part of the coalition, there will always be a border between those inside the security umbrella and those outside. As the security umbrella grows and strengthens, the borders will be more sharply delineated. In addition, those outside the security framework may increasingly become the focus of attacks themselves, or the assumed source of malicious actions. Neither condition would be considered very favorable and would lead to increasing pressure to join the coalition of those who have chosen to secure their territory in the domain.

4. Conclusions

This article sought to demonstrate the relevance of borders in issues regarding cyber security. As stated earlier, nation-states have borders. Domains, as merely mediums in which we interact, do not have borders. However, the relevance of a nation's borders in each domain is related to a nation's willingness and ability to assert their sovereignty in them. As long as threats are directed at nation-states, and legitimate response actions are retained by the state, they will remain important actors and their borders will continue to be relevant. Borders can be equally important in cyberspace because borders define boundaries of sovereignty regardless the domain and the ability to locate them physically. Even though borders have become less significant for all legal commerce, they have become even more significant for policing action against transnational threats. To explore their significance, I employed two dissimilar frameworks in an effort to broaden the discussion beyond purely technical and legal dimensions.

In comparing international drug trafficking with the national security elements of cybercrime, we see that there are several similarities. These similarities suggest that we can learn lessons from measures to secure borders against the shipment of illicit drugs. Whether the measures are largely symbolic or actually effective, they demonstrate national resolve and a determination to exert sovereignty. If the current drug wars teach us anything about national responses to transnational problems, the borders will become important in the fight to secure cyberspace if only for their political significance. More effective measures seem to center less on a tight control of the border itself, and more on improving the behavior of companies engaged in legitimate trade across the borders, and the behavior of surrounding countries within their territories.

The agent-based model of the IDS problem provided a theoretically different perspective. In this model, the exact location of the borders is not relevant. What is more important is the potential of, and benefit from coalition-building by actors responsible for securing their portions of cyberspace. If the IDS model teaches us anything, it is that we must work together on an international level. Unilateral efforts to secure borders are a losing proposition in today's interdependent reality. Nowhere is this interdependence more visible than in cyberspace.

The paper also included some considerations for public policy. Specifically, efforts can be made to induce better self-regulation of ISPs to avoid more intrusive border control measures. Also, measures to increase the visibility of national and private security measures can increase the incentives for others to make similar investments, and reduce the potential for a security dilemma arising between nation-states in cyberspace. Continuing to support and encourage neighboring countries to improve their legal regimes and law enforcement efforts is also an important step. However, some findings lead to further policy questions. For example, how do we change cultural attitudes to criminalize hacking behavior? Also, how and when do we increase visibility and share information with countries that may be the source of many threats in the domain? A final note of caution for policy consideration: sovereign powers must be careful not to use the concepts of borders to curtail the progress our nations have made to connect and better the world via this evolving and expanding environment. In other words, we can no more lock down the borders to counter malicious actors in cyberspace than we can lock down our nation's physical borders to fight terrorists and drug-traffickers.

If we accept that nations can play a role to improve cyber security in their country (within their borders) and influence others to do so as well, then there will continue to be an important role for borders as a physical and legal concept. However, if we find that it is not plausible for the state to affect security in its portion of cyberspace either technologically or conceptually, then the existence of borders in any sense becomes less relevant. Assuming this paper successfully demonstrated the former case is more true than the later, then regardless their exact physical location, the very existence of borders demonstrates a need for us to work together as an international community to develop transnational solutions.

Appendix 1: Agent-based Model Construction and Calibration

Most agent-based models are constructed in an object-oriented computer programming language and they interact in an environment. As described in the paper, the agents in this analysis interact within a network. Based on the IDS model equations, the significant parameters are as follows:

Agent Parameters

- c = Cost of investing in security to a level that defeats existential threats through cyberspace
- L = Loss of critical information from a successful attack
- p_{ii} = Probability of a direct, successful attack
- p_{ji} = Probability of an indirect, successful attack that occurs from within the network of highly-connected nations
- Network neighborhood
- Investment State (invest or not invest)

The next challenge is estimating the values of these parameters. All of the nations in the sample have made some level of investment, and the cost, at a national level of achieving an efficiently secure state of security can only be guessed in the absence of specific data on the threats to each. Therefore, another simple convention is employed. The agents are heterogeneous in that initial endowment of c, L, and p_{ii}, are randomly distributed amongst the agents. However, the possible values of these variables are normally distributed about a mean value. This convention allows us to assess the actions of the agents when changing the probability of indirect attack, p_{ji}, while holding other parameters within realistic bounds[15]. The value for p_{ji} for each agent is also normally distributed about a mean value, but it is a single variable for the agent's entire network neighborhood[16]. After consultation with cyber security experts regarding the potential costs and losses at the firm level[17], the remaining parameters were estimated by the following mean values:

$c = \$1,000,000$
$L = \$10,000,000$
$p_{ii} = 0.4$

Though the empirical data is not available, a mean value for p_{ii} was set at 0.4 based on a recent SANS Institute report [35] regarding attacks on firms in several industries. While holding these parameters constant, p_{ji} was varied from 0 to 1 to explore whether this sample network can behave in a manner predicted from IDS theory.

In addition to the agent parameters, there are rules that govern agent interaction. The agents make the security investment decision according to the interdependent security pay-off algorithms of the Kunreuther and Heal model[18]. In the current model, the behavior of the agents is determined by the following:

Interaction Rules

- Identify how many others in your network neighborhood have not yet invested in security
- Calculate the external risk created by their decision not to invest
- Determine if the external risk is less than or greater than the cost to invest
- If the cost is less, then decide to invest. If not, then decide not to invest
- Once all agents have made this decision, everyone changes their state as appropriate
- Repeat the above process until no one wants to change their state again

Initially, the agents are in a state where they have not invested. Since Kunreuther and Heal (2003) identified that the cost of the risk externality as the significant limiting condition, agents choose to invest in security when the inequality, $c < p_{ii}(L - X)$, is true. In this equation, X is the externality generated by others in the network that have not invested in security.

References

[1] B. Buzan, O. Wver, J.D. Wilde, O. Waever, Security: A New Framework for Analysis, Lynne Rienner Pub, 1997.

[15] In this assessment, 'realistic bounds' means relative to the other agents. The author did not attempt to estimate the actual, absolute costs of any of the parameters nor was it necessary to do so.

[16] In other words, the risk from each other agent, to which an agent is connected, is the same. This convention was necessary to generate a less complex decision algorithm for the basic model.

[17] For this analysis, it is only important that the relative values be appropriate and therefore it is assume that the values at the firm level can be scaled to the national level with the understanding that, just like the difference between small and large firms, the costs at a national level can vary widely as well.

[18] See Heal and Kunreuther [36] for a complete explanation of the pay-off matrix according to a two-person game.

[2] J. Lewis, Securing Cyberspace for the 44th Presidency, Center for Strategic and International Studies, Washington, D.C., 2008.
[3] Office of the Press Secretary, (2009).
[4] B. Kahin, C. Nesson, Borders in Cyberspace: Information Policy and the Global Information Infrastructure, The MIT Press, 1997.
[5] C.C. Joyner, C. Lotrionte, Eur J Int Law 12 (2001) 825-865.
[6] A. Wolfers, Discord and Collaboration: Essays on International Politics, The Johns Hopkins University Press, 1965.
[7] A. Streltsov, Diarmament Forum 3 (2007) 5-14.
[8] J.H. Herz, World Politics 2 (1950) 157-180.
[9] B. Buzan, People, States, and Fear: The National Security Problem in International Relations, University of North Carolina Press, 1983.
[10] D. Denning, in:, J. Arquilla (Ed.), Networks and Netwars: The Future of Terror, Crime, and Militancy, Rand, Santa Monica, Ca, 2001, pp. pgs 239-288.
[11] S.P. Gorman, in:, P.E. Auerswald, L.M. Branscomb, T.M.L. Porte, E.O. Michel-Kerjan (Eds.), Seeds of Disaster, Roots of Response: How Private Action Can Reduce Public Vulnerability, 1st ed., Cambridge University Press, 2006, pp. 239-257.
[12] F. Kramer, S. Starr, L. Wentz, eds., Cyberpower and National Security, Potomac Books, Dulles, Virginia, 2009.
[13] C. Wilson, Information Warfare and Cyberwar, Congressional Research Service, The Library of Congress, 2004.
[14] DHS, National Infrastructure Protection Plan, Department of Homeland Security, 2006.
[15] S. Gorman, A. Cole, Y. Dreazen, Wall Street Journal (2009) 3.
[16] Energetics, Roadmap to Secure Control Systems in the Energy Sector, 2006.
[17] O. Upton, Asserting National Sovereignty in Cyberspace: The Case for Internet Border Inspection, Master of Arts Thesis, Naval Post Graduate School, 2003.
[18] L. Shelley, Geo. J. Int'l Aff. 6 (2005) 5.
[19] P. Andreas, International Security 28 (2003) 78-111.
[20] J. Leyden, The Register (2009).
[21] T. Gibb, Bbc (2004).
[22] T. Thomas, Cyber Silhouettes, 1st ed., Foreign Military Studies Office, Fort Leavenworth, 2005.
[23] R. Deibert, R. Rohozinski, Tracking Ghostnet, Centre for International Studies, University of Toronto, Toronto, 2009.
[24] S. Flynn, in:, Beyond Sovereignty: Issues for a Global Agenda, Palgrave Macmillan, 1999, pp. 44-66.
[25] B. Krebs, Report: Russian Hacker Forums Fueled Georgia Cyber Attacks (2008).
[26] S. Flynn, Foreign Affairs 79 (2000) 57.
[27] W.S. Baer, A. Parkinson, Security & Privacy, IEEE 5 (2007) 50-56.
[28] H. Lafranchi, Christian Science Monitor (2009).
[29] P. Andreas, Border Games: Policing the U.S.-Mexico Divide (Cornell Studies in Political Econ, Cornell University Press, 2001.
[30] P. Bartram, (2009).
[31] J.E. Molini, Computers & Security 16 (1997) 189.
[32] T.C. Schelling, Micromotives and Macrobehavior, New, Norton, New York, 1978.
[33] H. Kunreuther, G. Heal, Risk and Uncertainty 26 (2003) 18.

[34] A. Barabasi, Scientific American 288 (2003) 9.
[35] SANS Institute, Top Ten Cyber Security Menaces for 2008, SANS Institute, 2008.
[36] G. Heal, H. Kunreuther, Journal of Conflict Resolution 49 (2005) 201-217.

The Virtual Battlefield: Perspectives on Cyber Warfare
C. Czosseck and K. Geers (Eds.)
IOS Press, 2009
© 2009 The authors and IOS Press. All rights reserved.
doi:10.3233/978-1-60750-060-5-106

Towards a Global Regime for Cyber Warfare

Dr Rex HUGHES
Cyber Security Project, Chatham House, London

Abstract. With two years having passed since the infamous cyber conflict between Estonia and Russia, international society still lacks a coherent set of principles, rules, and norms governing state security and military operations in cyberspace. For parties committed to promoting the cause of peace and stability in a multipolar world, this is a troubling notion since history shows that the likelihood of a new arms race is high when disruptive technologies dramatically alter the means and methods of war. As more nations aspire to project national power in cyberspace, a new digital arms race appears to be imminent if not already upon us. Thus, there is a central question confronting international society and more specifically the diplomatic community in cyberspace: What steps can be taken both today and into the future to forestall a major arms race and interstate competition in cyberspace? In order to begin addressing this complex question from the perspective of the Euro-Atlantic Community, this paper discusses both the challenges and opportunities of regulating 21st century cyber warfare. The paper is divided into three sections. Section 1 examines the evolution of the laws of armed conflict (LOAC) since the late 19th century. Section 2 examines how the LOAC apply to cyber warfare as viewed primarily from a US perspective (since US scholars have dominated the international regime discourse thus far). Section 3 examines what is needed to create a global regime for cyber warfare and specifically the role that NATO and the Euro-Atlantic Community can play.

Keywords. Law of Armed Conflicts, Cyber Warfare, Revolution in Military Affairs, Global Regime

> *Because the entire law of war regime has been built upon a Westphalian foundation, the transformative properties of cyber warfare are just as breathtaking. We are left pondering some fundamental questions - what constitutes force? What is a hostile act? When is self-defense justified in response to a cyber attack? Is the use of traditional means of force ever justified in response to a cyber attack? These are not easy questions and the international legal regime is lagging far behind the problems presented by the increasingly sophisticated technological possibilities in this area.*
>
> --Lt. Col Jeffrey K. Walker[1]

Introduction

When viewed systemically, the current generation of cyber weaponry demonstrates an enormous potential to alter the means of hostile attack and in turn of response. While our 21st century armed services are adjusting to the revolution in military affairs (RMA), the broader community of business, transportation, energy, research, health, academic, and social services look to their national leaders to provide plans and to conduct operations that will protect their domain of cyber space. Cyber defense for those old enough to remember may call to mind the home front nuclear alert drills plus the bunkers or bomb shelters constructed in the post WWII decades. In cyberspace both military and civilian networks are potential targets.

Overarching questions confront us: What is the current state of cyber warfare when viewed from an international affairs perspective? What options are available to policy makers that seek to fashion a global regime to govern 21st century cyber warfare? And more specifically to the theme of the first NATO CCD COE cyber war conference, what role can an international military alliance such as NATO play in advancing such a regime?

Since the enormous attack on Estonian digital networks, governments around the world have ordered their respective military branches to develop new offensive and defensive cyber capabilities. Some states have even gone as far as to create national cyber command authorities, as is evident in the United States [2]. However, as the attacks mount and more advanced 'cyber weapons' are introduced to the digital battlefield, there is little certainty or international consensus on the rules, or lack thereof, for governing modern cyber battles or larger warfare. Air Force Gen. Kevin P. Chilton, the head of U.S. Strategic Command (STRATCOM) issued a statement in May of this year that 'The Law of Armed Conflict will apply to this domain'[3]. STRATCOM defends the Pentagon's Global Information Grid at home and abroad through its Strategic Command Joint Task Force-Global Network Operations (JFT-GNO). Attempted penetrations of public and private systems number in the tens of thousands a day. As a commander who provides information for decisions by the US President and the Secretary of Defense, Gen. Chilton said that all combat options should be on the table for a US response to a cyber attack. He noted that many attacks thus far have been for the purpose of espionage, and that there can be an argument about the 'semantics of attack versus espionage and intrusion' [4].

1. REGULATING WARFARE

The Principles and Norms of Armed Conflict

The law of armed conflict (LOAC also commonly referred to as the 'laws of war' or *jus in bello*) as understood today originated in the mid-19th century, as did the humanitarian regulation of conflict and violence.[1] Since their early beginnings these laws applied primarily to interstate conflict as carried out by uniformed armed forces between two or more states. The principles, rules, and norms that guide today's LOAC can be found in a variety of sources: customary law, international treaties, judicial decisions, legal philosophers, and military manuals. Although the customs of the LOAC can be traced as far back to the 15th century medieval Europe, its more modern origins date back to the American Civil War of 1861-1865 [5]. Until that era Dale Stephens and Michael Lewis, note that 'there was no meaningful *jus ad bellum* because the right to resort to force was essentially unchallenged'. The ideals of knighthood in the Middle Ages provided some restraint against certain warfare cruelties [6]. Not until the 17th century was there a systematic legal code on war and peace. The Hugo Grotius

[1] Sections of the LOAC that deals explicitly with civilians are commonly referred to as *international humanitarian law*.

work of 1625, *On the Law of War and Peace* (*De jure belli ac pacis*), was based on the natural law. In the 18th century natural law and the Golden Rule were formulated by Emerich de Vattel in his 1758 work, *The Law of Nations* (*Droit des gens*). It was not until the American Civil War that the laws of armed conflict were codified and adopted by the leading world states. The first Geneva Convention was agreed upon in 1864 and employed the US Lieber Code (US War Department, General Orders No. 100, 24 April 1863) as a baseline. By 1868 in St. Petersburg a treaty, as noted below, was signed by leading nations/empires (excluding the US) that concerned regulating warfare methods and means.

Regulation of war and violence under humanitarian principles arose in 1863 with the establishment of the International Committee of the Red Cross (ICRC). During the 1876-1878 Russo-Turkish War the Ottoman Empire established the Red Crescent and there followed mutual agreement for respect from both sides whether honoring the principles of the Red Cross or the Red Crescent. The ICRC acts as a guardian and promoter of international humanitarian law.[2] The establishment of the First Convention of the Geneva Conventions in 1863 followed the formation of the ICRC and concerned the welfare of the wounded, civilians, shipwrecked, and prisoners of war. Four Conventions were adopted/revised through 1949. Amendment protocols about victim protections were added from 1977 through 2007. In 1899 and in 1907 peace conferences were held in The Hague for the purpose of regulating weapons in war and the customs and laws of war. The Geneva Protocol to the Hague Convention of 1925 prohibited other uses of gas and biological weapons. Later treaties of 1972 and 1993 covered the production, storage, or transfer of these weapons. A 1951 UN Convention Relating to the Status of Refugees covered post-WWII European refugees and was expanded in 1967 to cover other refugees without time or geographical limitation. While also referred to as the 'Geneva Convention' this UN treaty is actually not part of the Four Geneva Conventions [7].

New Weapons and LOAC

It was not until the 1868 ratification of the St. Petersburg Declaration that emerging military technologies were subject to any type of international legal review. This action is officially stated as the Declaration Renouncing the Use, in Time of War, of Explosive Projectiles under 400 Grammes Weight. A rash of new weapons introduced during the American Civil War, including armoured warships, submarines, as well as land mines, machine guns, projectiles filled with clear glass or explosives, exploding bullets, and dumdum bullets designed to flatten on impact. Weapons that caused unnecessary physical harm prompted delegates at St. Petersburg to issue a statement on this vexing issue in their declaration:

> *The Contracting or Acceding Parties reserve to themselves to come hereafter to an understanding whenever a precise proposition shall be drawn up in view*

[2] The ICRC explains its mission as 'an impartial, neutral and independent organization whose exclusively humanitarian mission is to protect the lives and dignity of victims of war and internal violence and to provide them with assistance'. The organization 'endeavors to prevent suffering by promoting and strengthening humanitarian law and universal humanitarian principles' [8]. The seven Fundamental Principles of the International Committee of the Red Cross are: humanity, impartiality, neutrality, independence, voluntary service, unity and universality. It was not until 1929 with the amendment of Geneva Convention (Article 19) that the Red Crescent was officially recognized as part of the ICRC.

> *of future improvements which science may effect in the armament of troops, in order to maintain the principles which they have established, and to conciliate the necessities of war with the laws of humanity.*[3]

In 1977, Protocol 1 of the Geneva Convention also cited the need to carry out international reviews of new weapons. The intent of Article 36 is to determine the lawfulness of new weapons before they are developed, acquired, or incorporated into a military's arsenal. Legal experts have also argued that scrutinizing the legality, means, and methods of new warfare is so basic that it applies to all nations, even those not a party to Protocol 1. Thus, states following international law must ensure that new innovations do not run afoul of international obligations. Article 35 states:

> *In the study, development, acquisition or adoption of a new weapon, means or method of warfare, a High Contracting Party is under an obligation to determine whether its employment would, in some or all circumstances, be prohibited by this Protocol or by any other rule of international law applicable to the High Contracting Party.*[4]

Article 6 is complemented by Article 82 of Additional Protocol 1, which requires that legal advisers be accessible at all times to battle commanders and that 'on the appropriate instruction to be given to the armed forces on this subject'. Thus, the basic aim of both the St. Petersburg Declaration and Protocol 1 of the Geneva Convention is to ensure that armed forces will carry out battlefield hostilities in strict accordance with the principles, rules, and norms as established by the LOAC. However, one problematic area is found in Article 36 that does not delineate specific means by which legal review of new weapons and methods of warfare will take place, leaving actual practice open to much interpretation. Few states have actually codified this mandate into state practice except for the United States and Sweden.[5] In 1999 the 27th conference of the International Conference of the Red Cross and the Red Crescent encouraged states to, 'to establish mechanisms and procedures to determine whether the use of weapons, whether held in their inventories or being procured or developed, would conform to the obligations binding on them under international humanitarian law'.[6] In 2003, in light of war in the Middle East, the ICRC reaffirmed this goal 'the legality of new weapons under international law' and 'in light of the rapid developments of weapons technology and in order to protect civilians from the indiscriminate effects of weapons and combatants from unnecessary suffering and prohibited weapons'.[7]

---

[3] Excerpt taken from the section renouncing the Use, in Time of War, of Explosive Projectiles Under 400 Grammes Weight, St Petersburg, 29 November / 11 December 1868.

[4] Article 35, Protocol, 1977

[5] The United States and Sweden both established formal review practices as early as 1974.

[6] Section 21, Final Goal 1.5 of the Plan of Action for the years 2000-2003 adopted by the 27th International Conference of the Red Cross and Red Crescent, Geneva, 31 October to 6 November 1999.

[7] Final Goal 2.5 of the Agenda for Humanitarian Action adopted by the 28th International Conference of the Red Cross and Red Crescent, Geneva, 2-6 December 2003 [hereinafter Agenda for Humanitarian Action].

2. LOAC in Cyberspace

Thus far, there is no international consensus on the application of the laws of armed conflict to 21st century cyber warfare, likely irregular warfare. This problem stems from both the loose definition of cyber warfare as well as the lack of precedent through which to guide present and future law. Another challenge is that not unlike other means of 21st century warfare, cyber warfare is coming of age in an era where the Westphalian state order is undergoing vast transformation. As the borderless realm of cyberspace both ignores and challenges state boundaries, a hands-off policy may disrupt or worsen cross-border transactions. As explained by Vida Antolin-Jenkins, USN JAGC (United States Navy-Judge Advocate General Corps), ‘Cyberspace operations for the most part do not meet the criteria for ‘use of force’ as currently defined by international law. Defining the parameters of proportional response through analogy is possible, but creates clear dangers of definitional creep into other areas of international relations that have long been the subject of long and contentious debate’ (Antolin-Jenkins 2005: 134). Let us remember that Gen. Chilton does not rule out responding with kinetic force to a cyber attack [2].

Disruptive Technologies on the Battlefield

Throughout history, there are numerous examples of disruptive technologies that have changed the way war has been waged, often resulting in enormous transfer of powers. Since hunter-gatherer tribes first wandered the earth, man has sought strategic advantage through the development and application of new technology. William Owens (US Admiral-Ret.), applying historian Martin van Creveld’s divisions of military history, reminds us that the ‘Age of Tools’ (prehistoric age to about 1500 AD) witnessed battles of the ‘muscular power of men and animals’ using ‘the wheel, the stirrup and iron weaponry’. In the ‘Age of the Machine’ (18th thru the 19th centuries) artillery weapons were eventually wielded by large units of fighting men [9]. The ‘Age of Systems’ developed between WWI and WWII with radar, long-range aircraft and radio coordinated ground-to-air attacks. We are presently living in a 21st century ‘Age of Automation’ where even older munitions and aircraft are deployed by sophisticated communications technologies as faster and more precise navigation and related digital devices are currently in development [9].

Using information and/or computer technology to impose one’s will upon an enemy is a form of warfare-- information warfare (IW). In 1995 Martin Libicki of the US National Defense University in classifying IW forms stated: ‘Seven forms of information warfare vie for the position of central metaphor: command-and-control (C2W), intelligence-based warfare (IBW), electronic warfare (EW), psychological warfare (PSYW), hacker warfare, economic information warfare (EIW), and cyberwarfare’ [10]. Robert Hanseman in the late 1990s anticipated how information warfare will blossom into a full-fledged “Revolution in Military Affairs” (RMA) [11]. Looking specifically to the US global role, Owens sees this RMA as an ‘opportunity to use the new information technology to change the very nature of our military, in a way that could reinvigorate American political, diplomatic, and economic leadership in the world for decades to come’. From a specifically US security perspective, Owens holds that in this new century the ‘changing world demands a new way of looking at war and the proper military force’ [9]. This paper maintains that in addition to this needed military strategy transformation there needs to be a new legal/treaty framework defining the cyber security regime.

Another question: What is the nature of the *force* deployed by cyber attackers or cyber terrorists and is this *force* an act of war? Michael Schmitt points to an International Court of Justice (ICJ) finding that 'supports a conclusion that a use of force need not be kinetic in nature' when the ICJ ruled (1986 Nicar. v. US) that 'although the funding of guerrilla forces was not a use of force, arming and training them was' [12]. Today cyber weapons may potentially be in the possession of varied and mixed-motive warriors, terrorists, or civilians. The legal assessment of cyber attack or cyber war, Schmitt offers, will come from the community. He concludes that a 'cyber attack that causes significant human suffering or property damage is obviously an armed attack justifying a response under the law of self-defence. Schmitt also speaks to the need for the law to mature since the 'global community finds itself at the cusp of normative change' in seeking a definition of aggression [12]. As this paper urges, the year 2009 is the time to begin constructing a global cyber security regime to bring clarity to the place of cyber attacks or cyber terrorism within the 1945 UN Charter.

Info Age Warfare

The transition through the phases of information technology with increasing dependency upon sophisticated devices and digital applications have led logically to phases --warfare involving information systems, hence new battlefields or rather, *battlespaces*. In one sense, modern information warfare can be traced back to the introduction of long-distance telecommunications. With the introduction of the telegraph, telephone, and radio, both civilian and military leaders gained an unprecedented command and control authority over troops movements and deployments. With a real-time link established between civilian and military headquarters, political leadership where possible could exert much more decision making authority, sometimes to ill effect. US President Abraham Lincoln was the first commander in chief to use the telegraph to issue orders to his Northern Generals in real-time. From across the Atlantic, Queen Victoria made extensive use of the telegraph to communicate her vision to overseas colonial viceroys.

During WWI, communications instruments were fully integrated into land, sea, and air campaigns. Later the sheer complexity and scale of WWII increased the need for cybernetic controlled weapons. As firepower increased again throughout land, sea, and air, the need for more precision and predictability increased. This was especially true in two areas of artillery and air combat. The Bletchley Park British code-breakers of WWII decrypted over 3000 daily German Enigma messages through the computational developments of Alan Turing [13]. The British and US through technology exchanges perfected their microwave and radar system and eventually were able to jam the German radars.

The atomic age only increased cybernetic development as the need for even more precise, reliable, and speedy command-and-control operations increased several orders of magnitude. The deployment of nuclear weapons required the most sophisticated air and space communications networks ever developed. Between the 1950s and 1960s, the space and nuclear race helped propel a host of new information communications technology (ICT) breakthroughs with the invention of the transistor and the microprocessor. From the 1970s forward, many of these dual-use military and space technologies found their way into the civilian sector. Historically, with each generation, the ability to hunt and destroy has increased.

Peace and Security and the Cyber Challenge

'A war of aggression is a crime against international peace. Aggression gives rise to international responsibility' states the UN Charter (Article 5, Paragraph 2 of the Definition of Aggression). This statement provides a basis for considering an attack on the information infrastructure of a nation as an act of aggression. Attacks have the potential to disrupt a nation's power grids, transportation links, health care service, emergency response, financial flows among many other venues. Under this Charter from 1945, a nation has the right to self defence. Lawmakers, diplomats, and military strategists need to confront the tasks for defining and framing the regime that will protect the cyber security for national defence and civil society functions. A major international military alliance such as NATO has a task suitable to its own mandate or charters in harmonizing accepted policy on aggression and protective national defence on the its own membership soil and their protectorates and in its global expanse of cyberspace.

Antolin-Jenkins sums up the predicament of a digital society: 'The strategic and economic power of the increased information awareness and connectivity are coupled with a hugely increased vulnerability to destruction and attack. This information network has created a new battleground'. She also reminds her readers that 'One estimate is that 95% of military information traffic utilizes civilian networks at some stage of communication' [22][14]. How or when could there be a separation of military from civilian networks? Is an attack on a military network also an attack on civilians, or *visa versa*? An advisor to the director of US national intelligence, Steven Chabinsky, reminds us that even if the laws of war 'would forbid targeting purely civilian infrastructure' we need to consider that 'terrorists, of course, don't limit themselves by the Geneva Conventions' [15].

Humanitarian Law for Cyber Weapons?

Does cyber warfare fall under the international humanitarian law (IHL)? Jeffrey Kelsey holds that IHL 'should evolve to encourage the use of cyber warfare in some situations and provide states better guidance in the conduct of these attacks' [16]. He argues that for a decade or more the 'potential threat and opportunity of cyber warfare' have confronted military planners while the 'international community has yet to reach consensus on the application of IHL'. This lack of consensus may be due to a variety of reasons, from holding that the 'current IHL framework can be applied to cyber warfare by analogy' to the realization that vast growth and fluidity of technology would make potential international agreements obsolete [16]. Kelsey further maintains that 'IHL applies to cyber warfare by analogy but contends that IHL must evolve to accommodate and, in some cases even encourage cyber warfare over conventional methods' [16].

The movement in a cyber attack across a neutral state becomes more than a 'mere communication signal', for cyber weapons can cause damage to states as have more conventional weapons. A weapon the 'size of a electron' could be a violation of the territory of a neutral state according to the Hague Convention that 'forbids the movement of weapons' across a neutral state which risks being drawn into a wider cyber conflict when its Internet nodes are engaged by a belligerent [16]. What obligations would a neutral state have as a conduit for cyber attack and mischief? Kelsey argues against establishing new treaties and in favour of states and their military commanders to follow established legal principles in cyber combat [16]. The question remains: What standards should emerge for a cyber security regime under established peace and war legal principles? Since cyber warfare is still in its infancy, some would argue that regulating it is a difficult if not an impossible challenge. However, the so-called *catastrophic* cyber attack is to be avoided, it would be foolish and impractical not to establish some type of international rules of the game as deterrence.

3. Towards a 21st Century Global Cyber Regime

Thus far, the diplomatic community has had little to say about the governance of cyber warfare. Two exceptions of major importance include former diplomats with knowledge of ICT have in recent months discussed with this author the major international relations quandaries from cyber threats and attacks plus their own concerns about diplomatic-level solutions. These former envoys are retired senior US Amb. Thomas Pickering who served over four decades in major posting for the US Dept of State, and Amb. David Gross, U.S. Coordinator for International Communications and Information Policy 2001-08. The latter observes that diplomatic silence may be attributed largely to a generational gap and a lack of technical understanding by policy makers since the Internet and associated networks are fairly recent developments; therefore, cyber security concepts in international affairs are still a nascent on the part of the diplomatic community (Gross to the author: April 2009). With wide-ranging diplomatic and corporate experience, Amb. Pickering sees an even larger problem in that forming an international treaty involves major, prolonged steps and major questions: What is the problem to be solved? How will the problem evolve in the future? (Pickering to the author: July 2009). Since cyber warfare is being conducted and developed during a period of wide interstate trade and general economic accord or agreement, there is an opportunity to design a governing framework before an actual global catastrophic attack takes place. Today the questions remains: Can governments be motivated to take action now before it is too late?

Special regimes have been formed for far-ranging interests or activities, such as treaties governing the Arctic, Antarctic, canals, international rivers, and outer space. While somewhat vague or undetermined there appears to be a consensus that outer space begins where airspace ends [17]. Examples of the treaties governing outer space include those governing the International Space Station (1998), Registration of Objects Launched into Outer Space (1975), INTELSAT or International Telecommunications Satellite Organization (1986), INMARSAT or International Maritime Satellite Organization (1976), as well as the ITU or International Telecommunications Union (1932; 1947 as UN agency) [17]. You may ask, 'Where does cyberspace begin?' It would appear that cyberspace begins with the keystroke to log on to a cyber network, whether from a mega terminal, a PC, a game console, or a mobile telephone.

Eventually, the ultimate venue for cyber warfare governance would be The Hague as the home of the world's first Peace Conference and for over a century as the international centre of justice and arbitration, as well as warfare governance. The Hague hosts several international organisations, including the UN International Court of Justice, the Permanent Court of Arbitration, the NATO Consultation, Command and

Control Agency (NC3A). As cyber warfare moves to the forefront of more government agendas, more questions arise as to how the Law of Armed Conflict and the Geneva Convention apply to cyberspace. Responses are likely to range on both sides of the fault line: those who see cyber warfare as fitting neatly under existing LOAC (as well as under the UN charter), and on the other side, those who see the need for an entirely new set of international laws and treaties to govern cyber warfare.

A Way Forward

While it is unlikely that these two countervailing diplomatic/legal views will be reconciled anytime soon, the time is now to begin having this debate in a more serious, focused manner. Again, because cyber warfare is a complex and dynamic issue, these debates will need to be hosted in many different venues and viewed from many different perspectives. NATO is already playing an important role in this debate by hosting conferences such as this June 2009 Tallinn gathering bringing together the relevant players from both member states and global partners. For the foreseeable future, this NATO Center of Excellence can play a critical role in bringing together the best experts both to analyze and to debate the problems from a number of unique cultural and disciplinary perspectives.

As the world's premiere military alliance, NATO is positioned to play a major role by facilitating significant interstate dialogue between civilian and military planners. Each year NATO hosts various fora where such engagements take place. The NATO Global Partnership Program provides a mechanism to reach out to other countries. NATO could also explore reaching out to other peace and security alliances, such as ASEAN Regional Forum (ARF) and the Shanghai Cooperation Organization (SCO), for the purpose of exploring confidence-building measures with that global hemisphere.

Although presently cyber warfare/defense is largely an ungoverned affair, the UN leadership has already acknowledged the severity of the problem and the need for governance. UN secretary-general Ban Ki-moon earlier this year announced that the UN Advisory Board on Disarmament Matters is to include cyber weapons in its arms list [18]. In his prepared February remarks to this Advisory Board the Sec-Gen stated:

> *This year you will be considering cyber warfare and its impact on international security. As you know, there have been many widely reported breaches of information systems in recent years. With both the public and private sectors growing increasingly dependent on electronic information, your work in this area is very timely. It will also complement the efforts of the panel of governmental experts that will be addressing information security later this year [19].*

The UN's International Telecommunications Union last year concluded its agreement with the International Multilateral Partnership Against Cyber-Terrorism (IMPACT) to conduct the ITU Global Cybersecurity Agenda (GCA) with headquarters in Cyberjaya, Kuala Lumpur. The GCA seeks international cooperation for governments, international law enforcement authorities, the private sector, international organisations, and civil society for the purpose of a secure cyberspace. Through five areas the GCA if focused on strengthening the legal framework, technical measures, organizational structure, capacity building, and international cooperation. [20] At its Cyberjaya headquarters inauguration the ITU-GCA was billed as 'public-private initiative' and a 'framework for cooperation aimed at finding strategic solutions to boost confidence and security' in a networked world [21].

Relevant committees beyond the UN should also begin to debate a proper a framework for cyber warfare while other international fora can and should play a role. Significant organisations such as the Institute of Electrical and Electronics Engineers (IEEE) and the Internet Society (ISOC) are positioned to use their technical legitimacy and their soft power to press forward with best practices for member states to follow in securing their own sovereign cyberspace. World Trade Organization (WTO) signatories should develop an agreement for pledges from nations not to promote or solicit mercenaries, or attack a member state's trade infrastructure. On the military side, the national leading military services should be encouraged to act with transparency where possible so as not to launch initiatives that unduly contribute to unnecessary cyber arms race between states.

One of the most difficult governance areas to reconcile will be in the area of police vs. military involvement in cyber security/defence. As the US-led global war on terror has shown, there is no clear line of authority when defending against threats where state involvement is murky at best. Due to the anonymous and secretive nature of cyber warfare, state involvement is often tricky for producing fool-proof forensics that can prove state involvement. While each country will ultimately need to sort through this problem in accordance to its national laws and constitutions, more global debate will be needed to clarify these issues.

Participants in Tallinn are witnesses to the positive leadership role assumed by Estonia and leading to the NATO CCD COE here at the scene of the first acknowledged, major interstate cyber conflict. There is potential for a cyber treaty to emerge should the North Atlantic Council embrace thoroughly the cyber warfare issue and engage the NATO Consultation, Control and Command Agency (NC3A) and the NATO Military Authorities (NMA). The Tallinn cyber convention questions and discussions are a start, but a protocol or treaty governing the conduct of cyber warfare needs serious consideration.

4. Conclusion

While cyber warfare is not an entirely new area of modern warfare (at least as viewed within an Internet world), its current evolution poses many challenges to international peace and stability. The increasing quantity and quality of online attacks threaten many parts of civil society that depend on reliable networks and information systems. Growing evidence of state-sponsored cyber attacks is especially alarming and could spark a serious arms race in cyberspace. Understandably, a number of countries have announced plans for full spectrum military cyber commands. As history has demonstrated, while international law cannot stop states from going to war with one another, it can go a long ways towards regulating their conduct should hostilities boil over into actual war. Some may argue that because cyber warfare is still in its formative stages, it is premature to begin work on a global regime to regulate it. However, it can also be logically argued that absence of some rules of the game, states will not feel constrained to develop and deploy cyber weaponry if the consequences are not understood by both military and civilian planners. While it is difficult to estimate the true potential for a catastrophic attack to spill over to kinetic warfare between states, the notion that the threat exists at all is cause enough to begin constructing a regime or legal framework through which to conduct cyber warfare.

History presents another lesson in that even with the best intentions and resources, a global cyber security regime will not transpire in short order. It will take many years to form an effective international consensus that might translate into a revision of the Law of Armed Conflict as spelled out by the Geneva Conventions. The operative concept is *regime*. And, the time to establish a global cyber security regime is *now*. As a proper follow-up to the innovative inaugural Tallinn CCD COE conference of 2009, NATO can and should play an important role by bringing together in short order the relevant stakeholders to outline a viable cyber security regime. Lt. Col. Walker quoted in the introductory epigraph to this paper, eight years ago properly challenge the legal community, and this writer has chosen to extend his challenges to the wider global policy community.

References

[1] Walker, Jeffrey K. (2001) 'The demise of the nation-state, the dawn of new paradigm warfare, and a future for the profession of arms' *Air Force Law Review* (51:2001).

[2] Stars and Stripes (2009) 'Pentagon Steps Up to Fight Cyber War' (30 May 2009) http://www.military.com/news/article/pentagon-steps-up-to-fight-cyber-war.html?col=1186032310810

[3] Schogol, Jeff (2009) 'Official: No options 'off the table' for U.S. response to cyber attacks', *Stars and Stripes* (Mideast edition, 08 May 2009). http://www.stripes.com/article.asp?section=104&article=62555

[4] Gertz, Bill (2009) 'Cyber warfare plans' *Inside the Ring* (04 June 2009) http://www.gertzfile.com/gertsfile/InsidetheRing.html

[5] Reynolds, Jeffrey. (2005) 'Collateral Damage on the 21st Century Battlefield', Air Force Law Review (56:2005).

[6] Stephens, Dale and Michael Lewis (2005) 'The law of armed conflict—a contemporary critique' *Melbourne Journal of International Law* (6:2005).

[7] 'Geneva Conventions: A Reference Guide' (2009) Society of Professional Journalists. http://genevaconventions.org (Accessed: 01 June 2009)

[8] ICRC-International Committee of the Red Cross (2009) 'The mission'. http://www.icrc.org/HOME.NSF/

[9] Owens, William (2001) *Lifting the Fog of War* (New York: Farrar, Straus & Giroux).

[10] Libicki, Martin (1995) 'What is information warfare?' *ACIS Paper 3*: August 1995; National Defense University Press. http://www.iwar.org.uk/iwar/resources/ndu/infowar/a003cont.html

[11] Hanseman, Capt. Robert G. (1997) 'The realities and legalities of information warfare' *Air Force Law Review* (42:1997)

[12] Schmitt, Michael N. (2003) 'The sixteenth Waldmar A. Solf lecture in international law' *Military Law Review* (176:2003).

[13] Keizer, Gregg (2006) 'British Museum unveils WWII computer replica' *InformationWeek* (08 September 2006). http://www.informationweek.com/news/security/government/showArticle.jhtml?articleID=192700296

[14] Antolin-Jenkins, Cdr.Vida M. (2005) 'Defining the parameters of cyberwar operations: Looking for law in all the wrong places?' *Naval Law Review* (51:2005).

[15] Tennant, Don (2009) 'The fog of cyber war' *Computerworld*, 27 April 2009.

[16] Kelsey, Jeffrey T.G. (2008) 'Hacking into international humanitarian law: The prince-les of distinction and neutrality in the age of cyber warfare' *Michigan Law Review* (106:2008)

[17] Aust, Anthony (2005) *Handbook of International Law* (Cambridge: Cambridge University Press).

[18] Marks, Paul (2009) 'Pentagon readies its cyberwar defences' *New Scientist* (16 arch 2009).

[19] Ki-moon, Ban (2009) 'Secretary-General's remarks to the Advisory Board on Disarmament Matter' (New York: 18 February 2009). http://www.un.org/apps/sg/sgstats.asp?nid=3717 http://www.un.org/apps/dsg/sgstatsarchive.asp

[20] ITU-International Telecommunications Union (2008) 'ITU's global cybersecurity agenda housed in Malaysia' (04 September 2008). http://www.itu.int/newsroom/press_releases/2008/27.html

[21] ITU-International Telecommunications Union (2009) 'ITU's global cybersecurity agenda housed in new centre in Malaysia: IMPACT headquarters in cyberjaya to focus on strengthening network security' (20 March 2009). http://www.itu.int/newsroom/press_releases/2009/08.html
[22] Knecht, Ronald and Ronald Grove (2001) *The Information Warfare Challenges of a National Information Infrastructure'* http://web.archive.og/web/2001110717440/http://invorwar.com/mil_c4i//iwchall.hyml-ssim

The Virtual Battlefield: Perspectives on Cyber Warfare
C. Czosseck and K. Geers (Eds.)
IOS Press, 2009
© 2009 The authors and IOS Press. All rights reserved.
doi:10.3233/978-1-60750-060-5-118

What Analogies Can Tell Us About the Future of Cybersecurity

David SULEK[a], and Ned MORAN[b]
[a] *Principal, Booz Allen Hamilton*
[b] *Senior Consultant, Booz Allen Hamilton*

Abstract. For more than a decade, leading experts in government and industry have warned of an impending Cyber Pearl Harbor, a surprise electronic attack with the potential to neutralize U.S. military power and cause massive disruptions in U.S. and global computer networks. This is a powerful historical analogy—but is it the right one? This paper articulates a framework to better explore and examine the use of historical analogies in their application to conflict in cyberspace. The resulting analysis does not seek to argue the Pearl Harbor analogy is a bad one. Quite to the contrary—our thesis is that while a cyber Pearl Harbor remains a possibility, is should not be treated by decision makers as an inevitability and that there may be equally powerful historical analogies to guide future cyber strategies.

Keywords: cyber conflict, cybersecurity, decision science, decision-making, historical analogies, public policy

Introduction

In their study Thinking in Time: The Uses of History for Decision-Makers, authors Richard Neustadt and Ernest May speak to the power and perils of making decisions through the use of historical analogies. They argue for a structured, critical inquiry to address an issue or crisis rather than leaping to a single analogy for which to formulate strategies and policy options (e.g., "Appeasement at Munich"). Systematic use of appropriate historical analogies can clarify the present situation, offer strategic insights, and inform policy options. On the other hand, incorrectly applying an analogy can muddy objectives, narrow policy options, and create blind spots for decision-makers.

One can debate when cybersecurity first emerged as an issue, but many consider the 1988 Morris Internet Worm a common marker. In the 20 years since this self-replicating program spread across the Internet at remarkable speed, attention has turned to countering fast-moving, continuously evolving cyber threats and vulnerabilities. In that time, a single historical analogy has appeared to dominate US Government thinking: the threat of a cyber Pearl Harbor.

This is a powerful, even seductive possibility. It connotes a bold stroke launched by an enemy without warning designed to neutralize US military power. This represents an imminent threat that is ignored only at one's own peril. The introduction of new weapons, strategies, doctrines, and tactics suddenly tilt the military balance toward the offense. Even those who do not directly advocate the Pearl Harbor analogy often employ similar imagery. For example, a number of cyber experts have suggested the potential for a "cyber 9/11"[1] or a "cyber Katrina.[2] While important distinctions exist between these analogies (e.g., Pearl Harbor centered on a state-based actor, 9/11 on a non-state actor, and Hurricane Katrina on an 'act of God'), the implications are clear. Drawing from history's lessons, experts warn of the potentially catastrophic dangers facing our cyber networks unless immediate, decisive action is taken.

If the Pearl Harbor analogy proves correct, one can argue the US and other countries will be better prepared. What if, however, the analogy proves erroneous or the wrong lessons are drawn? For example, could the focus on a single analogy ultimately create a self-fulfilling prophecy, something more akin to a modern *Guns of August?* This paper will not argue the Pearl Harbor analogy is a bad one. Instead, our thesis is that *while a cyber Pearl Harbor is a possibility, it should not be treated as inevitable.* To test this thesis, this paper will explore a range of historical analogies that might inform different options and courses of action available to decision-makers.

1. Thinking in Time

An inspiration for this paper is Thinking in Time, which outlined a systematic framework for policy practitioners to critically analyze key policy challenges and formulate well-reasoned strategies and options. Through case study analysis, Neustadt and May point to six problems that often negatively impact the quality of decisions:[3]

- A plunge toward action
- Overdependence on fuzzy analogies
- Inattention to an issue's own past
- Failure to think about key presumptions
- Stereotyped suppositions about persons or organizations
- Little or no effort to see choices as part of a historical sequence

To address these common shortcomings, the authors conclude that "better decision-making involves drawing on history to frame sharper questions [about a crisis or policy challenge] and doing so systematically, routinely."[3] Specifically, in response to a crisis or policy challenge, they recommend that decision-makers develop a detailed issue history. This history will enable them to clarify the overarching policy objectives and anticipate those conditions that are desired in the future after actions are taken. At the same time, an issue history provides the basis for determining which historical analogies might apply—and why. Neustadt and May then outline a process for developing issue history, summarized below:[1]

- **Determine the *Story* and *Timeline*.** The centerpiece of an issue history is a narrative story (what is happening today and why) accompanied by a timeline. The authors emphasize the timeline should begin at the earliest possible and relevant date of significance to ensure proper context for analysis.

[1] Authors note: we have taken the liberty of condensing and summarizing steps that are spelled out in detail in Chapters 6-14 of *Thinking in Time* [3].

- **Identify *Change Points*.** On that timeline, understand where significant changes altered the trajectory or thinking about the current issue.
- **Separate the *Known*, *Unclear*, and *Presumed*.** The authors point to the need early in a policy crisis to determine what is known (facts), what is unclear (absence of facts or evidence), and what is presumed (assumptions).
- **Challenge *Presumptions*.** Decision-makers must carefully review the core presumptions. Good presumptions are those that clarify and define a situation and surface concerns. Bad presumptions are value-laden, things that cannot be challenged save in its own terms by opposed values (e.g., the authors use the model of "communists are bad; market mechanisms good").

An issue history becomes the foundation from which decision-makers can judiciously compare current and past events to determine likenesses and differences. Embedded in these likenesses and differences are key insights that can shape future strategies. In concluding their analysis, the authors state, "Sensing that the present was alive with change, they knew the past would be outmoded by a future that had never been…but their image of that future could be realistic because [it was] informed by understanding its sources in the past"[3] In other words, no single historic event will prove a perfect analogy to the present moment—the underlying conditions will be different. However, thoughtful selection of historical analogies can offer decision-makers insights that enrich and inform the choices they must make while also enabling them to better anticipate the downstream implications of those choices.

2. The Cyber Issue History—In Brief

In April 2007, the Estonian government and many of that country's key lifeline infrastructures faced a barrage of coordinated cyber attacks. An unseen adversary launched sophisticated attacks to cause massive network disruptions—"a flood of bogus requests for information from computers around the world conspired to cripple the websites of Estonia banks, media outlets, and ministries for days."[4] Without delving into the specific causes, actors, and motives of the Estonian attacks, the entire event confirmed what many experts warned about cybersecurity. An adversary was able to employ cyber weapons and strike without warning. Directly attributing the source of the attack proved fleeting. Critical infrastructures appeared fragile in the face of withering DDoS attacks. Computer attacks were accompanied by social engineering and flash mobs to magnify effects. In response to this and other cyber events, the U.S. launched a comprehensive review of its national cybersecurity strategies.

Today's cyber issues, however, trace their lineage to the Superpower technology rivalry that was energized in 1957 with the launch of Sputnik. Following that psychological shock of this event, the US embarked on an ambitious program to ensure technological superiority for the foreseeable future. While the Space Program is often cited as the most significant post-Sputnik achievement, the seeds of the Internet were planted during this same era. The period beginning in the early 1960s and lasting through the late 1980s (see Figure 1) was dominated by government, industry, and academic researchers in their drive to develop internetworking technologies.

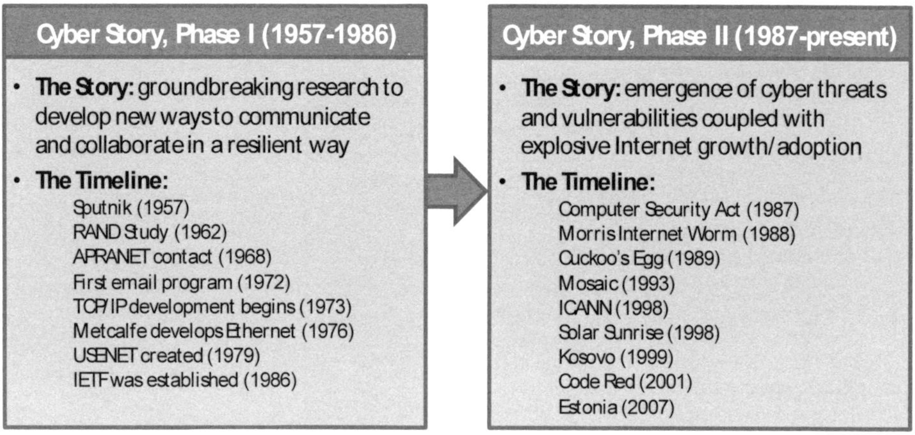

Figure 1. The Two Phases of the Internet's Issue History [5] [6]

Events in the late 1980s, however, would introduce vital change points—the emergence of cyber threats and vulnerabilities. After a raucous debate between the Executive and Legislative branches over roles and responsibilities for computer security, President Reagan signed the *Computer Security Act* in 1987. In 1988, the Morris Internet Worm is released and quickly self-replicates across the Internet, causing major disruptions. In response, the Defense Department creates the Computer Emergency Response Team at Carnegie Mellon University. In 1989, Stanford University professor Cliff Stoll publishes the *Cuckoo's Egg,* which detailed the real-life penetrations into US systems by a German hacker.

Two mega-trends dominate this second phase. The first is exponential growth in the number of hosts, users, computing power, and network capacity. For example, in 1984 there were 1,024 hosts worldwide; by February 2008, this number grew to more than 500 million.[6] Steep, explosive growth curves in these areas were accompanied by a growing military, economic, and societal dependence on the Internet and computer networks that permeate nearly every aspect of our lives. The second trend is the dramatic increase in cyber threats and vulnerabilities. During this period, cyber attacks grow in terms of *velocity* (speed of transmission), *volume* (attack frequency), *virulence* (impact, both direct and cascading), and *vector* (types of actors with the capability to launch attacks). The 2007 Estonian cyber attacks validated the dangers associated with the mix of growing dependencies, threats, and vulnerabilities. As one of the most wired societies in the world, Estonia was particularly vulnerable to this type of attack by a determined adversary employing hacking tools as the weapon of choice.

Before turning to consider historical analogies that might assist decision-makers formulate strategies and options for cybersecurity, Figure 2 outlines what we consider (at a high-level) known, unclear, and presumed about the cybersecurity issue of today.

3. A Framework to Explore Cyber Analogies

The use of analogies is rampant in cybersecurity—and this should come as no surprise. Neustadt and May note throughout their text that analogies are most often used when

What is Known	What is Unclear	What is Presumed
• Cyber threats, vulnerabilities, and risks continue to grow in terms of velocity, volume, virulence, and vector • Nation-states and non-state actors are investing in cyberwar capabilities • Decreased resources are needed to develop cyberwar capabilities • Internet access and network capacity will continue to grow—with Asia becoming a more influential actor • Attribution complicates response and deterrence	• How grave is the threat? • Will next generation Internet technologies and applications be more secure? • Is there sufficient political will (US, global) to address cybersecurity issues? • What types of policy approaches (regulation, market forces, international agreements, others) can change the current security conditions of the Internet?	• The United States will retain a key leadership role in governing and influencing the Internet • Nation-states are a more serious threat than non-state actors • Cyberwar is low risk and high reward • Increased public-private cooperation will improve security

Figure 2. For today's cybersecurity challenge, what is known, what is unclear, and what is presumed

issues are complex and decisions time constrained. Cybersecurity is an enormously complex issue with high tech threats and vulnerabilities, a community jargon that appears to layman as science fiction, attacks that appear with no warning, and a dizzying array of potential adversaries. Moreover, we live in a fast adapting socio-technology environment where users routinely change their favorite "killer applications" on an accelerated cycle, opening new doors of vulnerability.

While Cyber Pearl Harbor is perhaps the most prevalent historical analogy used to describe the cybersecurity challenge, others (such as Cyber 9/11 and Cyber Katrina) are being used with increasing frequency. Beyond these, some experts point to the need for a "Cyber Manhattan Project" or a cyber legal convention modeled after the Law of the Seas. Still others believe we are in the beginning phases of a "Cyber Cold War"[2] and most recently one cyber expert spoke of the need for a "Cyber Monroe Doctrine."[7]

Figure 3 depicts a framework to help sort through a plethora of analogies that might apply to cybersecurity. This model is built along two axes. The vertical axis divides those analogies motivated by *inspiration* (hope and possibility) versus those motivated by *desperation* (fear and danger). Consider, for example, the contrast between efforts to deploy the telegraph versus preparations for the Y2K software vulnerability. The former was motivated by a desire to speed communications across an unwieldy continent and to facilitate transatlantic communications; the latter driven by a time-certain fear of a major technological calamity. The horizontal axis divides analogies where change is *systematic* (linear, evolutionary) versus those where change is *disruptive* (transformative, revolutionary). For example, both the 9/11 attacks and the outbreak of the First World War were events that significantly altered the course of history. The former was a disruptive event occurring with little warning and of a scale not imagined, the latter a product of a system of mobilization and planning that made war unavoidable once a chain of events commenced (and of a scale not imagined).

[2] Panel at the 2009 RSA Security Conference entitled, "Is There A Cyber Cold War?"

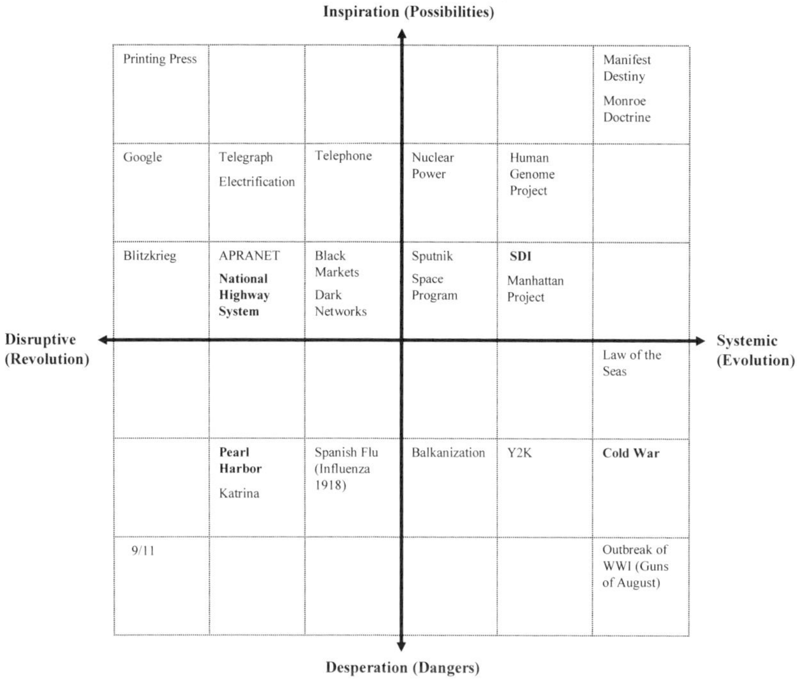

Figure 3: Framework to Analyze Cyber Analogies.

The remainder of this paper focuses on analyzing four analogies, one from each of the four quadrants of our framework: the Strategic Defense Initiative, the Cold War, the National Highway System, and Pearl Harbor. Each will be explored for its likenesses and differences to today's cyber issues.

4. The Strategic Defense Initiative (Inspiration, Evolution)

4.1. Overview

During the height of the Cold War, President Reagan proposed the developed of the Strategic Defense Initiative (SDI). Commonly referred to as the Star Wars program, SDI was envisioned as a system and capability to destroy Soviet Intercontinental Ballistic Missiles (ICBMs) while in flight in outer space. This proposed defensive shield held the potential of negating the carefully constructed logic of Mutually Assured Destruction (MAD) and conferring the United States a strategic military advantage over the Soviet Union. The mere threat of SDI forced the Soviet Union to respond by investing increased resources into its military programs in an effort to overcome the purported defensive shield constructed by the SDI. While SDI was never deployed or even proven to be technically feasible, the Soviets were compelled respond and many analyst believe the economic costs associated with this military buildup in response to SDI were a contributing factor to the downfall of the USSR.

While SDI provoked an offensive response from the USSR, the threat of cyber warfare has prodded the US to commit investments toward improving its cyber defense posture. The Bush Administration's Comprehensive National Cybersecurity Initiative (CNCI) is reported to have allocated close to $30B over the life of the program [8]; and this may, in fact, only represent a fraction of the resources required to protect the US from a cyber attack of national significance.

Despite the recent cyber attacks against Estonia and Georgia, the threat of large-scale cyber warfare between states is still theoretical. To counter the potential of this threat, the US has invested increased amounts of resources (both public and private expenditures) into cyber defenses designed to protect critical infrastructure—the purported targets of any cyber attack of national significance. As cyber defenses are not static and must constantly be monitored, evaluated, and improved in order to counter determined adversaries, resources must be committed over the long-term. This point is exacerbated by a fact of life in cyberspace—today, the balance strongly tilts toward the offense, where the ability to conduct offensive operations is cheaper, easier, and more effective in comparison to the high costs of mounting credible defenses.

4.2. Similarities and Differences

The similarities between SDI and cyber warfare lie in the responses to perceived threats. In both cases, the efficacy of the strategies and tactics were unproven. In the case of SDI, it was not clear the system would ever work—but the USSR could not take the chance that it might. Similarly, there may be open questions about whether a large-scale cyber attack might work over a sustained period of time against US military and infrastructure targets—but the US cannot take the chance that it might. Another similarity between these analogies is the relative costs of offense to defense. In the Cold War, the ability to produce nuclear missiles and other delivery systems was relatively inexpensive and certainly less expensive than trying to develop defensive systems that would be full-proof. Today in cyberspace, developing offensive capabilities is inexpensive, especially compared to the enormous costs of developing cyber defense-in-depth strategies.

The obvious differences between SDI and cyber warfare center on their application. SDI was inherently defensive in nature, whereas cyber warfare is perceived as primarily a stealthy, offensive weapon. Further, SDI held the potential to completely destroy the existing strategic paradigm of Mutually Assured Destruction and dramatically titling the global balance of power. Cyber warfare is still a new, yet-to-be-defined strategic paradigm where questions of balance of power are complicated by the roles and capabilities of governments, private corporations, and a host of non-state networks of actors (terrorists, organized crime, other dark networks).

4.3. Lessons That Can Be Drawn From This Analogy

SDI never actually worked or was deployed against the Soviet nuclear arsenal. And yet it has three important lessons for those who seek to develop cybersecurity strategies. First, it was a program largely motivated by inspiration. Always the eternal optimist, Reagan sought to find a solution that helped the world escape the horrors of Mutually Assured Destruction (MAD). Second and related, SDI did not accept the notion that the offense would always trump defense in the nuclear world. The entire MAD concept rested on the fact that during any nuclear exchange, both sides would retain sufficient offensive force to destroy the other. SDI was a bold move to change that paradigm. Third, in doing so, the US raised the costs for the offense. Today in cyberspace, the generally held view is the offense trumps defense with cost being the primary differentiator—it costs billions to erect defense-in-depth in cyberspace, and only thousands to attack it. However, SDI shows that it may be worth investigating strategies that seek to significantly increase the costs of the offense rather than trying to build the perfect defense.

5. The Cold War (Desperation, Evolution)

5.1. Overview

According to McAfee's 2007 Virtual Criminology Report, we are in the midst of a "cyber cold war." Specifically, the report states "attacks have progressed from initial curiosity probes to well-funded and well-organized operations for political, military, economic and technical espionage."[9] The analogy between the modern day cyber era conflict and the cold war conflict between the Soviet Union and the United States is primarily anchored in the idea that powerful nation-states are competing for influence and power without resorting to a direct conventional or nuclear war.

5.2. Similarities and Differences

The cyber as a Cold War analogy is ripe with similarities. The most obvious parallel between the Cyber and Cold War eras is the central role of espionage. The Department of Homeland Security's U.S. Computer Emergency Readiness Team received 37,000 reports of attempted breaches on U.S. Government and private sector systems, which included 12,986 direct assaults on Federal agencies in 2007 [10]. In addition, there were more than 80,000 attempted attacks on Department of Defense computer network systems. Countries such as China and Russia have been publicly implicated in many of these cyber attacks against US military cyber assets. In fact, Major General William Lord– the Commander of Air Force Cyberspace Command, has publicly stated, "China has downloaded 10 to 20 terabytes of data from the NIPRNet already." This appears to parallel efforts during the Cold War, where the Superpowers each invested resources into the creation and maintenance of rival spy networks. These networks were primarily designed to gather intelligence in an effort to gain a competitive advantage in diplomatic, economic, informational, and military confrontations.

Despite these similarities, this analogy is far from a perfect fit. First, the Cyber Era is multipolar as opposed to the bipolar structure of the Cold War. While the United States remains an unparalleled superpower, a number of other nation-states are quickly emerging as potential rivals to the US. In addition, there are a number of non-state actors (most notably terrorist groups) that threaten to acquire the means to launch cyber attacks of equal or greater capability that some nation-states. From a military perspective, this has occurred because the "costs of entry" are low—developing and maintaining a cyber capability is (in relative terms) remarkably inexpensive.

That stands in stark contrast to the Cold War, where the US and USSR needed to invest tremendous resources including time, treasure, and knowledge in order to become nuclear powers and to retain rough technological parity with respect to nuclear and conventional military forces. According to the Brookings Institute, the US spent approximately $5.5 trillion dollars on the construction and maintenance of its nuclear arsenal. The cost of becoming of nuclear power was high in part because of the tremendous capital investment required in the construction nuclear power plants, the physical weapons, and acquiring source materials.

In the Cyber era, organizations require only a fraction of these resources to become a "cyber power." According to a study conducted by the Naval War College and Gartner Inc. in 2002, it would require only five years and $200 million to execute a major cyber attack. [11] As the knowledge and the weapons, in the form of exploit code, required to conduct a major cyber attack has become increasingly available since the release of the Naval War College and Gartner study it is likely that such an attack could be carried out with less resources. The cost of a cyber warfare program is further reduced because there is very little capital investment required. Unlike nuclear weapons, cyber weapons are virtual and can be duplicated at very little cost.

5.3. Lessons That Can Be Drawn From This Analogy

The Cold War offers a powerful image, that of a protracted struggle between powers for political, military, and ideological supremacy. There are obvious similarities—the cat-and-mouse game of espionage the boils below the geopolitical surface; the proxy wars that may suddenly break out in cyberspace; and the importance of retaining technological superiority. There are obvious differences too—the Cold War was an ideologically-motivated, bipolar struggle between competing nation-states and the fear of MAD served as a governor on the actions of the two major actors.

However, one important similarity – and lesson to be drawn – is the close entanglement of economic, political, and security interests in devising a comprehensive strategy. In the Cold War, the US and USSR brought to bear all instruments of national power—economic, military, scientific and technological. In particular, the Mr. X telegram developed by George Kennan at the start of the Cold War outlined a comprehensive strategy where the US was able to bring all elements of its national power together toward a common objective, the containment of the USSR. A key predicate of that telegram was that conflict was inevitable between the two powers, and the U.S. required a proactive, comprehensive strategy to prepare for the characteristics of this new conflict. Given the new order being created in cyberspace – where the Internet touches all aspects of political, military, economic, and sociological life – perhaps one of the most important lessons from the Cold War is the idea of developing a Mr. X-like telegram for cyberspace that defines the boundary conditions for future conflict.

6. The National Highway System (Inspiration, Revolution)

6.1. Overview

In his 1955 State of the Union Address, President Dwight D. Eisenhower declared, "A modern, efficient highway system is essential to meet the needs of our growing population, expanding economy, and our national security." Nearly every aspect of this ambitious project sought to balance national security, public safety, and commerce as the country invested billions of federal dollars in road, bridge, and tunnel construction. The resulting National Highway System represented a remarkable achievement. More than 50 years after President Eisenhower's address, our automobile culture has fundamentally transformed America's way of life.

6.2. Similarities and Differences

A historian looking back 100 years from now might rightly proclaim the National Highway System and the Internet as two of the world's greatest cultural achievements, the Great Infrastructure Wonders of our age. There are numerous similarities. First, both represented significant step improvements in citizen access and mobility. Second, both were children of the Cold War. Eisenhower, fresh from his WWII experience in Europe where transportation and logistics were vital in achieving victory, the National Highway System was designed in part to ensure the US could quickly mobilize its forces and deploy them to Europe if the Cold War turned hot. The genesis of the Internet was, in part, efforts to build a resilient command and control system that could withstand a Soviet nuclear strike. Third, both infrastructures led to sudden—and unpredictable—cultural and societal shifts. In less than a decade, the National Highway System facilitated the growth of suburbs and accelerated the emptying of center cities. Within a decade, the Internet has created global online communities, perhaps a new form of "cyburbanization."

There are two additional, more subtle similarities. First, both infrastructures emerged from a combination of inspiration and desperation. In selling the idea of the National Highway System, President Eisenhower often changed his message to suit the audience. When speaking to war veterans, he emphasized security. When talking to Chambers of Commerce, he focused on the need to continue post-war economic growth. When talking to the Rotary Club, he stressed the need to reduce the number of highway fatalities. While the Internet has its roots in military resiliency, the research community helped guide and shape its development to promote greater collaboration. Second, both represented significant shifts in how the United States built infrastructures. Prior to Eisenhower, the States all developed their own roads and highways with differing standards and approaches. Eisenhower's approach "federalized" highways, leading to unchartered territory in terms of nationwide investments. Similarly, the Internet's development is truly unique as an infrastructure. Most US infrastructures began as regulated monopolies (telephony, power, banking, air travel) and were slowly deregulated. The Internet has never been regulated—and with its global reach and impact, has entered equally unchartered territories.

The differences between these two infrastructures are easily identifiable. One was US-centered; the other is global. The National Highway System was a top-down, Federally driven program; the Internet is, by its very nature, decentralized and governance is perhaps best described as ad hoc. Perhaps most germane, the National Highway System never could be viewed as the avenue of attack against the United States—it was a force multiplier and enabler. Cyberspace can and has offered some of the same features to the US as a force multiplier. But cyberspace also introduces (through technical vulnerabilities in networks) the means by which an adversary may attack and disrupt critical military and infrastructure operations.

6.3. Lessons That Can Be Drawn From This Analogy

Today, debates in cyberspace are far broader and more encompassing than cybersecurity. In the United States, we see a need to respond to our vulnerabilities to growing cyber threats and develop the ability to attribute attacks to better deter, prevent, and respond to them. At the same time, many of our citizens (and those of other countries) have expectations of online privacy. Beyond this, since entering office, the

Obama Administration has pushed for greater Internet access, online collaboration, open government, and transparency. At the international level, we are potentially entering a new era of Internet governance and influence where other nations share a common interest in limiting perceived U.S. dominance of the Internet and its governance structures.

While many of the analogies used today in the U.S. for cybersecurity have at their core a message of impending danger, President Eisenhower was able to demonstrate how to strike a balance in his messaging around the National Highway System. His message was large dose of inspiration—to lower highway fatalities, to create jobs, to improve the post-war economy—coupled with a tinge of danger, to remain vigilant should the US have to mobilize to Western Europe in the event of a communist invasion. Equally important, his approach blended the introduction of revolutionary concepts (e.g., a stronger Federal role in transportation) with evolutionary steps (e.g., the use existing State apparatus' to facilitate the flow of money). Any significant effort to address cybersecurity issues will require a similar approach—introducing new constructs, ideas, and strategies for cyber laws, Internet governance, etc., coupled with working within the existing confines of the system (at least initially).

The National Highway System analogy also offers a secondary lesson—that these types of decisions can carry a long tail and produce many unanticipated outcomes. Our transition en masse to automobiles changed our society (suburbs, summer vacations, an emphasis on automobile safety), changed our economics (dependence on foreign oil, rise of trucking), and our environment in ways President Eisenhower could never have predicted. Ultimately, this makes the case for adopting a balanced approach like that Eisenhower assumed in the mid-1950s: part inspiration, part desperation; part revolutionary, part evolutionary.

7. Pearl Harbor (Desperation, Revolution)

7.1. Overview

On December 7, 1941, the Imperial Japanese Fleet launched a surprise attack against the US fleet anchored at Pearl Harbor. The intent of this attack was to neutralize US military power in the Pacific as Japan continued to expand its empire.

7.2. Similarities and Differences

Many similarities exist between what happened at Pearl Harbor and what many experts believe theoretically could happen in cyberspace. Among the most obvious is the introduction of new strategies, tactics, and doctrine. During the First World War, aircraft were used to perform a variety of military missions, including bombing runs. As early as the 1920s, Navy's from across the globe began to recognize the potential of airpower as a new, offensive form of naval warfare. For example, General Billy Mitchell theorized that battleships could be sunk via an air bombing campaign. In the 1930s, Japan, the U.S., and Great Britain began to add aircraft carriers to their naval fleets. Theory was put to the test in November 1940, when the British launched the "first all-aircraft naval attack in history, flying a small number of aircraft from an aircraft carrier in the Mediterranean Sea and attacking the Italian fleet at harbor in Taranto. The effect of the British carrier-launched aircraft on the Italian warships foreshadowed the end of the 'big gun' ship and the rise of naval air power." [12] The Japanese studied this raid and built a war plan designed to strike at the heart of US military power in the Pacific, the US Naval Fleet anchored at Pearl Harbor. One can argue a similar progression has taken place in cyberspace, from the initial use of electronic warfare techniques during Operation Desert Shield/Storm through the spike in hacking attacks during the 1990s and 2000s witnessed in the US Government to the more coordinated and sophisticated cyber attacks launched against Estonia and Georgia (the modern day Taranto?).

As noted earlier in this paper, other similarities exist. This includes the notion of strategic surprise, with an enemy launching a no warning attack with devastating consequences and the desire to neutralize US military advantages. The general sense that intelligence and other information exists to point to the attack, information that may be overlooked or misunderstood if not placed in the correct strategic context.

At the same time, important differences exist in these two situations. First and foremost, in the case of Pearl Harbor, the enemy and its intentions were well known. For more than a decade (following the Japanese invasion of Manchuria in 1931), tensions between the US and Japan spiked. In previous conflicts, Japan had used strategic surprise to gain military advantage. The US was well aware of Japan's force projection capabilities given its large fleet of aircraft carriers. There would be no confusion about which adversary had determined to strike the United States in the Pacific. The same cannot be said in cyberspace, where the lack of attribution adds considerable complexity. Not only is strategic surprise possible in cyberspace, but it is also possible to veil the source of the attack. Complicating matters, there may be a number of actors (rival nation-states, rogue states, terrorist groups, and others) with an interest in not only launching a surprise attack, but potentially even attempting to stimulate conflict between the victim and a third party. For example, a rogue state might attempt to launch a large-scale cyber attack against the United States and make it appear the attacks emanated from another country.

Second, Pearl Harbor required a great deal of lead time and risk for Japanese planners, moving a giant fleet across half the Pacific Ocean while concealing their movements. At best, discovery of a large Japanese fleet would have removed plausible deniability about Japanese intentions. At worst, it could have resulted in military disaster, as it ultimately did at Midway. In cyberspace, concealment and plausible deniability are not only possible but relatively easy to achieve and the speed at which DDoS and other attack techniques can be produced eliminate much of the lead time that might result in an inadvertent discovery. In other words, strategic surprise in cyberspace may prove far easier to achieve than in other historic examples, such as Pearl Harbor or the Israeli attacks commencing the Six-Day War.

7.3. Lessons That Can Be Drawn From This Analogy

A critical lesson to be drawn from the Pearl Harbor analogy points back to a recommendation by Neustadt and May—understand the timeline and start it at the earliest possible point. The Pearl Harbor analogy is often used to describe either a successful surprise attack and/or the failure of a country to anticipate an attack despite weeks and even months of mounting evidence. Thus, with respect to cybersecurity today, the analogy is often used to create a sense of imminent danger. But when does the Pearl Harbor timeline begin? Should one think of the Pearl Harbor analogy beginning in the fall of 1941 as negotiations between Japan and the US begin to breakdown? Does the timeline begin with the Japanese invasion of Manchuria, which signaled larger imperial ambitions? In the 1920s with Billy Mitchell, or over the fields of Flanders during WWI when aircraft first played critical roles in military operations? Or does it begin in 1890 when Alfred Thayer Mahan published The Influence of Sea Power Upon History, 1660-1783, a book that was extraordinarily influential with two generations of Japanese naval strategists? From the Japanese perspective, does it begin with the arrival of Admiral Perry and his black ships in 1853?

Choosing the appropriate timeline for Pearl Harbor can greatly change the lessons cybersecurity strategists might learn from—particularly in identifying relevant historic parallels and how change points (Taranto, Estonia) altered decision-makers perceptions and actions.

8. Conclusion: The Power and Perils of Cyber Analogies

Analogies are powerful instruments in the hands of decision-makers. They can: create and cement an image in the public's mind about an issue; prove useful in sorting through the details of a crisis to find key insights; support efforts to develop sound strategies and well-reasoned policy options; and help leaders anticipate the future, ripple effects of decisions made today. The paper does not define the "best" historical analogy for decision-makers to consider in formulating cybersecurity strategies. A number of factors (e.g., new threats, breakthrough technologies, changing business conditions) can dramatically alter how the cyber issue may unfold—and, consequently, which analogies may best apply. The goal of this inquiry was to test our thesis by creating a framework that puts future cybersecurity events or crises in context—and to help decision-makers select the most relevant and applicable historical analogies. In doing this analysis, we've reached four conclusions.

First, no single analogy will suffice in considering the complex challenges of cyberspace. While the cyber Pearl Harbor analogy is rich in imagery, connoting urgency, it may not be applicable to the full range of cyber scenarios that may confront the US and the world. Moreover, it holds the potential to produce a dangerous blind spot—we may wait for the "big one," a large-scale surprise attack while suffering from "pinpricks" that, in the end, have a far more debilitating and degrading impact on our networks (and public confidence).

Second, our examination of historical analogies reveals those that strike a balance between inspiration and desperation tend to produce the most permanent, lasting results. The Strategic Defense Initiative, the Manhattan Project, the National Highway System, and even the creation of the Internet itself have their roots in a mixture of inspiration (e.g., to destroy incoming ICBMs, to end WWII quickly, to reduce highway fatalities, and to promote collaboration across a research community) and desperation (e.g., to counter a growing Soviet nuclear arsenal, to develop the Atomic bomb before Germany and Japan, to enable rapid mobilization should the Cold War turn hot in Europe, and the ensure command and control resiliency in the worst case event of a nuclear war).

Third, today many of the analogies used to describe cyber rest at the extremes of our model. Again, this is not a surprising outcome—these analogies present vivid images that grab the public's attention. At the same time, continuously planning only for the worst-case scenarios can erode public support when these events don't occur or their effects are far less than predicted. The Y2K issue offers an excellent illustration of this. The reality is that early vigilance and a date certain event helped the world prepare for this software glitch. The perception, however, was that Y2K was anticlimactic and created more skepticism around gloom-and-doom cyber warnings.

Fourth, one of the most important lessons from *Thinking in Time* is understanding the correct, appropriate timeline of an issue: when did this issue start? What are key change points or trends? The authors argued these timelines should start at the earliest possible date to fully understand the historical context. For example, one could examine the analogy of 9/11 in multiple contexts—the year leading up to the attacks; a timeline starting with the 1993 bombing of the World Trade Centers; or beginning with the 1982 US Marine barracks bombing in Lebanon. In this case, the same event can produce different timelines (and resulting historical lessons) that may ultimately change the focus and core meaning of an analogy. For leaders exploring the future of cyberspace or dealing with an active cyber incident, starting with the right timeline may prove critical in making good decisions.

References

[1] National Counterintelligence Executive Joel Brenner in a January 2009 CBS News television interview (http://www.dni.gov/interviews/20090118_interview.pdf) and by Michael Chertoff, Secretary of Homeland Security at 2007 RSA Conference (http://www.theinquirer.net/inquirer/news/1021392/cyber-attacks-significant)

[2] Remarks by Paul Kurtz on February 18, 2009 at the 2009 Black Hat conference in Washington, DC (http://www.blackhat.com)

[3] Neustadt, Richard E. and May, Ernest R., Thinking in Time: The Uses of History by Decision Makers, The Free Press, New York: 1986.

[4] Bruno, Greg, The Evolution of Cyber Warfare, The Council on Foreign Relations, February 28, 2008 (http://www.cfr.org/publication/15577)

[5] History of the Internet, http://www.davesite.com/webstation/net-history.shtml

[6] The Hobbes Internet Timeline (http://www.zakon.org/robert/internet/timeline/)

[7] Testimony by Mary Ann Davidson, Chief Security Officer at Oracle, to the United States House of Representatives Subcommittee on Emerging Threats, Cybersecurity, and Science and Technology, Source: http://homeland.house.gov/SiteDocuments/20090310143850 78976.pdf

[8] Andy Greenberg, Sketching Obama's Cyberplans, Forbes, Feb 20, 2009. http://www.forbes.com/2009/02/20/paul-kurtz-security-technology-security_kurtz.html

[9] Virtual Criminology Report – Cybercrime : The Next Wave, McAfee. Source: http://www.mcafee.com/us/research/criminology_report/default.html

[10] Jeff Bliss, "Dearth of Technical Experts Leaves US Open to Cyber Attack, Panel Says," Boston Globe, March 20, 2009.

[11] Margaret Kane, US Vulnerable to Data Sneak Attack, CNET News, August 13, 2002. Source: http://news.cnet.com/U.S.-vulnerable-to-data-sneak-attack/2100-1029_3-949605.html

[12] Source: http://wapedia.mobi/en/Battle_of_Taranto

The Virtual Battlefield: Perspectives on Cyber Warfare
C. Czosseck and K. Geers (Eds.)
IOS Press, 2009
© 2009 The authors and IOS Press. All rights reserved.
doi:10.3233/978-1-60750-060-5-132

The Information Sphere Domain Increasing Understanding and Cooperation

Dr. Patrick D. ALLEN and Dennis P. GILBERT, Jr
Johns Hopkins University, Applied Physics Lab
Booz Allen Hamilton

Abstract. Recent discussions regarding the emerging field of cyber warfare have focused on the term "cyberspace," and have included cyberspace as being considered its own war fighting domain, much like air, land, sea, and space. In this stage of the Information Age, the international community is grappling with whether it needs to define this information realm as a domain, similar to the air, land, sea, and outer space domains that already exist. History shows that there is always an advantage in a conflict to the side that *understands and operates* within a domain better than the opponent. In this paper, the authors propose a definition of a domain, define what constitutes a domain, posit how new domains are created over time, and describe the features of what is and is not a domain. These definitions and features lead to our proposal that the "Information Sphere" should the preferred international term, and it is this "InfoSphere" that qualifies as a new domain, with features both similar to and different from the four existing physical domains.

Keywords. domain, information, cyber, cyberspace

Introduction

Since classical times, two domains of operation dominated military and civilian operations: land and sea. The advent of powered flight in 1904 initiated the opportunity for a third domain. Shortly thereafter, actions by opposing elements in this airspace began during World War I. The Army and Navy each developed its own air capabilities, and at the end of World War II, the Army Air Corps became the US Air Force—about 50 years after the first powered flight. In a similar manner, the dawn of the "space age" in 1955 encouraged each of the U.S. military services to invest in their own efforts in the space domain. By the mid to late 1980's, with the advent of then US President Ronald Reagan's Strategic Defense Initiative (SDI), the US DoD acknowledged outer space as a fourth war fighting domain.

Based on the preceding observations, the historical trends for recognizing new domains tend to follow this sequence:

- First, the *capability* to begin to operate in that domain is developed, such as the first powered flight or the first space flight.
- Second, the capabilities to operate in that domain become relatively *commonplace*, such as air travel or Shuttle launches.

- Third, the capabilities in that sphere *to affect capabilities* in that domain and in the other domains become recognized and exploited.
- Fourth, sufficient recognition of the *unique and synergistic* nature of capabilities in the domain are recognized and further developed.
- Finally, *institutional and financial support* for the domain is developed.

1. Definition of a Domain

While considerable dialogue and research has been conducted on the subject, there does not appear to be an US military definition, NATO definition, or internationally agreed upon definition for a domain. Joint Publication 1-02, the Department of Defense Dictionary of Military and Associated terms, does not define a domain. In the absence of an internationally-accepted definition, we propose a definition of a domain, and describe features or criteria that distinguish one domain from any other domain in order to help frame the discussion about whether to define the Information Sphere as its own domain.

The Webster's New Collegiate Dictionary has two relevant general definitions of a Domain:

1. A territory over which rule or control is exercised.
2. A sphere of activity, interest, or function.

Webster further defines a sphere as an area or range over or within which someone or something acts, exists, or has influence or significance, such as the public sphere.[1] Note that while the definition of a sphere does include physical environments, it also includes non-physical environments. Following this train of thought that a sphere or domain does not have to be a physical domain, it is comprehensible that a sphere can also apply to area or range in which something acts, exists, or has influence or significance.

Using these definitions as a guide, we derived the following definition for a domain for consideration by NATO countries:

> *The sphere of interest and influence in which activities, functions, and operations are undertaken to accomplish missions and exercise control over an opponent in order to achieve desired effects.*

By breaking down this definition into its component parts, we can support that each of the existing four physical domains (air, land, sea, and space) qualifies as a domain, as defined above. The key components of our proposed definition of a domain are:

- It is a sphere of interest
- It is a sphere of influence in that activities, functions, and operations can be undertaken in that sphere to accomplish missions
- It is a sphere that may include the presence of an opponent
- It is a sphere in which control can be exercised over that opponent

Based on the above components, it is clear that the four physical domains of air, land, sea, and space each qualify as a domain. Each has its own sphere of interest and sphere of influence. Aircraft fly missions, ships navigate the waterways (both surface and subsurface), ground forces take and secure terrestrial objectives, and satellites orbit the earth. In each of these physical domains an opponent can be present and can interfere with friendly operations. Moreover, the NATO members have military capabilities in each of these domains, which can be used to control and dominate potential adversaries.

2. Features of a Domain

The authors offer for discussion what they consider are the six key feature of a domain:

1. Unique capabilities are required to operate in that domain
2. A domain is not fully encompassed by any other domain
3. A shared presence of friendly and opposing capabilities is possible in the domain
4. Control can be exerted over the domain
5. A domain provides the opportunity for synergy with other domains
6. A domain provides the opportunity for asymmetric actions across domains

The authors posit that if a domain has these six features, it qualifies as a domain, and if it does not have all six features, it should not qualify as a domain. This checklist of features can then be used as criteria to determine whether a new realm, such as the Information Sphere, qualifies as a domain. The following examples show how the four physical domains of air, land, sea, and space qualify as a domain according to these six features:

1. Unique capabilities are required to operate in that Domain. For example, aircraft are required to operate in the air domain, spacecraft for the outer space domain, ships for the sea domain, and land systems for the land domain. Note that each of these capabilities can readily differentiate themselves from capabilities in other domains.

2. A Domain is not fully encompassed by any other single Domain. For example, the air domain is not encompassed by the land domain, or vice versa. The capabilities, missions, and dominance techniques of the capabilities in each domain remain unique. A tank is not intended to operate in the air domain, while an airplane is not designed to operate underwater.

3. A shared presence of friendly and opposing capabilities is possible. Any domain can potentially be entered by opposing forces. This is not to say that every opponent is present in every domain, but that an opposing presence must be *possible* for the sphere of interest and influence to be considered a domain. *A potential shared presence* is an essential feature of a domain since dominance or control over the domain requires the possibility of an opposing presence or capability.

4. Control can be exerted. The presence of a potential opponent in the sphere of interest generates the need to influence or dominate such opponents in a domain. Since a domain is a sphere of influence as well as of interest, then it must be possible for one side's influence in a domain to dominate an opposing side's influence.

5. Provides opportunities for synergy. The capabilities in a domain must be able to provide synergistic opportunities with capabilities in other domains. The classic US Military's "Air-Land Doctrine" was an excellent example of how the capabilities of the land and air domains could be mutually supportive.

6. Provides asymmetric opportunities. Similar to synergistic opportunities are the opportunities for capabilities in a domain to gain an asymmetric advantage over opposing forces in other domains. For example, the US Army's Joint Fires Doctrine emphasizes the opportunity to use air assets as an asymmetric threat against opposing land and sea assets, while land or sea forces can be used to asymmetrically threaten enemy air assets. The principle of asymmetry must be a possibility for capabilities in a sphere of interest for it to be defined as a domain.

3. Proposed Definition for the Information Sphere's Domain

We will next address whether the Information Sphere qualifies as a domain, but first we have to provide a definition for the Information Sphere. Current DoD doctrine defines the Information Environment as "the aggregate of individuals, organizations, or systems that collect, process, or disseminate information; also included is information itself."[2] Regrettably, this definition puts the emphasis on the physical attributes of an information environment. In other publications, the Information Domain has been described as "the domain where information is created, manipulated, and shared," or "where information lives." These same authors have defined the Cognitive Domain as the "domain of the mind of the warfighter and the warfighter's supporting populace."[3] With this approach, the content, the connectivity, and the message have been segregated. We purport that these definitions diverge from the goals of the Information Operations mission area and the common understanding of Strategic Communication. Therefore, we first propose a definition of the Information Sphere, and second, for the Information Sphere's domain.

The definition we propose for the *Information Sphere* is:

The space defined by relationships among actors, information, and information systems.

To further elaborate on this definition, we also define actors, information, and information systems:

An actor may be a sender, liaison, modifier, transferor, or recipient (either intended or unintended) of information. Information is the data being passed among actors via information systems. An information system is any information perceiving system, information storage system, or communications system, (including couriers).

Based on these definitions, we propose a new definition for the domain we call the *Information Sphere*:

The space of relationships among actors, information, and information systems that form a sphere of interest and influence in or through which information-related activities, functions, and operations are undertaken to accomplish missions and exercise control over an opponent in order to achieve desired effects.

Note that the information by itself is not the domain, nor is the domain simply the information systems in which the information rides and resides. It is the space defined by the relationships among actors, information, and information systems that define the Information Sphere and allow it to qualify as a domain. We include the word "sphere" in the Information sphere to distinguish the fact that the domain we are proposing consists of more than just the information component, and calling it the "information domain" would encourage that misunderstanding. The other accepted domains do not use the term "the air sphere" or the "sea sphere," but we use "Information Sphere" to make the distinction from just the information component completely clear.

This definition is different from the previously referenced definitions for the cognitive, information, and cyber domains because the proposed definition of the Information Sphere explicitly includes the relationship among these three components. It is the *relationships* among these three components that define the meaning, context, and value of the Information Sphere, not the three components taken in isolation.

The US Military's Quadrennial Roles and Missions Review Report defines cyberspace as "a global domain within the information environment consisting of the interdependent network of information technology, infrastructures, including the Internet, telecommunications networks, computer systems, and embedded processors and controllers."[4] We believe that this is a good definition of cyberspace, but believe that cyberspace is still a subset of the larger Information Sphere domain. Just as naval surface actions and submarine actions are two components of the Sea domain, cyberspace, cognitive, and information are components of the more encompassing Information Sphere.

4. The Evolution of the Information Sphere as a New Domain

As mentioned above, there are five steps that capabilities in an environment tend to follow en route to becoming a new domain:

1) The *capability* to begin to operate in that domain is developed. From the advent of the PC and the birth of the public version of the Internet, communication and information capabilities have exploded. Combined with global transportations, these capabilities provide a global economy and social interactions to a degree previously unheard of.

2) The capabilities to operate in that domain become relatively *commonplace.* Fourth Generation cell phones, PCs, and Internet access are now commonly found almost everywhere in the world. Nations that have yet to develop their communications

infrastructures are jumping straight to fourth generation access that does not require the construction of expensive information infrastructures. Almost anyone in the world can achieve global communications via the Internet and cell phones.

3) The capabilities in that sphere *to affect capabilities* in that domain and in the other domains become recognized and exploited. Information has always been important to military and civilian operations. The Information Age has made the Information Sphere not only widespread but also *shared.* Opponents can reach our internet-connected networks without leaving their own country. There are few places in the world where the news media do not reach. Incidents in the remotest parts of the world often carry global implications beyond any time in previous history. As a result, conflict in the Information Sphere is becoming more prevalent and more important than even direct military action in many cases.

4) Sufficient recognition of the *unique and synergistic* nature of capabilities in the domain are recognized and further developed. As both the capabilities and threats in this sphere continue to grow, more and more resources are being allocated to the exploitation and securing of Information Sphere capabilities.

5) Sufficient *institutional and financial support* for the domain is developed. The US's efforts to create a new Sub-Unified Command for Cyber is one example of efforts toward developing the necessary institutional and financial support for operating and succeeding in the Information Sphere. The future may be an US Interagency organization, or even an international megacommunity, that represents the Information instrument of national power, along with the Diplomatic, Military, and Economic instruments. Note that the bringing together of Information Sphere capabilities from the other instruments of national power (including military) is the logical progression we would expect to see as the Information Sphere domain becomes institutionalized and supported financially.

5. What is Unique About the Information Sphere's Domain?

Now that we have described why the Information Sphere qualifies as a domain, this section will describe why the Information Sphere is also unique compared to the four physical domains. At the same time, we will describe why we believe that what makes the Information Sphere unique is yet one more reason why the Information Sphere should be treated as a new domain.

Since the definition of the Information Sphere includes actors, information, and information systems, it is evident that each of these three components must reside in a physical medium at any point in time. For example, an information server is located either on the ground, underground, in the air, in outer space, on the sea, or (potentially) under the sea.

In a similar manner, the information itself is either being stored on an information system, or is in some information conduit (including a portion of the electromagnetic spectrum) at any point in time. Finally, the human actors must be located in one of the four physical domains. Figure 1 gives an example of how one might consider the information sphere's domain with respect to the four physical domains.

Note that this figure shows that the Information Sphere is separate from each of the four physical domains, but is also accessible by, and provides access to, all four. Information may enter or exit via a physical medium, but that may or may not be relevant. For example, if an intruder is seeking an entry point into a network, the physical location of the entry point may matter to the intruder. However, once the intruder is in the network, the physical location of the entry point and any informational areas of interest are of less importance due to the degree of access provided by the network. What is important in this case is the relationship between security elements of the network (including people), the targeted information, and the intruder, rather than the physical location of the assets.

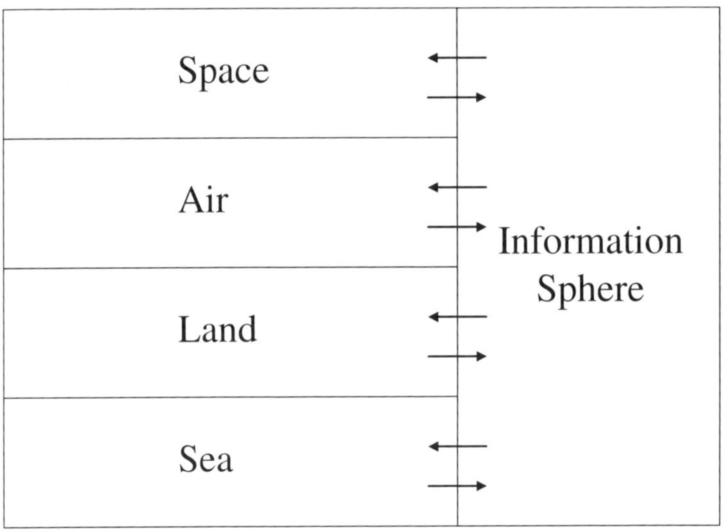

Figure 1. Information Sphere is a Unique Type of Domain[5]

Note that the concept of entry and exit points from one domain to another is prevalent in all domains. Aircraft and spacecraft land on the ground or at sea. Ships dock at land-based ports. The same is true for the Information Sphere. There will always be entry and exit points from the Information Sphere to and from the other domains, as the purpose of most activity in the Information Sphere is to affect something in the physical world. However, there are also actions and desired end states associated with operations *within* the Information sphere that are unique to the Information Sphere, irrelevant to and unaffected by the physical space in which the actors, information, or information systems actually reside.

Each domain has actions that are dependent and independent of each of the other domains. Similarly, *the Information Sphere is not completely encompassed by any physical domain*. For example, a distributed database that has elements either residing or in transit on land, in the air, on the sea and/or in outer space is not contained or fully encompassed by any of the four media in which it is located or passes through.

Moreover, *even the union of air, land, sea, and space physical environments does not fully encompass the Information Sphere*. The interactions we described for the Information Sphere often occur in a space of relationships where the physical location of the actual components is irrelevant once access has been achieved. For example, the ability for two actors to interact in some way does not depend on the medium or media within which the information exchange occurs. What matters are the interaction and the relationship between the actors, information, and information systems? Moreover, shared presence within all four physical domains does not equate to dominance in the Information Sphere, either in the control of information access, information systems, or in the beliefs and perceptions of groups of interest within those four domains.

It is these relationships between actors, information, and information systems that define the interest and influence mechanisms in the Information Sphere. Since these relationships can be satisfied by a wide range of paths into, out of, and through various physical media, the value, benefit, and vulnerability of elements within the Information Sphere are relatively independent from the four physical domains.

Another important distinction is that the *desired effects* of an information activity *eventually* reside in one or more of the four physical domains. For example, the information activity may be to bring down an enemy air defense system, which opens the way for the air operations, which shapes the upcoming ground or sea battles. However, there may be a significant delay between the initial information activity and any effect in one or more physical domains. For example, the placement of a back door on a target server does not have an immediate effect other than the opportunity for access at a later date. Until that access is exploited, there is no physical manifestation of a desired effect.

As another example, competition between opposing thoughts or beliefs frequently has a delayed reaction. The concept of freedom, for example, is often dormant until the opportunity to be free, or to achieve increased freedoms, becomes available. In the conflict among beliefs, a thought that is planted may blossom many years later after additional thoughts and physical events have occurred.

Therefore, the fact that the actors, information systems, and information that comprise the Information Sphere must reside at any instant in one of the four physical domains is either secondary or irrelevant to the functioning of the abstract relationships within the Information Sphere. Information easily transcends the barriers between the physical domains. The Information Sphere is a space where the understanding of relationships in that space can lead to dominance over opponents in that space.

6. The Information Sphere's Qualifications as a Domain

We argue that the Information Sphere qualifies as a domain according to our preceding definition for the following reasons:

- The space of relationships among actors, information, and information systems forms a sphere of interest
- It is a sphere of influence in that activities, functions, and operations can be undertaken in that sphere to accomplish missions
- An opponent to friendly operations may function in that sphere
- Control can be exercised over that opponent in or through that sphere

Besides meeting the four preceding criteria described above, the Information Sphere also meets each of the six features required of a domain.

1. Unique capabilities are required to operate in that Domain. Information capabilities are required to operate in the information realm. The Information Sphere requires unique equipment and personnel skills to function effectively, accomplish missions, and dominate any enemy presence. Information capabilities operating in the Information Sphere are both unique and differentiable from the capabilities designed to operate in other domains. For example, a computer system (and associated software/code) optimized for hacking into enemy computer networks is a unique asset different from air, land, sea, and space platforms. Hacking skills are unique from the more traditional set of pilot, sailor, soldier, and astronaut.

As information capabilities become more specialized, the uniqueness and differentiability of these capabilities will continue to grow. For example, the Information Sphere now has a set of unique equipment (materiel) and personnel skills required to effectively operate in, defend, and attempt to dominate the domain. With these new capabilities comes a range of unique support structures, such as doctrine, organization, training, leadership development, facilities, and policy.

2. A Domain is not fully encompassed by any other single Domain. The Information Sphere is not fully encompassed by any combination of land, sea, air, or space domains. The Information Sphere has capabilities and functions that are meaningful only in this information environment.

3. A shared presence of friendly and opposing capabilities is possible. Until recently, the Information Sphere rarely allowed for *a shared presence*. A shared presence was not feasible primarily due to physical and geographical separation and the inherent time delays. With the birth of the Information Age, however, the Information Sphere is frequently shared. Examples of this sharing include the range of information media, including the Internet, local and wide area networks, television, radio, print media, video and audio recordings, and other capabilities. Due to the explosion of information and information capabilities, the Information Sphere allows for a shared presence more than ever before. As a result, dominance and control in this domain have become much more important than in the past.

4. Control can be exerted. For the Information Sphere, control can refer to the control of the information systems in a region of the information sphere, control to the access of the information in that information sphere, or even the dominance of one belief over another in a region of the information sphere. As an example, air-to-air radars on fighter aircraft may try to jam or spoof the radars of opposing forces in the air domain, thereby attempting to control the information sphere. The recent spate of alleged nation-state sponsored hacks into sensitive but unclassified US military and contractor information systems is an example of the type of temporary but useful control our opponents can undertake in the Information Sphere.[6] Influence over population groups is a constant competition in the Idea Battlespace among ideas vying for dominance over other ideas.[7]

5. Provides opportunities for synergy. The Information Sphere provides synergistic support to all the other domains, and vice versa. The ability to gather information directly from an enemy information source can assist air, land, sea, and space operations. Conversely, the ability to take out an enemy information system from the

air can force the enemy to use an information system already compromised by our side.[8]

6. Provides asymmetric opportunities. Information capabilities can provide an asymmetric threat against enemy capabilities in other domains. In his book, *The Next World War*, James Adams[9] describes a case where a computer virus was entered into a printer that was supposed to be delivered to an enemy site in order to neutralize an enemy air defense system. In a similar manner, physical assets can be used to disrupt and destroy opposing information systems.

7. Benefits of Treating the Information Sphere as a Domain

First, there is always an advantage in a conflict to the side that *understands and operates* within a domain better than the opponent. This is true on land, air, sea, and space, and can also be true in the Information Sphere. Obtaining dominance in the Information Sphere will likely lead to continued dominance in the four physical domains via asymmetric effects. By defining the Information Sphere as a domain, a body of knowledge or military science of operating in the Information Sphere will be more thoroughly developed to improve understanding and consensus on the subject.

Second, *representing* the relationships of information among actors and information systems in a manner useful to planners and decision makers will help improve the effectiveness and efficiency of operations in and through the Information Sphere. For example, the ability to readily visualize relationships in a common format will facilitate a unity of effort and common understanding of objectives and constraints. Defining the Information Sphere as a domain should lead to an investigation and experimentation on a number of methods to represent these relationships, and the best-of-breed methods should emerge to enhance our capabilities in this domain.

Third, focusing and *preparing enhanced capabilities* in the Information Sphere will enable superior *influence and control* in this domain. The side with better personnel, equipment, doctrine, organizations, and leadership will have a significant advantage over the opposition. If the military departments of NATO countries choose to define the Information Sphere as a formal war fighting domain, then the resourcing to more effectively and efficiently function in that domain should follow.

Fourth, defining the Information Sphere as a domain allows for increased emphasis on *planning and employing* all instruments of national power—diplomatic, informational, military, and economic—in a common, coordinated endeavor. Since information is a common element in the use of all instruments of national power, the ability to function effectively in this domain will encourage the coordination and synchronization of effects among all these instruments.

Fifth, defining the Information Sphere as a domain will help increase the emphasis on improved *information assurance and cyber security*, which can and should lead to improved economic and national security. Defining the Information Sphere as a domain will help define the common areas of interest between these sectors, and eventually lead to common, or at least coordinafted, resourcing in the areas of information security.

Finally, defining the Information Sphere as a domain can help *focus international efforts* on the important conflicts already ongoing in this domain. In addition to the skirmishes in cyberspace, the battle for the hearts and minds of many groups of actors worldwide has been raging since the birth of philosophies and political systems. In a battle of the minds, the physical location of the people believing in something is less important than the dominance of that belief over other competing beliefs. Defining the Information Sphere as a domain will help highlight the need for renewed effort and capabilities in this cognitive realm.

8. Summary

This paper presented the definition and features of a domain, a definition for the Information Sphere, and why the Information Sphere qualifies as a domain along with the four physical domains of air, land, sea, and space. The paper also presented why the Information Sphere has some distinct differences from the four physical domains, and the benefits of treating the Information Sphere (which includes cyberspace) as its own domain. Lastly, the paper describes why the Information Sphere is a comprehensive domain that encompasses the areas of cyberspace, cognition, personnel, and information itself, which is why we include the term "sphere" in the definition. Referring to this new area as the "Information Domain" would imply that the focus of this domain is focused just on the information component, which does not adequately capture the full scope of the new proposed domain: the "Information Sphere Domain."

References

[1] http://www.merriam-webster.com/dictionary/sphere; accessed 14 January 2009.

[2] Joint Publication 1-02, Department of Defense Dictionary of Military and Associated Terms (Washington D.C.: U.S. Government Printing Office, 12 April 2001), page 203.

[3] Alberts, David S., Gartska, John J., Stein, Frederick P., Net-Centric Warfare-Developing and Leveraging Information Superiority (DoD C4ISR Cooperative Research Program, 2nd Edition (Revised) August 1999.)

[4] Gates, Robert M., Quadrennial Roles and Missions Review Report, Department of Defense, January 2009, http://www.defenselink.mil/news/Jan2009/QRMFinalReport_v26Jan.pdf, accessed 1 February 2009.

[5] Allen, Patrick D., Information Operations Planning (Artech House, New York, 2007), p. 298

[6] Wortzel, Larry M.,Chairman, et al., 2008 Report to Congress (U.S.-China Economic and Security Review Commission, Washington, 27 October 2008), http://www.uscc.gov, accessed 25 November 2008.

[7] Allen, op cit., p. 114

[8] Koch, Andrew, Information warfare tools rolled out in Iraq, (Janes' Defense Weekly, Washington, 6 August 2003)

[9] Adams, James, The Next War, Simon and Schuster, New York, 1998.

The Virtual Battlefield: Perspectives on Cyber Warfare
C. Czosseck and K. Geers (Eds.)
IOS Press, 2009
© 2009 The authors and IOS Press. All rights reserved.
doi:10.3233/978-1-60750-060-5-143

Sun Tzu was a Hacker: An Examination of the Tactics and Operations from a Real World Cyber Attack

Billy K. RIOS[a]
[a] *GreyLogic., LLC*

Abstract. This text will cover the operational and tactical techniques used in a "real world" cyber-attack and includes an analysis of the planning, command, control, execution, and outcome of these cyber-attacks. The text focuses on the cyber-attacks against the nation state of Georgia in 2008, as the author was in a unique position to observe the communications, execution, and responses from both attacking and defending entities. The various aspects of the attacks will be described and linked back to traditional concepts of Maneuver Warfare as described in Marine Corps Doctrinal Publication 1 (MCDP-1).

Keywords. cyber warfare, cyber attack, offensive cyber capability, defensive cyber capability, Georgia, maneuver warfare

Introduction

"*Doctrine must evolve based on growing experience, advancements in theory, and the changing face of war itself*"

General C.C Krulak, *MCDP-1*

The computer system sitting on your desk brings a new dimension to modern warfare. Just as the machinegun and airpower changed the face of warfare, so does offensive cyber capabilities. The ubiquity of computer systems in our military, government, and civilian infrastructure have solidified the importance of offensive cyber capabilities to the point where packets will be the "bullets" that will be fired in future conflicts. Software loaded on computer systems will be considered the "terrain" on which cyber warfare is waged. The use of packets as weapons presents a novel approach to warfare and will inevitably cause changes in military doctrine within our lifetime; however the employment of cyber capabilities continues to abide by many of the traditional concepts and principles of warfare. Associating the concepts of cyber war in the terms of conventional warfare and known doctrine can make analysis less daunting and provides a new perspective when measuring the impact and deciding how best to employ cyber capabilities. In this text, the author will examine the execution of a cyber-attack and correlate the principles of the attack to traditional concepts studied in maneuver warfare. The specific scenarios examined in this study involve the cyber-attacks conducted against the country of Georgia in 2008. Although the author uses a specific cyber-attack to illustrate his points, the author hopes that the concepts presented in this text can be universally applied to any cyber-attack.

1. Maneuver Warfare

The principles of maneuver warfare will be referred to extensively in this text. While a complete description of all the concepts associated with maneuver warfare are beyond the scope of this text, the author begins by providing a fundamental description of maneuver warfare and its foundations. According to the Marine Corps Doctrinal Publication 1 (MCDP-1), maneuver warfare is described as a:

"war fighting philosophy that seeks to shatter the enemy's cohesion through a variety of rapid, focused, and unexpected actions which create a turbulent and rapidly deteriorating situation with which the enemy cannot cope"

The focus on disrupting enemy cohesiveness makes maneuver warfare unique from other military doctrine. In maneuver warfare, entire enemy units and strongholds are bypassed in order to reach a "decisive opportunity" to exploit a "critical vulnerability" in the enemy's position [1]. Exploitation of critical vulnerabilities provides a pathway to attacking the enemy's "center of gravity" [1]. It is important to understand that despite its name, maneuver warfare is not limited to the maneuvering of units or spatial operations. Temporal actions such as psychological and technological disruption are also key elements of maneuver warfare. In maneuver warfare, well timed combat power brought to bear on strategic points on the battlefield, preemptive strikes geared towards the elimination of the enemy's decision making ability, surgical strikes on communication systems, elimination of the enemy's logistical chains, and suppression of enemy combat power are all more highly valued over high body counts or gained geography [1].

While the physical destruction of enemy forces and equipment is not the primary focus of maneuver warfare, physical destruction and firepower do play a central role at decisive points in battle, especially when destruction of enemy forces results in the degradation of the enemy's overall cohesion. Advanced weapon systems and technical superiority such as superior weaponry, stealth technology, and highly trained special operations forces (SOF) can increase the aggressor's opportunities to deliver decisive firepower on the right targets at the right moments, disrupting the enemy's normal operating rhythm and decision making ability. For example, in the lead up to Operation Desert Storm in 1991, SOF and air power disabled and destroyed a significant portion of the Iraqi command and control systems, disrupting Iraqi command and control at the highest levels [2]. While the success of the air power and SOF units would not have "won the war" in isolation, they disrupted the enemy's cohesion and decision making, allowing for more effective follow on operations which would ultimately win the war. It is this "disruptive capability" that is the quality that makes offensive cyber capabilities so attractive. The ability to disrupt the enemy's tempo, rhythm, and decision making from afar, in a lighting fast manner, while exposing very little, is extremely appealing to many commanders.

Although offensive cyber capabilities offer a novel approach to disrupting the enemy's normal rhythm and decision making, the prudent commander understands that much like air power, naval power, intelligence, and other individual military capabilities, offensive cyber capabilities cannot "win a war" by itself [3]. Instead, these offensive cyber capabilities must be used as a component in the overall combined arms effort focused on disrupting the enemy's cohesion and exploiting critical vulnerabilities. Once the enemy's battle rhythm and decision making is disrupted, the disruption must be exploited via follow on actions. Disruption creates opportunities that should be ruthlessly exploited. This exploitation often leads to additional opportunities, which eventually leads to a decisive opportunity to launch a decisive attack against the enemy [1].

The author will present specific scenarios where offensive cyber-attacks were used in a manner that was consistent to the principles of maneuver warfare. The author chooses to focus on the 2008 cyber-attacks launched against the nation state of Georgia. These attacks are chosen due to the resources and vantage points the author held whilst the conflict progressed [4]. These resources and vantage points gave the author an insight into both sides of the conflict; however most of the examples and scenarios will focus on the aggressor and offensive actions. Before diving into the specific attacks that occurred against the Georgia infrastructure, it is important to define several terms which are used regularly in describing maneuver warfare. The author chose to focus on the following terms throughout the course of the text.

- Decentralized Command and Commanders Intent
- Combined Arms
- Initiative
- Centers of Gravity and Critical Vulnerabilities

1.1. Commanders Intent

The commander's intent provides the means for subordinates to exercise judgment and initiative. MCDP-1 states that each mission has two parts: (1) the task to be accomplished and (2) the reason or intent behind it. Commander's intent provides the reasoning and intent behind the assigned tasks and missions. Commanders intent is crucial for as the situation changes on the battlefield, the specific tasks assigned to the subordinate may become obsolete, but the intent is lives beyond the assigned tasks and continues to guide the subordinate's actions [1]. If the subordinate understands the commander's intent, they will be able to execute actions without the presence of direct orders and those actions will be in line with the commander's desires (which should ultimately advance strategic objectives).

In addition to the promotion of initiative, effective use of commander's intent allows for decentralization of command, pushing decision making to the lowest level. It is at these levels where forces are able to react and exploit opportunities in the most effective and efficient manner. Decentralized command and asynchronous execution is essential in the success of conventional operations in today's "small wars" [5] as well information based campaigns.

1.1.1. Target Lists and Commanders Intent

In August of 2008, the Grey Goose project commenced. Grey Goose was a pure open source intelligence initiative aimed at gathering and analyzing intelligence related to the Georgia cyber-attacks. During Phase I of the Grey Goose project, a number of

Russian hacker forums were mined for data detailing over 29,000 separate forum events with correlation of those events to status of Georgia cyber infrastructure [4]. One of the first items discovered on the various Russian hacker forums were target lists, providing the domain names of various Georgian servers to be attacked. A portion of the target list is shown in figure 1.

Сайт	Доступ с РФ (есть/нет)
www.parliament.ge Парламент;	-
http://occupation.tspteam.com	+
www.cec.gov.ge Избирком;	+
www.mdf.org.ge Муниципальный фонд развития;	-
www.mfa.gov.ge МИД;	+
www.corruption.ge Anti-Corruption Program;	-
http://smr.gov.ge/en/home	+
http://stoprussia.org/	+
www.insurance.caucasus.net Страхование;	-
www.mc.gov.ge Минкультуры;	-
www.nsc.gov.ge Совет безопасности;	-
www.supremecourt.ge Верховный суд;	+
www.iberiapac.ge Минтранс;	
www.court.gov.ge Department of material service;	+
www.civil.ge Ассоциации ООН в Грузии;	-
http://www.forum.ge/	+

Figure 1. Target list from Russian hacker forum

The target lists discovered on the various forums were simple and provided no specific direction or instruction as to how the various sites were to be attacked. The targets lists were simply lists of servers that should be targeted and attacked by forum members. Forum members were not assigned specific tasks, the specific techniques to be used were not defined, and the definition of a "successful attack" was broad and conceptual. Instead of publishing specific actions, the publisher of the target list allowed the forum members to decide the best course of action to carry out the attacks. Soon after the target list was posted, the forum was filled with chatter related to the most effective means of attacking the various servers. The communication was not directed to "higher" (to the forum administrator) asking for guidance, but instead focused on "lateral communication" (to other forum members), updating each forum member on newly discovered vulnerabilities and weaknesses [6]. As the lateral communications increases, each forum member cherry picks the data most appropriate to their interests and skillsets. This prejudicial filter helps maximize the impact of each individual forum member by allowing them to focus on applying their specific skillsets to attacking the servers on the target list, quickly identifying those servers that are most vulnerable to those specific skills possessed by the individual. An individual contribution to the forum is shown in figure 2.

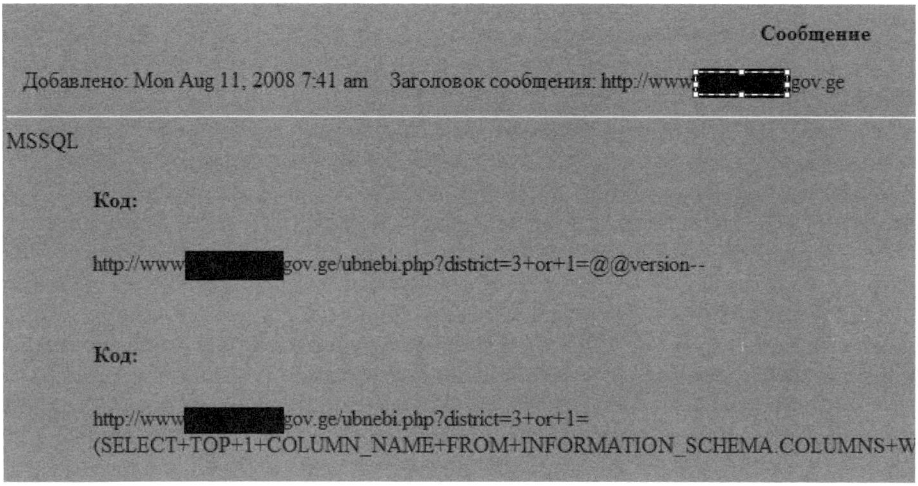

Figure 2. Various Forum Members Discussing Vulnerabilities

Despite its lack of detailed instruction, the target list established the framework for the most effective use of forum member skillsets. The target list and broad concepts of “success” are prime examples of an effective use of commander’s intent [7]. As opposed to identifying individual forum members, attempting to determine their skillset, and assigning the forum members specific tasks/servers/exploits, the forum administrator establishes the overall intent of the attacks and essentially publishes the intent. The forum administrator publishes the target list (task to be accomplished) and broad guidance (the reasoning/intent behind the task). The forum administrator provides little/no guidance as to how the tasks are to be accomplished. Instead, the forum administrator relies on the forum members to develop their own methods to accomplish the tasks within the overall intent, exploiting weaknesses as they become evident, and relaying updates through lateral communications. As the situation changes (new vulnerabilities discovered, new defenses encountered, new tools released…etc.) the forum members proceed within the original intent and do not wait for further instruction. This allows maximum flexibility and effectiveness in attacking, a flexibility that simply cannot be matched in a highly centralized, top down command and control structure [1].

1.2. Combined Arms

Combined arms are utilized to maximize combat power. The term “combined arms” refers to making use of all the available resources to the best possible advantage. Combined arms are typically achieved through the complementary use of different weapon systems [1]. The weaknesses of one weapon system are supplemented by the strengths of a different weapon system. The classic example of combined arms automatic direct fire weapons (machine guns) and indirect fire weapons (grenade launchers). If the enemy infantry becomes pinned down by the automatic fire, they become vulnerable to grenade attacks. If the enemy maneuvers to avoid the grenade attack, they expose themselves to the automatic weapons fire. The ultimate goal of combined arms it to utilize a full integration of various arms to achieve a situation so that when the enemy counteracts one arm, they are making themselves more vulnerable to another [1].

1.2.1. SQL Injection, DDOS Tools, and Combined Arms

In order to extract the maximum effect from offensive cyber strikes, the strikes must be used as part of a combined arms effort. This combined arms effort can involve the leveraging the exploitation a single cyber related vulnerability to accomplish successful exploitation of another cyber related vulnerability. The combined arms effort could also involve the exploitation of cyber related vulnerabilities in conjunction with the use of kinetic weapons or conventional forces. During the investigation of data made available to the Grey Goose project, exploitation of several SQL injection vulnerabilities against various Georgia applications were discovered [4]. SQL injection attacks in the logs of a Georgia server are shown in figure 3.

```
[Tue Jul 01 05:32:43 2008] [error] [client 91.XXX.XXX.XXX] File does not exist:
/home/virtual/XXX/public_html/small4, referer: http://www.XXX.gov.ge/full_text.php?nid=-
6038%20union%20select%201,2,3,4,5/*
[Tue Jul 01 05:33:19 2008] [error] [client 91.XXX.XXX.XXX] File does not exist:
/home/virtual/XXX/public_html/small4, referer: http://www.XXX.gov.ge/full_text.php?nid=-
6038%20union%20select%20unhex(hex(version())),2,3,4,5/*
[Tue Jul 01 05:33:35 2008] [error] [client 91.XXX.XXX.XXX] File does not exist:
```

Figure 3. Examination of log files show evidence of SQL Injection attacks

These SQL injection vulnerabilities were initiated from Russian IP addresses and log data provides many of the exact SQL injection queries that were used in the attacks. These targeted SQL injection attacks began in July of 2008, months before the high profile attacks against Georgia in August of 2008. The SQL injection attacks started with simple fingerprinting of the backend database servers being used by the vulnerable applications. An examination of the log files shows that once the fingerprinting of the backend database was complete, the Russian hackers extracted the usernames and passwords associated with the vulnerable applications. The usernames and passwords are valuable because they provide the foundation for attacks against other systems. For example, once the usernames and passwords are extracted, the hacker can test those username and password combinations against other, better protected information systems (password reuse) [8]. If any of the users of the compromised application have reused their password on other systems, the hacker can now masquerade as a legitimate user on that other system. Password reuse can also lead to the compromise of personal and business email accounts, providing a stream of intelligence that can be used in conjunction with other attacks (both cyber and conventional). If a hacker has gained access to the business and personal email systems of those employees, the hacker will be in a prime position to collect intelligence on those individuals, feeding the captured data into traditional intelligence analysis and fusion.

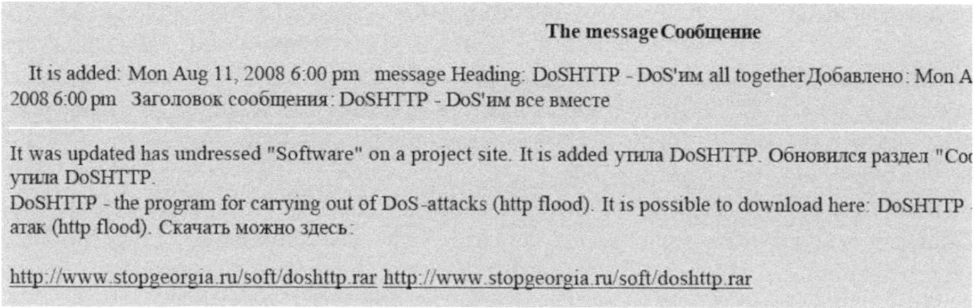

The message Сообщение

It is added: Mon Aug 11, 2008 6:00 pm message Heading: DoSHTTP - DoS'им all together Добавлено: Mon A
2008 6:00 pm Заголовок сообщения: DoSHTTP - DoS'им все вместе

It was updated has undressed "Software" on a project site. It is added утила DoSHTTP. Обновился раздел "Со
утила DoSHTTP.
DoSHTTP - the program for carrying out of DoS-attacks (http flood). It is possible to download here: DoSHTTP
атак (http flood). Скачать можно здесь:

http://www.stopgeorgia.ru/soft/doshttp.rar http://www.stopgeorgia.ru/soft/doshttp.rar

Figure 4. Forum Members Make DDoS Tools Available to Novice Hackers

Once the attacks against Georgia became public and conventional action was imminent, many of these SQL injection vulnerabilities were posted on various hacker forums, allowing others to take advantage of the SQL injection vulnerabilities. In addition to the SQL injection attack vectors, tools were developed and posted to enable the flooding of Georgia servers, creating a distributed denial of service attack against various parts of the Georgia infrastructure. The automated tools gave even novice hackers the ability to disrupt communications while also spreading the attacking surface. An example of the tools being distributed is shown in figure 4. Both highly skilled, targeted attacks coupled with unskilled, broad DDoS attacks were used against the Georgia infrastructure, forcing the defender to address both issues simultaneously. Once attention and resources were dedicated to defending the information systems, the Russian military began the conventional campaign [9].

The exploitation of a single application level vulnerability leads to further compromise and exploitation, long after the initial vulnerability is fixed. Pilfered information leads to more pilfered information, which in turn leads to even more information theft and swells into enormous amounts of sensitive data being stolen. This data is fed into traditional intelligence analysis efforts, helping to paint the picture of the lives of government employees involved with various projects on separate, seemingly unrelated servers [10]. As the situation changes, some vulnerabilities can be repurposed to fit new situations and scenarios. Various attacks (SQL injection, Command Injection, DDoS…etc.) are launched simultaneously, forcing the defender to addresses multiple scenarios and skillsets simultaneously. Eventually, the Russian military initiated contact though conventional means, adding yet another dimension onto the defenders dilemma [9]. This chain of events began with the targeted exploitation of a single vulnerability in a single application and grew into multiple attacks launched simultaneously, along with the beginning of a conventional campaign.

1.3. Initiative

MCDP-1 describes two states that all actions in war are based upon; "initiative" and "response". MCDP-1 describes initiative as the ability to "dictate the terms of the conflict and force the enemy to meet us on our terms" and response as "resistance to initiative" [1]. Taking the initiative is considered the more preferable of the two states as, "it is through the initiative that we seek to impose our will on the enemy" [1]. In traditional military operations, initiative is established by forcing the enemy to assume a reactionary stance against active actions. These actions typically gain initiative though the effective use of surprise, tempo, concentration, and audacity. Offensive cyber capabilities offer tremendous opportunities to gain the initiative [11]. The very nature of cyber-attacks brings about the elements that are commonly associated with initiative: surprise, tempo, concentration, and audacity. Initial contact can be initiated with little risk as attacks can be launched from a variety of locations, including non-state sponsored educational and commercial networks. The nature of today's networking makes uncovering undisputable links to State sponsorship an extremely difficult task [12]. These attacks have the ability to disrupt conventional systems and decisions making from long ranges, helping shape the battlefield well before any rounds are fired.

An example of how cyber-attacks can be used to help establish the initiative for conventional forces is presented in the attacks against Georgia in 2008. In July of 2008, before a single shot was fired by conventional forces, Russian based hackers had already penetrated Georgia government applications with SQL injection and other application level attacks [13]. During the pre-emptive cyber strikes, several high profile systems (such as the website of the President of Georgia) were compromised. The pre-emptive attacks undoubtedly captured the attention of the Georgia government, forcing parts of the Georgian government to utilize a "decision making cycle" in order to determine the appropriate response [14]. By initiating contact and forcing the Georgian government to enter the decision making cycle, the Russian hackers gain the initiative, forcing a reactionary stance by the Georgian government. Much like conventional attacks, a single, un-sustained attack is insufficient in maintaining the initiative, so the Russian hackers followed with sustained attacks against a wide range of government systems. Eventually, these cyber-attacks were followed by conventional ground and air attacks. Each phase of the attack is meant to keep the enemy off balance and in a reactionary state. Agility, tempo, and surprise continuously disrupt the defenders decision making, allowing the attacker to dictate the terms of engagement. As the defenders observe, orient, decide, and attempt to act upon targeted attacks, the attackers launch broad denial of service attacks against the entire infrastructure. As the defenders rush to observe, orient, decide and act to defend the wide scale cyber-attacks, the attacker changes the terms of engagement and initiates the conventional ground and air campaign. These offensive cyber-attacks were not the ultimate end state; instead they were used to augment the achievement of initiative in conventional warfare in support of the true main effort.

1.4. Centers of Gravity (CoG) and Critical Vulnerabilities

Building a combat capability is not sufficient to win a war; to win a war, the built up combat capability must be directed towards a decisive objective. Although several interpretations for Centers of Gravity (CoG) exist, MCDP-1 considers CoG as the "sources of strength for the enemy". These sources of strength need not be physical and can encompass "intangible characteristics such as resolve or morale". MCDP-1 states that centers of gravity are to be attacked (although not directly, if well-defended). While CoG focuses on the enemy's strengths, critical vulnerabilities focus on the enemy's weaknesses. While the enemy is likely to have several vulnerabilities, some of these vulnerabilities will result in greater damage than others. Some these vulnerabilities may "contribute significantly to the enemy's downfall while others may lead to only minimal gains" [1]. Those vulnerabilities which offer the greatest impact are known as critical vulnerabilities. These are the vulnerabilities that are to be pursued by attacking forces and should be the focus of efforts.

The ubiquity and prevalence of information systems increases the overall attack surface and number of exposed vulnerabilities. Finding, classifying, and determining which of these vulnerabilities are "critical vulnerabilities" is crucial in the effective employment of offensive cyber capabilities. Once again, planners must not silo cyber capabilities, as information weapons can be used to create opportunities for maneuver against conventional critical vulnerabilities and ultimately CoG. Much effort has already been placed in determining the CoG and critical vulnerabilities in planning for conventional warfare, these CoG should be reevaluated to find avenues where offensive cyber capabilities can help maneuver against critical vulnerabilities and create opportunities for attacks against CoG. Critical infrastructure is one such conventional CoG that has already been identified where offensive cyber capabilities can bring new avenues of attack and exploitation [15]. The enemy's critical infrastructure has always been a prime target for conventional and kinetic weapons, information weapons simply bring new avenues to reaching and disrupting this critical infrastructure. As planners begin to understand the capabilities of offensive information weapons, some CoGs and critical vulnerabilities will change. Not all of these new CoGs and critical vulnerabilies will be purely military, civilian infrastructure will be affected as well. As CoG to blur from military objectives to civilian objectives, war will truly become an "extension of politics" as opposed to a struggle between two different military forces.

2. New Weapon Systems, Classic Principles

"Over time, perhaps as little as in twenty years, and as the leverage provided by technology increases, this threshold will finally reach its culmination - with the ability of one man to declare war on the world and win."

John Robb, *Brave New War*

Conventional weapons (rifles, indirect fire weapons, explosives…etc) have physical limitations. These physical limitations cannot be overcome even if the individual employing that weapon system is highly skilled or experienced with that weapon system. These physical characteristics provide the basis for the development of tactical guidance for effective employment of the weapon system. For example, the M4 carbine is designed to be employed as a short/medium range weapon in force while the M24 Sniper Weapon System is designed for precision, long range fire from trained marksmen [16]. This foundation for the employment guidance for the M4 and M24 are based on the specific characteristics and physical capabilities of the weapon system. The skillset of the individual employing the conventional weapon system may stretch the weapon systems capabilities in a small way, but the ultimate capabilities of the weapon system remained tied to the physical characteristics of the weapon.

The weapons (information weapons) used in cyber-attacks are different from conventional weapons. With conventional weapons, the physical weapon system represents the capabilities being brought to the battlefield. Information weapons enjoy a different type of relationship. With information weapons, the attacking power of the weapon system is directly correlated to the skillset of the individual using the weapon system. The capabilities of conventional weapons systems are bound to the physical characteristics of the weapon system. Two identical rifles fired by two differently skilled operators will continue to fire with the same muzzle velocity and rate, as the physical characteristics constrain the ultimate capabilities of the weapon. Information weapons on the other hand, are bound directly to the skillset of the individual employing the information weapon [17]. Two identical laptops employed by two differently skilled operators will have completely different capabilities. As the individual's skillset increases, so does the striking power and effectiveness of the information weapons employed by that individual. Creating an offensive cyber capability is less about finding the right hardware and more about finding the right people and skillsets.

The importance of the individual brings about unique challenges for intelligence organizations when attempting to understand and estimate the enemy's offensive cyber capabilities. With conventional weapon systems, capability can be tracked via procurement, tracking of the physical location of the weapon systems, and active surveillance of the weapon system. Physical movement and logistical operations associated with the employment of conventional weapons are eagerly watched by intelligence organizations and is used in all types of intelligence analysis. Developing a nuclear capability for example, require distinctive materials, specialized knowledge, and distinctive facilities for development. Non-proliferation is tracked through safeguards which monitor materials, inspections of facilities, and surveillance. Each safeguard has a threshold that indicates when a nation state may be attempting to create a nuclear capability. Nation states developing an offensive cyber capability can do so in a much more subtle way. There are no specific thresholds, distinctive materials, or facilities that indicate an offensive capability is being developed [18]. Individuals with a solid understanding of offensive security can be trained or recruited from both academia and corporate environments putting impressive offensive cyber capabilities within reach of every nation, regardless of size or economy. These individuals are the capability. With information weapons, the capability rests with the operator of the information weapon, not the equipment itself. The commercially available laptop available at any major retail outlet can be used to conduct attacks against any nation in the world. The striking power of this attack is measured not by the hardware, but the skillset of the operator. This makes tracking via procurement and logistical operations impossible. The wide spread availability of sufficient hardware coupled with the lack of distinctive, easily tracked characteristics not only lowers the barrier for entry for establishing an offensive cyber capability, it makes determining the true source of the attacks virtually impossible. Intelligence organizations must now shift focus from identifying physical equipment and logistical actions to identifying key capabilities and specific skillsets possessed by individuals. This is an extremely difficult and daunting task, making determining the true capabilities of nation's offensive cyber capabilities difficult.

In August of 2008, the Grey Goose project kept a Russian hacker forum under surveillance watching the interaction of the various forum members. Investigators determined that the forum had over 600 registered members (users were required to register in order to read/write posts). A sample of the forums member list is shown in figure 5.

It was impossible to immediately determine which of the forum participants represent a legitimate offensive capability and which forum members are simply "script kiddies" (unskilled participants). Analysts for the Grey Goose Project were forced to analyze thousands of forum events, learning about the various topics being discussed by the forum members. Each forum post was analyzed for technical sophistication and technical leadership. Relationships between forum members were mapped using

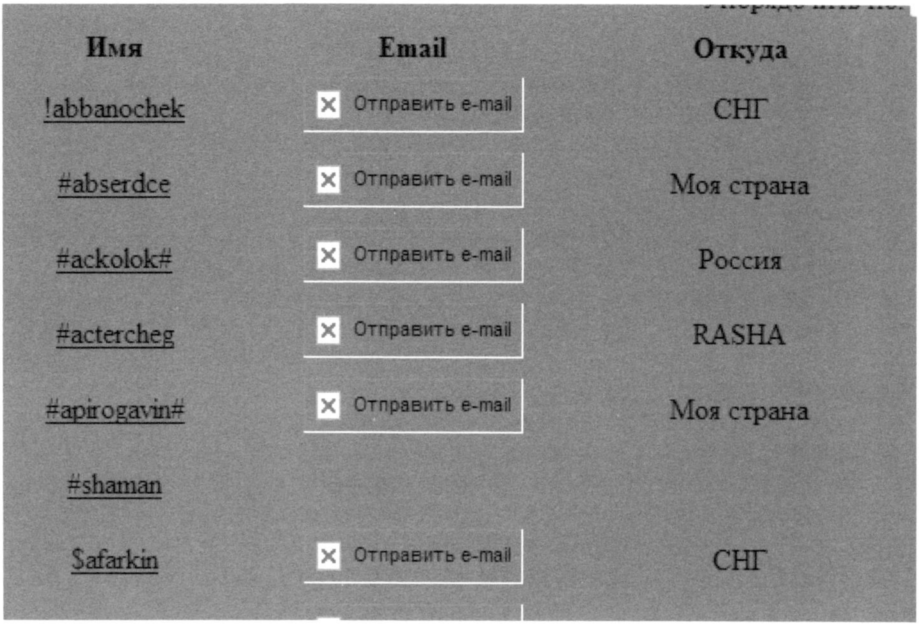

Figure 5. Member List from a Russian Hacker Forum

technically sophisticated Palantir analysis platforms [19]. Only after extensive analysis could the Grey Goose investigators determine which members represented the true offensive capability of the forum. Once these individuals were identified, surveillance focused was focused on these key individuals. Each individual was noted and ranked using forum participation as the key indicator of their technical sophistication (individual contributed vulnerabilities, contributed tools, provided advice for exploitation…etc.). This approach allowed the analysts to focus on the handful of individuals driving the offensive capabilities of the entire forum. It was these individuals that were offensive capability, not the tools, hardware, or even the forum [4].

3. Conclusions

"*Preparing to win in combat must be the highest priority in the allocation of time, dollars, and rewards, at every level and under all circumstances*"

William S. Lind, Maneuver Warfare Handbook

3.1. Employment

The focus on using cyber capabilities to "win wars" must be at the forefront when developing doctrine. It is easy to become enamored with the seemingly magical displays of exploitation and technical jargon; however planners must recognize that cyber capabilities represent one of many dimensions of warfare. Planners must not silo cyber capabilities, employing them in isolation, but must consider how best to augment conventional capabilities with cyber capabilities. Having a robust cyber capability is important, however winning the "cyber-battle" while losing the conventional war is unacceptable in any scenario. In time, information weapons will be the weapons of force, perhaps establishing themselves as the "main effort" in campaigns with conventional forces designated as "supporting efforts". Until that time, information weapons are to be employed much like other conventional weapon systems as supporting efforts, helping shape the battlefield in support of the main effort (typically conventional forces). Planners must strive to integrate cyber capabilities into conventional warfare as a supporting arm. Information weapons, much like other supporting arms, must be cognizant of the main effort and should strive to shape the battlefield in support of the main effort. Commanders must be acclimated as to how to request supporting cyber capabilities and understand what gaps offensive cyber capabilities can cover.

3.2. Command

Rigid, highly centralized command makes the development and effective employment of offensive cyber capabilities difficult. Commanders must be careful not to impose rigid requirements or artificial constraints onto cyber capabilities. Judicious use of commander's intent is essential in establishing the decentralized operational command necessary for the development and effective employment of offensive cyber capabilities. A top down, micromanaged effort will kill the speed, tempo, and most importantly the creativity required in effective cyber-attacks. Hierarchical, centralized ***administrative*** chains of command are essential for the good order and discipline in military ranks; however ***operational*** chains of command should strive to push decision making down to the lowest level, using commanders intent to guide decision making and initiative. This forces a more decentralized approach to employment of cyber capabilities, allowing for the need flexibility needed to successfully employ offensive cyber capabilities. Without this decentralized approach, employment of offensive cyber capabilities will ultimately fail.

3.3. The Individual is the Weapon System

The leverage technology brings coupled with the increasing ubiquity of information systems builds upon the power wielded by individuals with the right skillsets. As operational commands become more and more decentralized and the impact of the individual becomes more and more powerful eventually, a tipping point will be reached and the individual will represent the offensive capability. The concept of the measuring a nation state's striking power and capability through the surveillance and tracking of ground forces, air, and naval equipment will eventually succumb to the identification of highly skilled individuals that represent the offensive cyber capability. As our reliance on technology continues to evolve, the ability of the lone individual to disrupt conventional operations also increases. Eventually, the power of the lone individual will grow until a lone, highly skilled individual (or a small team of highly skilled individuals) will be able to impose their "political will" on other individuals, corporations, and even nation states.

References

[1] United States Marine Corps, *Marine Corps Doctrinal Publication 1 - Warfighting*, United States Marine Corps, Quantico, VA, 1997. http://www.dtic.mil/doctrine/jel/service_pubs/mcdp1.pdf
[2] Department of Defense, *Conduct of the Persian Gulf Conflict*, Department of Defense, Washington, DC, 1991. http://www.dod.gov/pubs/foi/reading_room/305.pdf
[3] Correlli Barnett, *Victory Through Air Attack? It's a Pie in the Sky*. TimesOnline, 2009. http://www.timesonline.co.uk/tol/comment/columnists/guest_contributors/article5469051.ece
[4] Jeffery Carr, *Grey Goose Phase I*. GreyLogic, 2008. http://www.scribd.com/doc/6967393/Project-Grey-Goose-Phase-I-Report
[5] Kevin Benson, *Tactics for Small Wars*. Institute of Land Warfare, 2008. http://www.ausa.org/SiteCollectionDocuments/ILW%20Web-ExclusivePubs/Landpower%20Essays/LPE08-1.pdf
[6] Marine Corps Combat Development Command, *Offensive Fundamentals I*. Marine Corps Combat Development Command, 2009. http://www.usna.edu/USMCInfo/Documents/Pubs/b0354.pdf
[7] Lt Col. Lawrence Shattuck, *Communicating Intent and Imparting Presence*, Military Review, 2000. http://www.au.af.mil/au/awc/awcgate/milreview/shattuck.pdf
[8] Ives, Walsh, Schnieder, *The Domino Effect of Password Reuse*, Communications of the ACM, 2004. http://portal.acm.org/ft_gateway.cfm?id=975820&type=pdf&coll=GUIDE&dl=GUIDE&CFID=38680500&CFTOKEN=56506836
[9] Kenneth Corbin, *Lessons From The Russia-Georgia Cyberwar*, Institute of Communications Studies, 2009. http://ics.leeds.ac.uk/papers/vp01.cfm?outfit=gdr&folder=442&paper=750
[10] Whitaker, Evans, & Voth, *Chained Exploits: Advanced Hacking Attacks from Start to Finish*, Pearson, 2009.
[11] Office of the Secretary of Defense, *Military Power of the People's Republic of China 2006*, Office of the Secretary of Defense, 2006. http://www.defenselink.mil/pubs/pdfs/China%20Report%202006.pdf
[12] William Jackson, *US Already at War in Cyberspace*, Government Computer News, 2009. http://www.gcn.com/Articles/2009/04/22/RSA-cyberwar.aspx
[13] Toomas Hobermagi, *Estonia Promises Georgia Help in Fighting a Cyberwar*, BalticBusinessNews.com, 2008. http://balticbusinessnews.com/Default2.aspx?ArticleID=4743f52b-6f71-4ebd-ad1c-ca4ab13ad921
[14] Wikipedia, *OODA Loop*, Wikipedia.com, 2009. http://en.wikipedia.org/wiki/OODA_Loop
[15] Cordesman, *Cyber-Threats, Information Warfare, and Critical Infrastructure Protection*, Greenwood Publishing 2002.
[16] Department of Defense, *Basis of Issue Plan – M24*, Department of Defense. http://www.fas.org/man/dod-101/sys/land/docs/bnI063AA.htm
[17] Killgannon & Cohen, *Cadets Trade the Trenches for Firewalls*, Nytimes.com, 2009. http://www.nytimes.com/2009/05/11/technology/11cybergames.html?_r=2&ref=technology
[18] Wikipedia, *Nuclear Proliferation*, Wikipedia.com, 2009. http://en.wikipedia.org/wiki/Nuclear_proliferation
[19] Palantir Technologies, *Palantir Government Services*, 2009. http://www.palantirtech.com/

The Virtual Battlefield: Perspectives on Cyber Warfare
C. Czosseck and K. Geers (Eds.)
IOS Press, 2009
© 2009 The authors and IOS Press. All rights reserved.
doi:10.3233/978-1-60750-060-5-156

Belarus in the Context of European Cyber Security

Fyodor Pavlyuchenko[a]
Translated from the Russian language by Kenneth Geers[b]
[a]*www.charter97.org*
[b]*Cooperative Cyber Defence Centre of Excellence*

Abstract. During the first decade of the 21st century, Internet censorship in Belarus has evolved into a government tool used to combat political dissent. State-sponsored denial of service (DoS) attacks against civil society have become a domestic crisis that threatens not only freedom of expression in Belarus, but also the integrity of Internet resources throughout Europe. The ongoing cyber conflict between state and non-state actors in Belarus is analogous to the struggle between the Russian government and its internal adversaries in cyberspace. In this essay, we recount the history of cyber censorship and attacks against Charter '97, a popular Belarusian website, and discuss the effectiveness of countermeasures.

Keywords. Belarus, Internet, censorship, charter '97, denial of service, cyber crime

Introduction

For over a decade, the Charter '97 website in Belarus has been a leading venue for Belarusian public policy discussion. Since its founding, however, the site has been forced to defend itself against practically all existing Internet censorship strategies. Following a disputed political referendum in Belarus in 1996, Alexander Lukashenko alone has governed the country. His government has suppressed freedom of speech, and Charter '97 is well-known for siding with Belarusian dissidents.

While modern technology has offered the world significantly improved communications, it also creates new threat vectors. Nation-states can abuse their power over telecommunications, creating dangerous precedents and fostering long-term political instability.

In cyberspace, the activities of the Belarusian secret services are reminiscent of their colleagues in Russia. It is critical that democratic states in Europe strengthen independent Internet resources in all European countries, and strive to extend the rule of law in the whole of European cyberspace.

1. Background: Internet Censorship in Belarus

For over a decade, there has been virtually no independent traditional media in Belarus. Under pressure from the authorities, the popular newspapers of the 1990's have ceased to exist or have seen their circulation greatly reduced. The independent FM radio stations have been closed, and an independent Belarusian television channel has never existed. It is therefore unsurprising that, despite its high cost, the Internet has been and remains the only source of objective information for the majority of Belarusians, and the number of Web users has grown to nearly one-quarter of the country's population.

09.09.2001

On September 9, 2001, at 1200, Belarusian Internet users came into conflict with the authorities for the first time – on the day of the national presidential elections. The national telecommunications firm Beltelecom – a monopoly provider in the country – intentionally blocked access to a range of popular political websites. By the following afternoon at 1600, all Internet censorship had ceased.

From a technical perspective, such information blocking is easy for a telecommunications monopoly to perform. The data packets can be filtered on the ISP's primary router based on their source or destination Internet Protocol (IP) addresses. With the help of the 'tracert' (traceroute) command, the point of network interruption was easy enough to find. During this event, the prohibited sites continued to be accessible outside Belarus, but not within the country.

Some of the popular but blocked sites, including *.home.by, *.minsk.by, *.org.by, *.unibel.by, *.nsys.by, and *.bdg.by were physically hosted on servers in Belarus, within the .by network domain. These sites were disabled by altering their Domain Name Service (DNS) records. In that context, it is important to note that the management of the .by Top Level Domain (TLD) is the responsibility of a special state agency, the Operations and Analysis Center, which falls under the direct control of the President of Belarus.

Some websites, including www.charter97.org, created multiple mirror sites in an effort to stay online. All such mirrors were promptly blocked by the government. Furthermore, popular 'anonymizer' websites and pages that published lists of free proxy servers were also blocked. In all (including the sites that were directly and indirectly blocked), over 100 websites were inaccessible.

It must be emphasized that there were no legal grounds to perform Internet censorship in this case. In fact, such censorship directly violated the Belarusian constitution. The official explanation from the Ministry of Communications and Beltelecom was that too many Belarusians were trying to access the sites in question at one time, and that this led to a self-inflicted denial of service. From a technical point of view, that explanation does not hold water. For its part, the Belarusian government had no comment, even though Internet censorship – in this case, computer sabotage – is an offense under Belarusian law. An official investigation into the facts of the case was never undertaken.

24.10.2001

The Charter '97 website was completely deleted from its server by unidentified hackers. A few days after this incident, under pressure from the Belarusian secret services, the hosting company was forced to break its contract with Charter '97, and the site was no longer allowed space on its server.

20.01.2004

In January 2004, Charter '97 was the target of a massive distributed denial of service (DDoS)-attack for the first time, which lasted more than 3 weeks. The attack followed the publication of a journalistic investigation into a possible connection between high-ranking officials from the Belarusian Interior Ministry (from a department responsible for investigating computer crimes) and the trading of child pornography on the Internet. In a strange coincidence, Natalya Kolyada, a human rights activist working with our website at that time, was also convicted on misdemeanor charges.

The DDoS attack was supported by a botnet that included more than 55 thousand active IP addresses. This network of infected computers spanned the globe, and included compromised machines in Latin America, the United States, South-East Asia, China and India. The geographic dispersion of the botnet allowed the power of its attack to be spread across a 24-hour period. Further, the power and focus of the attack changed several times, which indicated an active command and control (C2) over the activity.

While it is impossible to affirm that this attack was politically motivated, a simultaneous campaign of harassment was organized against the employees of our site on official Belarusian television. Our employees were, among other things, accused of trading in online pornography.

14.07.2004, 21.07.2004

On July 21, 2004, there were mass protests in Minsk to mark the 10th anniversary of the Lukashenko government, and our website again came under a DDoS-attack. Charter '97 had planned to host a webcast to cover the protests. One week prior, on July 14, a 'test' cyber attack had paralyzed our server for 2 hours. On July 21, the main attack began at 1400 – 4 hours before the demonstrations began – and lasted until the political protests were over. The technology and power of the attack were similar to the attack in January of that year.

10.10.2004

The next large-scale attempt to block Charter '97 and other independent websites occurred in the autumn of 2004, during parliamentary elections and a national referendum on the lifting of presidential term limits in Belarus. On the day before the election, correspondents were unable to access our website and could not call us by mobile or landline telephone. At the same time, various opposition websites were again blocked by a filter on Beltelecom's primary router.

This time, many Belarusian network users were better prepared to combat censorship, and immediately switched to Internet proxies and anonymizers. But the Belarusian Internet authorities had an effective new weapon in their arsenal: the artificial stricture – or "shaping" – of Internet bandwidth. The use of this tactic meant that, in principle, forbidden sites were still available. However, it took anywhere from 5 to 10 minutes for the censored webpages to load in a browser. Thus, most Web users were unable to gain full access to Charter '97, as well as a range of other targeted sites. All other Internet resources were accessible as normal.

There was no announcement from the Ministry of Communications or Beltelecom regarding this incident, nor was any official investigation undertaken.

19.03.2006

The next time that websites were blocked in Belarus was on March 19, 2006 – the day of Belarusian presidential elections. Well before the election took place, in an initiative called 'Free Internet', Charter '97 began to offer site visitors information regarding various ways to circumvent censorship. This strategy ensured that all attempts to block information using IP-filtering failed. However, network 'shaping' – or the intentional stricture of specific streams of network bandwidth – was again the method employed.

On March 18, the day before the election, a website filtering 'test' was conducted from 1600-1630. On election day, sites belonging to opposition presidential candidates and their political party websites, leading independent news sources, and www.livejournal.com (an international blogging site very popular with Belarusians) were blocked.

Representatives from Beltelecom announced that the technical interruptions were caused by the overloading of particular circuits, and no formal investigation was ever undertaken.

25.04.2008

On the eve of massive street protests in Minsk, the government changed its strategy. At 1420, a test DDoS-attack on Charter '97 lasted 30 minutes. The following IP addresses were used in the attack: 89.211.3.3, 122.169.49.85, 84.228.92.1, 80.230.222.107, 212.34.43.10, 81.225.38.110, 62.215.154.167, and 62.215.117.15. For about 10 minutes, our site was difficult to access, but we were able to restore normal traffic before the attack ended.

On April 26, the main DDoS-attack took place. It began five hours before the start of the demonstration. Charter '97 had intended to conduct a live webcast of the protests, but the attack paralyzed our server. Our hosting company, www.theplanet.com, attempted unsuccessfully to mitigate the attack. Its hardware was designed to defend against an attack only up to 700 Mbps. The volume of this DDoS reached over 1 Gbps. We had no choice but to turn off the site and simply wait for the attack to end. The perpetrators were apparently satisfied with their achievement, and the attack was over by the next day. It is important to note that other independent online media were subjected to similar attacks at the same time, especially 'Belarusian Partisan' and the Belarusian-language version of 'Radio Liberty'.

The server hosting 'Belarusian Partisan' crashed as a result of the DDoS-attack. Further, system administrators even temporarily lost control of the site, and for several days, unknown hackers used the website to publish fabricated, scandalous news stories. The Belarusian Partisan editors were forced to denounce the false reports on other websites. The level of expertise required for this attack was high enough that there is no doubt the Belarusian special services were involved in the incident.

The technical capabilities of the Radio Liberty (RL) server – home to the Belarusian, Albanian, Azerbaijani, Tajik, and Russian RL services – were sufficient to contain a similar attack for more than 3 days. However, the site was nonetheless difficult to access until 1500 on April 28, and that was enough to cause a minor diplomatic scandal. The U.S. mission to the Organization for Security and Cooperation in Europe (OSCE) issued a statement on the cyber attack. The Belarusian Ministry of Foreign Affairs publicly denied any official involvement.

Lamar B. - Saturday April 26th, 2008; 8:11 AM CDT

Dear Customer,

The block is to protect our networks integrity. 1G of traffic would cause latency for customers on that switch. I will removed the null once it below 700Mps of traffic and place it back in the guard. I have also informed our next shift of your situation. I am sorry for the inconvenience.

Lamar
NOC Technician

Current Traffic:
treebeard.organichost.com
69.93.158.34 1.1 G (14.6%)

Figure 1. Explanatory email from the Internet Service Provider (ISP)

Increasingly, experts believe that various political DDoS attacks share some common characteristics, and that there may be important links between discrete attacks such as those which hit Estonian and Georgian websites.

08.06.2009

The most recent example of a politically-motivated DDoS attack on Charter '97 occurred in June 2009. The incident may have been related to the recent conflict between the governments of Russia and Belarus, which resulted in the imposition of economic sanctions against Belarus and a worsening political situation inside the country.

The DDoS attack lasted more than a week, and for a while it paralyzed our site completely. The strength of the DDoS-attack in this case had not been particularly high; about five thousand IP addresses took part in it. With the support of our ISP, the Charter '97 technical support staff was able to neutralize the attack.

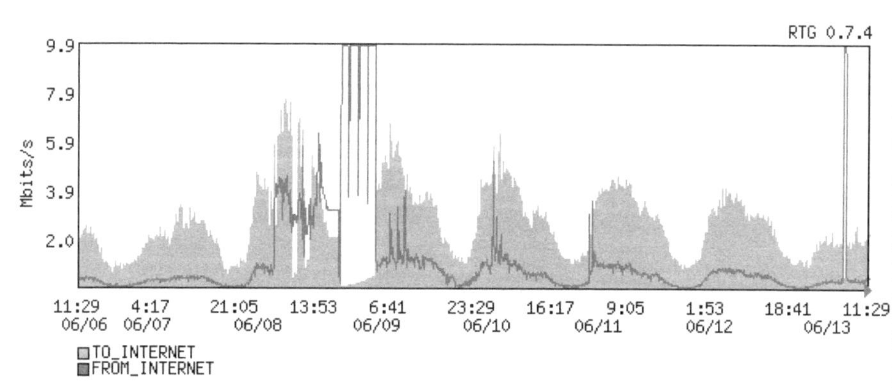

Figure 2. Timeline for the June 2009 attack against Charter '97

2. Countermeasures and their Effectiveness

Throughout the history of these DoS attacks, we have been looking for ways to counter censorship by the Belarusian authorities. We have tried many different methods, but none of them has been completely effective. Our situation might best be described as a competition to outmaneuver an opponent who has more resources than we do.

Initially, our strategy was to strengthen our technical capabilities and to increase our expertise in computer security. The site was moved to a relatively powerful, hardened server, and we created a system for monitoring vulnerabilities and attempted intrusions. We employed encryption, not only for access to the server itself, but also for access to our content management system. We created a multi-tiered system of access to the server and to the site, as well as the ability to quickly replace all passwords in the event administrators and/or journalists were arrested. We developed a distributed system for creating server data backups. Based on our experience so far, it is best to use simple, open-source technologies such as UNIX, PHP, and MySQL to help with site mobility (i.e. the rapid transfer of the site to another hosting platform). Firewall and caching technologies are sufficient to repulse DDoS-attacks of average strength. Combined, these efforts helped to prevent subsequent compromises.

Separately, Charter '97 launched the 'Free Internet' project. This website provides recommendations to site visitors for what to do in case the site is blocked, and explains how to use an Internet proxy, anonymizer, Virtual Private Network (VPN), and software such as Tor. Visitors are encouraged to disseminate information independently through their own blogs, forums, chat rooms, social networking sites, e-mail, and instant messengers. Site information is rebroadcast, for example, via RSS or email to mirror sites and partners. The successful use of these measures can overcome the simple blocking of IP addresses. However, with our limited resources, there is no current, solid countermeasure to DDoS.

We believe that the authorities have concluded that their most effective methods of censorship are DoS attacks and information manipulation. In support of the latter, specially-trained 'visitors' insert themselves into discussions on popular websites, and monitor or help 'guide' the course of discussion. During politically sensitive times, or whenever political dialogue becomes excessive, the site can be temporarily blocked by DDoS.

3. Government Power and Cyber Crime

The current power structure in Belarus not only attempts to suppress political dissent on the Internet, but also flagrantly violates the Belarusian constitution. In Belarus, there is no legal basis for Internet censorship, much less for computer hacking and DoS attacks on websites. There is only one case of censorship in Belarus that was officially acknowledged and at least somewhat justified: in 2005, Beltelecom blocked two Russian gay sites for possession of pornography, at the behest of the Ministry of Culture's Republican Commission for the Prevention of Promoting Pornography, Violence and Cruelty. All other cases of Internet censorship involved the use of blatantly criminal methods.

DDoS-attack techniques against Belarusian independent Internet media could easily be used to block all sites covering current affairs in Belarus. Further, the lack of the rule of law within Belarusian Internet space could facilitate the growth of organized cyber crime on the international level.

There is active cooperation between Belarusian and Russian special services in cyberspace: the Agreement on Cooperation of the Commonwealth of Independent States (CIS) in Combating Cybercrime was signed in 2000. Finally, there are similarities in the cyber attack methods used against Estonia, Georgia, and the websites of human rights organizations in Belarus and Russia, which suggest that these crimes have common roots.

4. What is Required?

The challenge of DDoS attacks threatens civil society throughout Eastern Europe. In Belarus, Ukraine, Russia, Georgia, Armenia, and Azerbaijan, the government may already have used DDoS attacks as a tool for countering dissent on the Internet. To mitigate this threat, an international, collaborative approach is required. A good start would be the establishment of an international hosting platform designed to support freedom of speech throughout Europe, built by a team of international experts. They should investigate cyber crimes based on aggregate data, and work toward the development of effective defense methods and technologies. The mere creation of such a platform would be a helpful step, and could enhance the level of cyber security and freedom of expression throughout Europe.

The Virtual Battlefield: Perspectives on Cyber Warfare
C. Czosseck and K. Geers (Eds.)
IOS Press, 2009
© 2009 The authors and IOS Press. All rights reserved.
doi:10.3233/978-1-60750-060-5-163

Politically Motivated Denial of Service Attacks

Jose NAZARIO
Arbor Networks, United States

Abstract. Cyberwarfare has been waged for well over a decade, utilizing methods such as website defacement, data leakage, and distributed denial of service attacks (DDoS). This paper focuses on the latter, attacks that are easily carried out and designed to overwhelm a victim's network with wasted traffic. The goal of a DDoS attack is to make the use of the network impossible for internal or external users. Through a brief examination of the history of these attacks, we find they previously were designed to inflict punitive damage on the victim but have since grown into sophisticated censorship tools. Our approach measure such attacks by looking at Internet backbone traffic, botnet activities, BGP routing changes, and community chatter about such attacks to provide a robust picture of politically targeted DDoS attacks. Our analysis indicates that most of the attackers are non-state actors but are able to fluidly utilize a growing botnet population to launch massive denial of service attacks. This finding has broad ramifications for the future of these attacks.

Keywords. DDoS, botnets, Estonia, Georgia, China, Russia

Introduction

Internet attacks take on many forms, including system compromises and information theft, as well as denial of service attacks designed to disrupt services. Motivations for cyber-attacks include frustration, fun, and extortion, especially against gambling and pornography sites. Anger and frustrations appears to be the major motivation in attacks against gaming sites and forums, where player-on-player attacks happen quite frequently. Politically targeted attacks are extremely rare in the sphere of daily attacks. The types of attacks launched depend on the attackers' skills and motivations.

A distributed denial of service (DDoS) attack is nothing more than a coordinated effort that instructs PCs to send a victim a flood of traffic designed to overwhelm their servers or consume their bandwidth [[1]]. Regardless of the attacker's underlying motivations, the attacks are designed to disrupt the normal flow of the site for internal or external users. The PCs used in the attacks can be the bots in a botnet or a zombie army, or it can be tools willingly installed on peoples' computers. A simple form of a DDoS attack is when individuals work together and to continuously reload a website in a browser such as Internet Explorer. In each case, the purpose is to aggregate the PCs' bandwidth together to overwhelm an adversary who is usually superior in their bandwidth resources, and to do so from a large enough number of locations to make source-based filtering unmanageable.

DDoS attacks are among the most visible and disruptive of cyber-attacks. When coupled to political motivations, they can be seen as an extension of politics in the 21st century, to borrow a phrase from von Clausewitz. Currently, researchers infer a political motivation for various attacks based on internal information, such as the nature of the victim and the attack commands seen. Investigators may also use external sources to validate this finding by looking at news reports and website conversations discussing diplomatic grievances and their redress through online attacks. In this context, cyber-attacks are sometimes referred to as "fifth generation warfare".

Arbor Networks' Peakflow products are used by many Internet service providers to detect and defend against DDoS attacks [[2]]. Independently run Peakflow deployments collect data on such attacks and provide a distillation of the events to the Arbor Networks ATLAS portal. Attack data is gathered in three different ways to provide a nearly complete picture.

The first data source is direct traffic measurement using Arbor Peakflow deployments around the world using data collected as part of the ATLAS project. Customers can share attack information with each other through the Fingerprint Sharing Alliance. Some of this data is made available in ATLAS and can be analyzed by network or country affected or launching the attack. Peakflow counts attacks based on the traffic types (e.g. TCP SYN, ICMP echo request) and destination networks for a time period using dynamically learned baselines or static thresholds. Therefore, Peakflow may register multiple concurrent attacks if they target the same destination but use different traffic, such as a simultaneous TCP SYN flood and ping flood.

The second way attacks are measured is to look at commands sent to botnets to launch attacks. Malicious software analysis can be used to discover botnets and infiltrate them by communicating with the botnet's command and control (C&C) server by mimicking legitimate bot clients, enabling a record of the botnet's activities for later analysis. This data is valuable to understand the attack's root origins for disruption but also for post-event analysis to understand the nature of the attackers. Most of the attacks tracked are against inconsequential targets, but sometimes they target victims such as financial firms, major e-commerce sites, or government assets.

The third form of continuous measurements is to look at border gateway protocol (BGP) routing data used to provide Internet backbone routing. Sometimes the paths may change during an attack as a direct result of the attack, such as BGP session drops during congestion, or through attempts to mitigate the attack. Changes to the BGP routes for a victim can indicate an attack.

This paper focuses on the confluence of DDoS attacks with political targets and political or ideological motivations. DDoS attacks are crippling because they are designed to make the networks they target unusable, either to inflict damage to the victim or, in the case of many recent events, to silence their opponents by making their resources inaccessible. This paper does not analyze information on attacks such as website defacements or compromises through malicious software that may be a part of these attacks.

1. Major Events in Political DDoS Attacks

DDoS attacks became widely popular in the late 1990s following the development of toolkits such as Tribe Flood Network and Trinoo [[1]]. These methods were quickly adapted for political targets. Major attacks from the past 10 years can be used to highlight changes and illustrate how sweeping this problem can be.

Very early events in this field include attacks on NATO computers in the former Yugoslavia during the campaigns in the late 1990's, and also the attacks from Chinese hackers on US military sites following the bombing of the Chinese embassy by a US plane in the former Yugoslavia during that NATO mission [[3]]. This list of attacks shows how many different regions are affected and how many different motivations exist for these attacks. It also shows how these attacks have evolved over time. This section also shows that such attacks didn't start with Estonia in May 2007 and didn't end with Georgia in the summer of 2008.

1.1. Hainan Spy Plane Incident

In April 2001, a US Navy spy plane was on a reconnaissance mission off the southern coast of China when multiple Chinese fighter planes intercepted it, and one of the planes clipped the US Navy plane, causing damage to both planes. The pilot of that Chinese fighter plane was lost after his plane broke up in mid air, and due to the damage it sustained during the accident, the US Navy plane had to make an emergency landing in Chinese territory on Hainan Island. The crew was held for several days before diplomatic efforts released them.

During this time, tensions between the US and China ran high. Among the events that occurred were multiple attacks, including DDoS attacks and probes on US military Internet sites. The Chinese hacking group "Honker Union" is believed to have been behind the attacks [[4]].

The attackers in this situation appeared to see these attacks as acts of patriotism. The public outrage was undeniable and bubbled over to Internet forums. Multiple groups and parties appeared to take part in these actions.

1.2. Estonia, 2007

Beginning in late April 2007, the European nation of Estonia was hit by a series of coordinated denial of service attacks. Ethnic Russians make up a significant percentage of Estonia's population, and by many accounts Estonians and the ethnic Russians co-existed peacefully [[5]]. As is commonly found throughout Russia and much of the former Soviet Union, Estonia has a statue of a Soviet soldier commemorating the end of World War II.

The statue has been a sore point in Estonian politics for many years and was moved in April 2007, leading to civil unrest within Estonia and complaints by the government in Russia [[14]]. Coinciding with the street protests, online DDoS attacks began to target Estonian government and private sector sites, including banking institutions and news sites.

The attacks seen in Estonia built up over the course of a few weeks and peaked on Victory Day, May 9. On this day, Peakflow systems around the world measured attacks lasting 10 hours each with a peak bandwidth utilization of 95 Mbps. This data comes from multiple Peakflow sensor sources that are aggregated into ATLAS via ISPs that provide transit for Estonian ISPs [[6]].

The attackers used multiple attack methods. They used Russian language forums and blogs to spread tools such as ping flood scripts and to coordinate their efforts, and they also recruited botnets into the effort. For example, they worked hard to take their collective tools, botnets, and activities and fire them at the same time (e.g. 11pm in Moscow).

The attacks in Estonia hit many parts of the infrastructure, including the websites for the prime minister, parliament, various ministries, and even government name and mail servers. News reports contained information about slowdowns with some banks and financial transactions. All of this is consistent with a nation that makes heavy use of the Internet for daily life suffering from systemic flooding.

Most of the attacks measured in ATLAS died out after Victory Day, although reports from first-hand accounts within Estonia indicate that they continued for several weeks.

1.3. China and CNN

In April 2008, the CNN news personality Jim Cafferty commented on air about the Chinese preparations for the Olympics in Beijing, China. Many Chinese found these remarks offensive, and this sentiment quickly brewed into anti-CNN hacking events.

A number of hacking groups activated and worked to coordinate their activities. The attacks included website defacements and many probes to try and disrupt the CNN.com website. Peakflow and ATLAS also monitored the flows for the site as well as for botnet attack activity [[7]].

During the investigations, a number of Windows tools developed to target CNN specifically were discovered, in addition to a few botnets that were targeting CNN more generally. The first tool was dubbed "Supper DDoS" is a simple flooder usable by an average computer user, with only an input for the victim's address together with "Attack" and "Stop" buttons. This tool was distributed on Chinese language forums by an unknown number of authors.

Researchers also discovered a botnet apparently operated by "Ice Kernel", using a bot dubbed "KernelBot". KernelBot is a flexible DDoS attack system, supporting common attack types, as well as full control of the victim's PC. Commands for this botnet targeting CNN's website appeared during this event. Another tool released in late April 2008 to target CNN was a specialized version of the NetBot Attacker tool, a general purpose DDoS tool that is usually deployed on a victim's PC using standard malware infection methods. This particular version of NetBot Attacker is hard coded to target CNN.com and provides the user with some basic control over their PC. This kit includes the flooding portion of the bot and the attacker's UI for control, something not normally seen. Typically, the bots run without any UI for the victim.

1.4. Georgia and Russia

In July 2008, the website for Georgian President Mikheil Saakashvili was hit with a DDoS attack. In this case the botnet was based on a codebase that is only seen in Russian-language botnets. The command and control server for this botnet was located in a regional ISP, PaeTech, and had been under surveillance by ATLAS and other researchers for some time. This was the only attack launched by this botnet and lasted from July 18-20, 2008 [[9]].

These attacks were corroborated together with ShadowServer, a volunteer botnet monitoring team. We attempted to reach the site during the attack and found that the Georgian President's website was unable to load from a number of North American vantage points, consistent with a major attack [[10]]. When asked by the press, spokespeople for Saakashvili's office said that no such attacks had occurred, however.

A message was included in the attacks that read "win love in Russia", consistent with the ongoing tensions in the region. A few days before the DDoS attacks began in July 2008, the ITAR-TASS news agency from Moscow ran a story with the translated headline "Withdrawal of Georgian troops only way out of Abkhazia conflict - Medvedev". At the time, Russian president Dimitry Medvedev had been in power for a few months. There had been ongoing, minor skirmishes between Georgia and Russia over two regions within Georgia. South Ossetia and Abkhazia, two semi-autonomous areas have historically stronger ties to Moscow than does the rest of Georgia. These two regions had been seeking more independence and closer ties to Russia than Tbilisi would allow. The diplomatic flames going back and forth were substantial and included reports of gunfire between Georgian and Russian forces. After the shutdown of the PaeTech C&C server, the July 2008 attacks stopped [[9]].

A few weeks later, in early August, a large-scale shooting war between Georgia and Russia broke out with Russian tanks entering Georgian territory. Almost immediately, very substantial DDoS attacks began to flood into Georgia and were caused by multiple botnets and ping flood scripts. Targets included the Georgian president's site, various ministries, news agencies, and others [[6]]. Furthermore, ATLAS monitors recorded some attack commands into Russia at the same time, suggesting that someone - either Georgian or possibly a Georgian sympathizer - tried to counter attack.

Arbor Peakflow and ATLAS live traffic monitors on the Internet showed that the peak size of the attack was substantially larger than the attacks in Estonia the year before. The peak bandwidth recorded during the attacks was over 800 Mbps, and the attack were much more intense [[9]].

Key Georgian properties were quickly relocated to various countries with better defense capabilities. The president's website, for example, was moved to Atlanta Georgia and Tulip Networks. Other sites were moved to Estonia, which had experience and tools after the previous year's attacks [[11]].

This was the first time in nearly 10 years that a military conflict and a cyber conflict coincided, the most recent being attacks between Israel and Palestinian militias. These attacks on Georgian websites, especially after what happened in Estonia, have raised concerns around the world by governments concerned about an apparently growing trend of politically motivated attacks on government networks. This is discussed later in this paper.

1.4.1. Investigating Active Routing Attacks

One unique aspect of the attacks is that Georgia gets nearly all of its Internet access from two main countries: Russia and Turkey, with some additional connectivity from Europe. Analysis by Bill Woodcock at Packet Clearing House shows that nearly all of the major connectivity routes go through Turkey or 0Russia. This provides a high bandwidth connection for Russian bots, if they are located in Russia, to flood Georgia. Turk Telecom, the main upstream for Georgia in Turkey, is also a major source of bots.

Russian ISPs were accused during the fighting of filtering or blocking Georgian sites, which would have been possible for some routes but not all. In our analysis, we have found no data that suggests that Russian ISPs performed such filtering [[12]].

Routeviews monitors did see some unexplained BGP announcements via Turk Telecom but we attribute those to fighting the DDoS traffic or drops due to congestion, rather than active attempts to disrupt normal Georgian traffic.

Our measurements indicate that approximately 100 BGP updates per day occurred for Georgian prefixes immediately before the onset of ground fighting with Russia. After ground fighting began, less than 10 BGP updates per day were seen. The August cyber attacks began within 24 hours of Russian tanks rolling into Georgia, making the data hard to decipher conclusively. Any BGP disruptions could be due to fighting on the ground, DDoS attacks (and congestion leading to drops), or active disruptions by upstream peers.

1.5. Democratic Voice of Burma

Starting in the summer of 2008, DDoS attacks were launched against the Burmese dissident site the Democratic Voice of Burma (DVB) and its sister sites. Many of the attacks were website defacements and the attackers got in through a poorly configured and poorly secured site. ATLAS monitors recorded some packet flooding to the sites, as well [[14]].

Most of the attacks were apparent attempts to censor the sites and to thwart planned 8-8-2008 protests around the world. 20 years before on August 8, 1988, a significant protest occurred in Burma against the ruling Junta centered on the 8-8-88 date. The number 8 is very significant in Chinese and Burmese society, providing the protests on 8-8 are a powerful rallying point. The Burmese government is believed to be behind the attacks, although no such evidence has been provided.

1.6. Russian Elections, 2007

In the lead up to the Russian elections in late 2007, the website for the dissident politician and well-known chess Grand Master Gary Kasparov and his political party were both hit with substantial DDoS attacks. Kasparov has been a very vocal counterpoint to the powers in Moscow, specifically former Russian president Putin's administration, for many years. During the attacks, Kasparov's site was inaccessible, and so was his political party's [[13]].

The attack command activity traced back to botnets possibly run by Russian or pro-Russian hackers. The botnets have been used in the past to strike political targets among other targets.

1.7. Radio Free Europe/Radio Liberty

In April 2008, Radio Free Europe and Radio Liberty (RFE/RL) websites were hit with DDoS attacks [[15]]. It is thought that the attacks were in retribution for the reporting that RFE/RL made to cover the anniversary of the Chernobyl disaster.

The attacks started on April 26 and first targeted the website of RFE/RL's Belarus Service and quickly spread to other RFE/RL sites. Within a few hours, eight different RFE/RL websites serving Belarus, Kosovo, Azerbaijan, Tatar-Bashkir, Radio Farda, South Slavic, Russian, and Tajik-language listeners were all affected by such attacks.

The botnet behind the attacks was a Russian-language botnet that had been active in other politically motivated attacks in the recent past.

1.8. Ukraine Anti-NATO Protests

In March 2008, various Ukrainian newspaper sites were hit with DDoS attacks due to internal political tensions. The C&Cs behind the attacks were located in the Ukraine [[13]], although it is possible that outsiders or parties operating within the Ukraine used these botnets.

Also in 2008, the website for '5.ua', a news website for Ukraine, came under attack with the message "NATO go home" in the HTTP request as part of the flood. These attacks coincided with street protests against NATO expansion into the Ukraine. ATLAS monitoring tracked the C&C behind the attacks in this case to the hosts 'my-loads.info' and 'ultra-shop.biz', a BlackEnergy botnet controlled located (at the time) in China that uses multiple names for the same IP address [[14]].

1.9. Kazakhstan Government Criticism, MSK Forums

In early 2009, the forums for the Russian website MSK came under denial of service attacks. It is believed that these attacks were in retribution for the MSK site posting a PDF copy of a newspaper that was censored through the Kasakh government by pro-Moscow forces. The newspaper published an article written by the Kasakh president that was critical of the Russian government. When no other newspaper would carry the article, MSK offered to host it online and came under attack shortly thereafter [[16]].

The MSK site forums, in response to the DDoS attacks on site in conjunction with the Kazakhstan newspaper, hosted a poll on who people thought were responsible for the DDoS attacks. The poll, dated March 2, 2009, asked, "Who do you think organized DDoS-attack on forum.msk?" The results speak very significantly at the amount of distrust in the region:

Kremlin (185)
FSB (121)
Pro-Kremlin youth organizations (68)
MIA (4)
Administration of the Moscow region (3)
Administration Himok (11)
Communist Party (14)
Simple network hooligans (21)
Anti-power (23)
Neo Trotsky Fighters (22)
Other (15)

At this time it is still unclear what group launched the attacks, although ATLAS data indicates the attacks were lead by the botnets hosted on the sites 'candy-country.com', '22x2x2x22.com', and 'sexiland.ru'. All three of these are identified BlackEnergy-based botnet controllers.

1.10. Russian Opposition Websites

In late December 2008, a related attack struck the newspaper sites 'grani.ru', 'ikd.ru,' (which publishes news about demonstrations going on around Russia) and 'nazbol.ru' (the website of the banned National Bolshevik Party) [[17]]. All of these attacks are consistent with the basic premise that the opposition is routinely censored by

DDoS. Data gathered by Arbor Networks indicates that some of the same botnets behind the MSK attacks (above) participated in these attacks.

1.11. Israel-Gaza/Hamas

During the Israeli-Hamas fighting in Gaza in January of 2009, multiple cyber attacks were launched both from Israeli hackers and Palestinian (and pro-Palestinian) attackers. The bulk of the attacks were website defacements, although we did see some DDoS attacks [[18]]. This is not the first time such cross-border cyber-attacks have occurred. In fact, the long-standing Israeli-Palestinian conflicts are the source of many such attacks and the cause for many website defacements on both sides of the conflict.

One of the tools distributed during these attacks was the "Patriot DDoS tool" from the website "Help Israel Win". The tool was loaded onto a number of websites and domains and was routinely shut down by various groups. It had also undergone a number of iterations to fix bugs and evade any antivirus detection. This is another example of the voluntary cyber attacks sometimes observed in the wild during diplomatic conflicts and shooting wars.

1.12. Kyrgyzstan, January 2009 – False Positive?

In mid-January 2009, reports started appearing that the small former Soviet Bloc nation of Kyrgyzstan was under a cyber attack. The data so far consists mainly of a few NetFlow logs and some web server logs of a few sites in Kyrgyzstan, but very little else. The main site reporting this attack, in a blog posting by Secure Works researcher Don Jackson, blamed the Russian government for the attacks [[19]]. This was followed up on the IntelFusion blog with some analysis and speculation as to the causes behind any such attacks [[20]].

In a posting on January 30, the author at IntelFusion made a case that the Kyrgyzstan government itself launched the attacks [[21]], basing this on some speculations that are consistent with the events in the region. While many researchers' attention in the United States was drawn to the threats at the time to close the Manas airbase (vital to NATO and US efforts in Afghanistan), events within Kyrgyzstan reveal another story. Instead, IntelFusion's analysis suggests that it was an effort to silence critics, since the Kyrgyzstan government is already very pro-Moscow and will happily comply with any offers that Moscow wields. Indeed, Moscow did openly offer Kyrgyzstan money if they closed the Manas air base.

ATLAS data was unable to discover independent data to suggest attacks came through the usual routes such as botnets and coordination via forums [[22]]. ATLAS data also did not show any Internet backbone flow data that suggests that the attacks crossed the normal channels.

1.13. Kommersant, 2008

On March 14, 2008, The Kommersant newspaper had complained to police and prosecutors about a massive hacker attack on its web site, which it suspected was orchestrated by the pro-Kremlin youth group Nashi. Nashi is one of several youth groups in Russia that has been involved in street protests and highly organized activities. They are also suspected in several online attacks including the ones against Kommersant. At the time, the Kommersant paper had published articles critical of

Nashi and the government and came under fire, possibly in retaliation for this reporting. ATLAS data tracked several botnet C&C servers issuing commands to their BlackEnergy-based botnets to launch attacks against the Kommersant servers [[23]]. During the attacks, the Kommersant website was moved to the UK for improved hosting, although the attacks continued after the relocation.

1.14. Kazakh opposition websites allegedly under DDoS attacks

In February 2009, a Kazakh newspaper website came under attack for publishing material critical of the government in Astana [[24]]. The newspaper's site, ‘zonakz.net’, had published articles and recordings of several government officials purportedly committing crimes and acts of corruption. The site was first shut down in Kazakhstan and then moved overseas where it came under a DDoS attack.

In a report titled “The Contradictory State of Kazakhstan” that appeared on the site EurAsia.net, reporter Bruce Pannier wrote about the attacks [[25]]:

> Critics claim there is ample evidence of increased scrutiny of media outlets -- whether traditional or Internet-based.
>
> The owner and editor in chief of the independent weekly "Almaty-Info" is currently on trial for divulging state secrets in a November 2008 article, and is also being sued for defaming a businessman.
>
> Also this week, the head of the zonakz.net website complained that Kazakh law enforcement agencies were blocking access to the website, which is known for having carried material critical of, and at times potentially damaging to, the government.
>
> After a shutdown of zonakz.net’s domestic servers that followed its posting of purported recordings and transcripts of senior Kazakh officials’ phone conversations, the site was registered abroad only to find access blocked by a new distributor-denial-of-service program known as DDOS-attack

Additionally, various political parties have described DDoS attacks against news outlets in Kazakhstan as a means of silencing political opponents [[26]].

1.15. Iranian Elections, 2009

Beginning in mid-June, 2009, Arbor Networks began to see signs of Internet attack activity following the disputed presidential elections in Iran [[32]]. Street protests were organized using online forums and especially the Twitter service, and DDoS attacks against Iranian media and government sites began almost immediately. Most of the attacks used simple “page reboot” scripts, which are websites that construct a repeatedly reloading web page for an attacker that can be used by just browsing to the website. To maximize their effect, attackers coordinated the timing of their efforts using Twitter. However, attackers just as quickly suggested the attacks stop due to bandwidth consumption issues in light of the country’s Internet traffic filtering. It is unclear if the attacks had any significant impact on the target sites’ availability.

1.16. Coordinated South Korean-US Attacks, July 2009

Beginning on July 4 2009, a series of DDoS attacks began to strike first South Korean and then both South Korean and US government and commercial websites [[33]]. Sites targeted included the Korean Assembly, the US and South Korean presidents' websites, the US State Department, the public websites for the US stock exchanges NYSE and NASDAQ, and popular sites in South Korea such as 'naver.com'. Investigations revealed a botnet that was apparently built using a variant of the MyDoom worm from early 2004 together with rudimentary DDoS attacks such as HTTP request floods, UDP and ICMP floods. The attacks continued from July 4 until July 10, when the infected PCs were programmed to encrypt files and render themselves unbootable.

The targets, the US and South Korea, together with the timing between a North Korean missile test launch on July 4 and the 15th anniversary of North Korea's Kim Il Sung's death on July 8 lead some to suggest that North Korea was behind the attacks. To date, we have not seen any evidence of this. The real motivations for these attacks remains a mystery, but it is widely considered a political attack.

2. Attackers' Motivations

In many of the above cases, classic right-wing sentiments are apparently behind the attacks. In most cases, we appear to see attackers using DDoS attacks to express support of an official government position, either against external or internal foes. This is analogous to street protests organized by a political party to stifle opposition through a show of force. Increasingly, we are seeing DDoS attacks used to silence opposition sites, such as in the Kommersant attacks, the attacks on MSK, and the recent attacks in Kazakhstan. A notable exception is the Iranian attacks in June 2009, where anti-Iranian government protesters apparently organized a series of DDoS attacks to protest the election results. The July 2009 attacks on government sites in South Korea and the US may have been a protest, but it is unclear at this time.

In many of these situations, the attacker is able to employ classic guerilla warfare tactics to grow their size and power through the use of propaganda that appeals to an ethnic or national base. In these conflicts the attackers first answer the rally call at the beginning of diplomatic or military hostilities to begin their attacks. They then extend this force by providing easy to use tools through an extensive network of social forums and media including blogs, bulletin boards, and specialized information sites (often dubbed "inform" sites by the Russian hacker underground). Materials posted and re-posted here encourage new recruits to seek retribution against their enemies and join the fight. What starts as a small, core group is can grow into a massive force. Propaganda effects can be so strong, and long lasting, that Estonia still watches for renewed attacks every year on Victory Day. They have seen some attacks but nothing that rises to the level of the 2007 attacks.

By using cheaply and widely available technology, the enemy can leverage IP protocols, botnets, and applications as a force multiplier. That is to say that by using such tools attackers have a reach and power significantly beyond their normal capacity. The techniques to launch these attacks are commonly discussed; fortunately any advance in the sophistication of these techniques is much slower. However the attackers are able to codify their methods into easy to use tools that can be shared freely. There is an increasing emphasis on the ease of use for these tools by outsiders or non-technical parties. An example is the appearance of websites that use dynamic HTML methods to launch HTTP floods simply by loading a specific website. These tools were popular in the recent DDoS attacks on the Iranian government following a disputed national election, commonly using the website 'pagereboot.com'.

3. Attackers' Aims and Goals

Historically, these DDoS attacks have been aimed to cause the victim some punitive damage or register their dissent with the victim's actions. These are the apparent motivations in the attacks from Chinese hackers in retaliation for the embassy bombing in the late 1990s, and the 2007 Estonia attacks, the 2008 Georgia attacks, and the 2009 attacks on Iranian websites. We have seen changes with recent attack activity. Lately, the apparent goal of the attacks is to censor the opposition, either a dissident populace within the country, or dissidents outside the country, or an adversary elsewhere in the world. These are the kinds of attacks we see in the Russian elections of 2007 and subsequent attacks.

The Internet has become a major communication tool for news media, governments, political parties, the opposition and dissidents. Striking at their voice, their printing press, and their Internet channels makes perfect sense. This is apparently the main motivation of the attacks against the Democratic Voice of Burma, where a coordinated series of website hacks and defacements, as well as some DDoS attacks, were used to disrupt global protests against the ruling military in Myanmar.

The cheap and easy availability of the tools and weapons - botnet armies, hacker groups, and the like - have caused governments around the world to eye this approach as a means of silencing enemies. Even when there is no direct tie to the government, such actions can benefit the ruling party's aims. However, in every case we have been unable to conclusively say that the government has been behind the attacks. If governments use such tactics and tools in modern information warfare, then these attacks, by using independently operated botnets, make an excellent attack tool with plausible deniability for the attack director.

4. Attribution

Many have accused government actors or sponsored actors of carrying out these sorts of DDoS attacks. It is important to note that we cannot attribute any of these attacks to a specific group or agency with our data. We simply do not have the evidence to confirm it. All analysis of the data we have suggests non-state actors, however. This comes from observing the attack through three major means: direct data observations, community discussions encouraging and organizing the attacks, and analyzing the botnets and tools used to conduct the attacks.

In a LiveJournal account that we spotted we read representative during the denial of service attacks on Estonia in 2007 [[27]]. The post contains a simple DOS batch script that lists Estonian servers and IP addresses to be ping flooded and enters an infinite loop. The messages around the posting, and in similar forum postings, describe the Estonians as "fascists", "amateurs", and saying that they must be attacked.

Based on flow data from one of the attacks during the Estonian incident, we mapped where the traffic origins to geographic coordinates. The result quite clearly shows how widely distributed the attacks were sourced, namely from all over the world. In this case this particular attack was from a botnet. We do not think that this attack used source spoofing as all of the IP addresses in question mapped back to allocated netblocks and not unallocated IP address space, as is commonly seen when the attacks used spoofed or forged source IP addresses.

Some of the attacks were from far more discrete sources and likely came from the ping flood scripts that were in circulation. These were run by far fewer people and therefore had a smaller base of hosts to come from. We identified these attacks by their traffic type, ICMP echo request, and by the networks the traffic sources aggregated to, network allocations in Europe and Russia.

During the investigations into who launched the attacks, a 20-year-old Estonian student was charged and fined for his part in the attacks [[28]]. His fine was very small, only about $1650. Based on our data showing botnets, ping flood scripts, and the attackers' discussion, we conclude that it is unlikely that Dmitri Galushkevich is the only person responsible for the attacks, however.

Attribution continues to a significant challenge in this problem space when retaliatory measures are considered. In the July 2009 attacks on South Korean and US websites, the South Korean intelligence services stated through the press that they suspected North Korean hackers were behind the attacks. This was picked up and used as a call for retaliation on North Korea by a US lawmaker a few days later. Clearly, these kinds of attacks can spiral into significant diplomatic incidents if great care is not taken.

4.1. Role of Russian Youth Groups

An examination of recent attacks shows that in many cases there are political skirmishes with Russia at the core of the attacks. In these scenarios, one commonly fingered segment of the Russian hard-line community is political youth groups. These organizations are partially state-sponsored and used to hold pro-Kremlin rallies, but have also been accused in various physical attacks over the years. As noted earlier in this paper, they have been accused of the Kommersant attacks, among others. The Russian youth group Nashi claimed responsibility for the Estonian attacks of May 2007 in a news report from mid-2007 [[29]].

Claims about who was behind the Estonia attacks in 2007 were renewed during a 2009 videoconference between Moscow and Washington, and was described in a news report [[30]]. The participants talked about the methods and technologies of information warfare in the 21st century, based on examples of the "Inform Campaign" model that accompanied the military and economic conflicts in recent years (the five-day war in Georgia in August 2008, Israeli military operation in Gaza in early 2009, the gas delivery conflict between Ukraine and Russia, etc.). "Inform campaigns" are routinely used to coordinate such attacks and are widely thought to be government assisted if not outright sponsored.

> Sergei Markov, a State Duma Deputy from the pro-Kremlin Unified Russia, claimed in a March 3, 2009, discussion that his assistant was responsible for the attacks. Said Marvov, "They did not know what to do next. There were feasts, to whom they could not reach. They call to me and say: Sergey, what to do now? Here, we have disabled Estonian sites. I do not

know what to do! I say: So what? Let's let this information that is learned." Markov reportedly said ominously, "and, incidentally, such things will happen more and more." Nashi, the Russian youth group, renewed their claim of a role in the attacks as well.
"In this way, the boys expressed their protest against the policy of the state of fascism carried out by the leadership of the Republic of Estonia", - quoted Commissioner movement Webplanet.ru.

4.2. Hainan Island incident

The Chinese hacker group "Honker Union" took credit for the 2001 hacking incidents in relation to the Hainan Island incident, including the DDoS attacks and the probes on US government computers. This claim is widely believed to be accurate [[4]]. Honker Union is now merged with another Chinese hacking group. Such groups appear to operate openly in China and can sometimes organize such political attacks.

4.3. Botnets behind Georgia-Russia Cyber War

Many of the botnets we listed above, and more, actively participated in attacks against Georgian websites. We recorded well known as well as new BlackEnergy-based botnets striking Georgian targets, most launching generic flood attacks. We identified only a few botnets launching attacks into Russia.

One of the sites set up to coordinate cyber-attacks on Georgia as well as to share ongoing information about the war was the site 'OSInform.RU'. The website contained imagery of death and skulls, and also claims of genocide, material seen consistently in sites set up by Russian hackers detailing attacks on Georgian sites. Multiple blogs begin sharing a simple ping flood scripts targeted Georgian sites, a very similar scripts to the ones seen in Estonia.

A "Stop Georgia" site was set up to coordinate cyber attacks on Georgian web properties. Self appointed representatives of the Russian hacker underground claimed to be behind the site, and it was hosted in multiple locations (via mirroring). The translated comments on the site were:

> *Our response to aggression by Georgia*
>
> *We - the representatives of Russian hacker underground 0 will not tolerate provocation by the Georgian in all its manifestations. We want to live in a free world and exist free from aggression and lies space. We do not need the guidance from the authorities or others, but act according to their convictions based on patriotism, conscience and belief in the virtue of justice. You can call us criminals and cyber-terrorists, continuing with war and killing people. But we will fight and unacceptable aggression against Russia in cyberspace.*
>
> *We demand the cessation of attacks on information and government resources on RUNET, as well as appeal to all media and journalists with a request to cover events objectively. Until the situation has changed, we will impede the dissemination of false information by the Georgian government and information resources. We did not launch an information war, we are not responsible for its consequences.*

We call for the assistance of all who care about the lies of Georgian political sites, everyone who is able to inhibit the spread of false information.

StopGeorgia.ru

P.S. There is one formal mirror project - www.stopgeorgia.info. All other resources have nothing to do with the movement StopGeorgia.ru.

The "Stop Georgia" site also contains a list of sites belonging to Georgia government agencies or Georgian properties abroad. The exhaustive list provides victim IP addresses for targeting and shows their status.

Russian attackers had significant coordination to their activities that was quickly set up, many within a day of the ground offensive beginning. We are not clear on the timelines of the buildup of border tensions or any propaganda campaigns by Russia against Georgia, although a significant lead up to the shooting war could have allowed attackers to establish their operations in time for the ground hostilities.

5. Official Responses Since Estonia

The spring 2007 events in Estonia have served as a clear wake up call to governments around the world about the power of cyber attacks and the damage they can inflict. The events in the summer of 2008 against Georgia were a forceful reminder of the attacks and added great urgency to this analysis. Many governments are reviewing their own vulnerability to DDoS attacks or more common infiltrations. A small handful of nations are investigating active cyber attack programs of their own.

5.1. Defensive Responsibilities

Especially since May 2007, but even more after the 2008 Georgia attacks, governments and groups around the world are worried about being a victim of a cyber attack. NATO, the EU, and other groups have been investigating their role in responding and their responsibilities and obligations. To date neither the EU nor NATO has articulated clear strategies for countering such attacks on member states.

The IMPACT alliance (http://www.impact-alliance.org/) has been founded in Malaysia to combat cyber terrorism and has been working to become a UN of cyber security, in part with the help of the ITU.

5.2. Role of Attribution in Response

Attribution is a key aspect for any large-scale response including retribution attacks or seeking redress via the international community, such as in the UN or via diplomatic channels. These kinds of attacks give a nation-state clear plausible deniability if they are actively sponsored, and an even bolder claim if these are simply run out of the civilian populace but tolerated or even tacitly controlled.

Some have claimed that the use of subtle language cues is commonly employed by the Chinese to direct such attacks. Phrases that seem innocent can have a sweeping impact on how the populace responds, either in street protests or in online attacks. If this is the case then we should expect that these kinds of attacks would continue and become a tool for managing opposition or foes in the 21st century. Their impact - bandwidth, durations, victims - is likely to grow and their frequency, scale, and the number of origins is likely to grow as well, as we have seen in the past several years.

6. Recommendations

Recent history has shown that packet flooding attacks are increasingly a favorite weapon of politically motivated attackers regardless of their geographic region. These attacks threaten communication mechanisms, the integrity of elections, and the freedom of an independent press, the activities of dissident groups and politicians, and may, in the future, grow in sophistication and disrupt normal daily life. In this time we have seen investigations and defense measures spawned from independent parties, the commercial sector, and the government sector through mostly ad-hoc means. While this has been marginally effective so far, this has quickly become an untenable situation.

A number of recommendations follow based on the author's experience in a number of the conflicts described above.

6.1. Broad Defensive Contributions Must be Possible

If we are to successfully defend national infrastructure against the sorts of attacks that affected Estonia and Georgia then we must be open to all forms of assistance. In both cases the public were firmly on the side of the victim (Estonia, Georgia), a sentiment that must be harnessed more effectively in the future. This must be turned into *Schwerpunkt* - a unity of purpose and goals - which will make us effective in our mission of defending the Internet.

Commercial tools from various vendors, including the author's employer, exist to detect and filter DDoS attack traffic and have been deployed to help thwart some of the attacks reviewed above. The technology in these tools is commonly available and the only barrier to their deployment is budget. However, as a total solution to the political DDoS problem this is insufficient from a cost or management perspective. We must think about how to utilize new methods to defend critical and civilian infrastructure as well as government infrastructure.

The enemy, attackers, uses public sentiment on his side to grow an organic legion of supporters to aid in their cause. Their aim is more amorphous than the defenders' role but the principle applies: by utilizing propaganda campaigns and nationalist and ethnic sentiment, he grows his army of volunteers. This is exactly analogous to the enemy in guerilla warfare.

Defenders do not use organic support for their mission of stopping these attacks, however. Outside support has been used to some extend in the recent past, with Tulip Networks in Atlanta, Georgia, in the United States providing bandwidth and connectivity for some of the Georgian infrastructure under attack. This was made possible through a direct, personal friendship that enabled this help. This kind of assistance is rare and no formal agreements are in place, leaving victims at risk.

For the victims, successes in defending an online presence usually come when a group or an individual acts on his or her own with the best interests in mind. Many more individuals or groups who could help are usually blocked from providing assistance. More outsiders are willing to help in these cases through meaningful ways, and we must enable them to provide aid if we are to defend these networks and this infrastructure. One challenge that will have to be addressed is to discover which offers are credible or worthy. However, a network of professionals to defend against these sorts of attacks exists in the commercial Internet service provide realm.

Governments must be open to assistance from the private, commercial sector for dedicated DDoS-resilient hosting for public facing Internet properties. At this time the targets of these attacks mainly consist of information-only sites, but in the future will surely include key infrastructure equipment such as VoIP exchange points, DNS servers, and email systems which, if targeted, could impact the ability of a government to communicate internally. Governments and other likely political targets such as newspapers must identify how they can migrate their infrastructure to a third-party's systems to ensure continuity.

Furthermore, governments and targets must be trained and willing to accept a rapid deployment of commercial tools to defend against these kinds of attacks. All members of the government's information technology staff should be able to receive an offer of help and determine its credibility, and route that offer to the appropriate internal party for follow up. We have seen this work in limited cases in the past but too often we find that government victims in these attacks do not know how to accept an offer of assistance in a timely fashion.

6.2. Improved Efficiency in the Decision Making Process

A review of the OODA loop, or the Boyd cycle, provides ample areas to review and seek improvement in our current posture [[31]]. The cycle is built of four core steps that provide feedback to each other: observe, orient, device, act. The faster and more accurately one side can complete the loop - and begin the cycle again - the bigger an advantage he has.

Our observation points are currently piecemeal and hampered by competing business interests. This is nothing new, but it means we have a poor foundation on which to base our decisions. Because we lack a complete overview of Internet activity about the origins of attacks and how we may stop them, we often waste valuable time defending against attacks when we could stop them at their root. Information collection, sharing, and recall are woefully ignored and falling behind.

As a community of defenders we are usually able to orient at the broader goal - defend a specific country's assets (e.g. Estonia), identify the attackers behind it - but our more specific tactics to achieve that goal are unfocused and lacking. We fail to communicate what we need, what we find, and what the next steps are.

Our decision making process is often mired in consensus building and dogged by second-guessing. We are ineffective in many cases because we fail to make decisions for fear of making the wrong one. Committees with the wrong stakeholders and people who have no value to the process hijack and derail the process.

Finally, our actions are bound by laws and jurisdictions but also by seeking the permission of too many parties. In short, we move too slowly, too blindly, and too ineffectively, if we move at all. We are not consistently effective.

Moving forward, governments and coordination centers must be given the authority to act without requiring a consensus of all parties but rather act quickly in the best interests of the group. This should be treated as an authority akin to a military command authority and should coordinate public-sector, private-sector, and military efforts at combating attacks. Careful balance must be taken to work with carriers, for example, to avoid disruptions to the infrastructure, a key facet to ensuring the carriers will accept outside leadership in such events.

7. Conclusions

DDoS attacks provide a simple, easily available mechanism to disrupt the Internet presence of a group or a small nation. Previously, they have been confined to retaliatory attacks seeking punitive damage to the victim, but in recent years the role of the Internet in publishing newspapers or organizing dissident efforts has grown. The growing importance of the Internet to potential victims has not escaped cyberwar practitioners. DDoS attacks will continue as a tool of censorship as long as the Internet remains a communications medium.

Cyber-warfare takes on different forms in different areas of the world. Political targets and motivations in DDoS attacks are most popular in Russia and the region, less so in China, Asia and the Middle East. China favors more surgical, infiltration events for serious cyber warfare. We have seen an explosion of DDoS tools from Chinese hackers, although most of their targets are commercial sites located in China, but many are in Korea or Japan. These sites are the targets of bullying or extortion attacks that do not yet rise to the level of political warfare. Burma benefits from website defacements and destruction. Israel and Palestine often use website defacements to challenge each other. At this time we expect to see DDoS attacks continue to be a political weapon in the Russian power sphere, particularly for former Soviet bloc nations.

These attacks will continue to provide the nation-state benefits from their actions as well as plausible deniability should they actively engage in such actions. Because of this we expect their frequency to grow in the Russian region, together with their sophistication as victims begin to develop improved defenses. Furthermore we anticipate that other nations may begin using DDoS attacks as a simple, blunt force political weapon to silence critics or opponents.

Much of the theory of cyber-warfare remains to be written, but may borrow from other warfare theories. Specifically theories on guerilla and asymmetric warfare need to be reviewed to understand the enemy's tools and tactics, as well as to understand responses. While governments and private industry control the communication's fabric, they have yet been unable to muster a unified, consistent defense. Instead, defenses have largely been ad-hoc and at the mercy of generous outsiders. Responses must be cohesive if not unified in order to be consistent, an approach that would be well informed with an understanding of defense tactics learned from studying theories of cyber-warfare.

References

[1] Mirkovic, J. and Reiher, P., A taxonomy of DDoS attack and DDoS defense mechanisms, in *ACM SIGCOMM Computer Communication Review*, 2004.

[2] Arbor Networks Website, http://www.arbornetworks.com/en/products.html.

[3] ACTIVISM, HACKTIVISM, AND CYBERTERRORISM: THE INTERNET AS A TOOL FOR INFLUENCING, Denning, D.E., in *Networks and netwars: The future of terror, crime, and militancy*, 2001.

[4] Cyber Protests: The Threat to the U.S. Information Infrastructure, National Infrastructure Protection Center, 2001. Available online at http://www.au.af.mil/au/awc/awcgate/nipc/cyberprotests.pdf.

[5] US State Department Website. Available online at http://www.state.gov/r/pa/ei/bgn/5377.htm.
[6] Estonian DDoS Attacks - A summary to date, by Jose Nazario, on Security To The Core weblog, May 17, 2007. Available online at http://asert.arbornetworks.com/2007/05/estonian-ddos-attacks-a-summary-to-date/.
[7] CNN Attack Summary, by Jose Nazario, on Security To The Core weblog, April 21, 2008. Avaiable online at http://asert.arbornetworks.com/2008/04/cnn-attack-summary/.
[8] Cyber Attacks Against Georgia: Legal Lessons Identified, by Eneken Tikk, Kadri Kaska, Kristel Rünnimeri, Mari Kert, Anna-Maria Talihärm, and Liis Vihul, 2008.
[9] Georgia On My Mind – Political DDoS, by Jose Nazario, on Security To The Core weblog, July 20, 2008. Available online at http://asert.arbornetworks.com/2008/07/georgia-on-my-mind-political-ddos/.
[10] The Website for the President of Georgia Under Attack - Politically Motivated? by Steven Adair, in Shadowserver Foundation Calendar, July 20, 2008. Available online at http://www.shadowserver.org/wiki/pmwiki.php/Calendar/20080720.
[11] Estonia hosts Georgian Web sites to halt hackers, on FoxNews.com, August 26, 2008. Available online at http://www.foxnews.com/wires/2008Aug26/0,4670,EstoniaGeorgiaHaltingHackers,00.html.
[12] An In-Depth Look at the Georgia-Russia Cyber Conflict of 2008, Jose Nazario and Andre M. DiMino, at the Botnet Task Force meeting, October 2008.
[13] Political DDoS? Ukraine, Kasparov, by Jose Nazario, on Security To The Core weblog, December 13, 2007. Available online at http://asert.arbornetworks.com/2007/12/political-ddos-ukraine-kasparov/.
[14] Political DDoS: Estonia and Beyond, Jose Nazario, in a presentation give at Usenix Security, 2008. Available online at http://www.usenix.org/events/sec08/tech/slides/nazario-slides.pdf.
[15] Radio Free Europe DDoS, by Jose Nazario, on Security To The Core weblog, April 29, 2008. Available online at http://asert.arbornetworks.com/2008/04/radio-free-europe-ddos/.
[16] MSK Forum, February 28, 2009, available online at http://forum.msk.ru/notice/.
[17] Russian Opposition Websites Shut Down By Attacks, December 25, 2008, on The Other Russia. Available online at http://www.theotherrussia.org/2008/12/25/russian-opposition-websites-shut-down-by-attacks/.
[18] The Effects of War: Gaza and Israel, by Jose Nazario, on Security To The Core weblog, January 5, 2009. Available online at http://asert.arbornetworks.com/2009/01/the-effects-of-war-gaza-and-israel/.
[19] Kyrgyzstan Under DDoS Attack From Russia, by Don Jackson, on SecureWorks Research Blog, January 28, 2009. Available online at http://www.secureworks.com/research/blog/index.php/2009/01/28/kyrgyzstan-under-ddos-attack-from-russia/.
[20] The Kyrgyzstan DDoS Attacks of January, 2009: Assessment and Analysis, Jeff Carr, jart Armin and Greg Walton, on IntelFusion blog. Available online at http://intelfusion.net/wordpress/?p=516.
[21] Why I believe that the Kyrgyzstan Government hired Russian hackers to launch a DDOS attack against itself, by Jeff Carr, on IntelFusion blog. Available online at http://intelfusion.net/wordpress/?p=520.
[22] Kyrgyzstan DDoS Attacks, by Jose Nazario, on Security To The Core weblog, February 2, 2009. Available online at http://asert.arbornetworks.com/2009/02/kyrgyzstan-ddos-attacks/.
[23] Russian DDoS Attacks: Kommersant, by Jose Nazario, on Security To The Core weblog, March 19, 2008. Available online at http://asert.arbornetworks.com/2008/03/russian-ddos-attacks-kommersant/.
[24] Quick Notes on Cyber Warfare News, by Jose Nazario, on Security To The Core weblog, February 19, 2009. Available online at http://asert.arbornetworks.com/2009/02/quick-notes-on-cyber-warfare-news/.
[25] THE CONTRADICTORY STATE OF KAZAKHSTAN, by Bruce Pannier, in EURASIA INSIGHT, March 6, 2009. Available online at http://www.eurasianet.org/departments/insightb/articles/pp030609d.shtml.
[26] Kazakhstan: Five political parties report about the information terrorists to the public prosecution office, February 25, 2009, on the website Ferghana.ru. Avalable online at http://enews.ferghana.ru/news.php?id=1024.
[27] "Load quickly on chuhonofilam", in a posting on a LiveJournal blog by w8kl8dlaka. Available online at http://w8lk8dlaka.livejournal.com/52383.html.
[28] Student fined for attack against Estonian Web site, Jeremy Kirk, InfoWorld, January 24, 2008.
[29] Nashi, Russia's new militant nationalist movement, Rediff India Abroad, May 21, 2007. Available online at http://www.rediff.com/news/2007/may/21nashi.htm.
[30] Behind the Estonia Cyberattacks, Radio Free Europe/Radio Liberty, March 6, 2009.

[31] Osinga, Frans. Science, Strategy and War: The Strategic Theory of John Boyd. Abingdon, UK: Routledge, 2007.

[32] Iran DDoS Activity: Chatter, Tools and Traffic Rates, by Jose Nazario, on the Security to the Core weblog, June 19, 2009. Available online at http://asert.arbornetworks.com/2009/06/iran-ddos-activity-chatter-tools-and-traffic-rates/.

[33] Korean/U.S. DDoS Attacks – Perplexing, Disruptive, and Destructive, by Steven Adair, on the Shadow Server Foundation Calendar blog on July 10, 2009. Available online at http://www.shadowserver.org/wiki/pmwiki.php/Calendar/20090710.

The Virtual Battlefield: Perspectives on Cyber Warfare
C. Czosseck and K. Geers (Eds.)
IOS Press, 2009
© 2009 The authors and IOS Press. All rights reserved.
doi:10.3233/978-1-60750-060-5-182

A Brief Examination of Media Coverage of Cyberattacks (2007 - Present)

Cyrus FARIVAR
Freelance Technology Journalist (NPR, PRI, CBC, The Economist)

Abstract. As cyberattacks become more frequent, they draw new attention in the media. Indeed, there has been a significant spike in journalistic coverage of cyberattacks and cybersecurity in the last year alone, making this particularly relevant now. The aim of this paper is to provide an overview of coverage and make suggestions for future journalists and policymakers to work better together to better understand this new threat.

Keywords. Media studies, media, journalism, cybersecurity, cyberattacks

Introduction

In the last decade, there have been countless cyberattacks against various political, military and economic targets in the United States, Europe and elsewhere. Some have been made public, and others remain classified. Many of these cyberattacks have been against various American military targets and some have have overlapped into cyber-espionage territory. (For the purposes of this paper, I will focus only on political-oriented direct cyberattacks, not cyber-espionage.)

Since 2007, the world has seen three major politically-oriented cyberattacks (denial of service attacks) against three former Soviet Union countries, Kyrgyzstan (January 2009), Georgia (August 2008), and Estonia (April-May 2007). All three likely originated from within Russia, and may have implicitly involved the Kremlin, despite official denials. As such, this increase in cyberattacks has resulted in a corresponding increase in the amount of coverage this issue receives in the English-language print media. In the case of the 2007 attacks against Estonia and the 2008 attacks against Georgia, both made the front page of *The New York Times*. However, while there has been more attention paid to this issue, some of it has been misleading at best and false at worst. Therefore, it is in the interests of the cybersecurity community and the media that cover them to better understand how the media has treated cyberattacks, and to improve public understanding of this phenomenon.

1. Not all Cyberattacks Are Created Equal (Kyrgyzstan)

On January 28, 2009, The Wall Street Journal ran this headline: "Kyrgyzstan Knocked Offline."[1] However, the six-paragraph article, which relied on two sources, only one of which was named, described how a denial-of-service attack hit the country's two main ISPs, accounting for nearly 80 percent of the country's bandwidth. While such a tactic would seem like major news, it was treated as a minor, largely unimportant story. The Journal relegated it to page A10 of the newspaper, indicating that the news was only moderately important. The attack was also covered by a few industry publications, including Computerworld and The Register. The New York Times, ignored the story in print and only wrote about the event on its blog, The Lede.[2]

This lack of attention shows that when a minor, obscure country gets hit, it's difficult to develop much interest in such a story – particularly when it's a country that doesn't have an active online presence, nor that is accompanied by any kind of corresponding real-world action, nor is it an active member of a multi-national organization like the European Union or NATO. This is not to say that the attack against Kyrgyzstan should not have warranted more coverage. If any North American, E.U., or East Asian nation suddenly had 80 percent of its online capacity knocked out, it likely would have made international headlines, as it did in late 2008 when an undersea cable near Egypt was cut by accident, and not as a result of a cyberattack.[3] This is an unfortunate example of a double-standard in the media should be rectified the next time something like this happens.

2. When a Cyberattack Accompanies Real-World Events, People Take Notice (Georgia)

In August 2008, when Georgia suffered a cyberattack that accompanied its invasion by Russia, the world sat up and took notice. *The Wall Street Journal* reported: "Georgia States Hit By Cyberattack," while *The New York Times* noted: "Before the Gunfire, Cyberattacks."[4] Most media outlets sat up and took notice that a cyberattack element corresponded with actual physical attacks. Even though these attacks again took the form of "hacktivism," and denial-of-service attacks, these media outlets tended to analyze the online component in more straightforward and plain terms. The *Times* noted that the attacks simply "overload and effectively shut down Georgia servers."

As the second major cyberattack in recent memory, the Georgia attack was notable as the cyberattack was squarely set in the context of the events on the ground. Perhaps one of the reasons why the attack against Kyrgyzstan never captivated the attention of reporters and editors in the same way was because there was no clear narrative of why it happened – competing theories about obscure political disputes in far-off countries perhaps don't work. In Georgia, like in Estonia before it, there was a clear example of a former occupying power asserting its dominance, like a bully beating up a little kid. This attack was also notable as it was the first (and possibly only) cyberattack where a journalist became an active participant in the war – albeit in a very minor way. Evgeny Morozov, a Belorussian journalist now living the United States, in his *Slate* piece "How I became a soldier in the Georgia-Russia cyberwar," showed how easy it was for an average Russian-speaking Internet used to quickly acquire the tools necessary to throw an "e-Molotov Cocktail."[5] Morozov was likely the first journalist who quickly understood how such an attack could emerge so quickly. In essence, nationalist fervor plus an Internet connection could rapidly constitute a "cyberwar." He concluded his piece this way, noting:

In less than an hour, I had become an Internet soldier. I didn't receive any calls from Kremlin operatives; nor did I have to buy a Web server or modify my computer in any significant way. If what I was doing was cyberwarfare, I have some concerns about the number of child soldiers who may just find it too fun and accessible to resist.

My experiment also might shed some light on why the recent cyberwar has been so hard to pin down and why no group in particular has claimed responsibility. Paranoid that the Kremlin's hand is everywhere, we risk underestimating the great patriotic rage of many ordinary Russians, who, having been fed too much government propaganda in the last few days, are convinced that they need to crash Georgian Web sites. Many Russians undoubtedly went online to learn how to make mischief, as I did. Within an hour, they, too, could become cyberwarriors.

3. The First Cyberwar (Estonia)

2007 was the first time that *The New York Times* ever used the word "cyberwar." In its May 29, 2007 article, "Digital Fears Emerge After Data Siege in Estonia," the American newspaper of record followed the tone that had already been set for much of the worldwide English-language media coverage of the event.[6] The *BBC*, which was one of the first major news outlets to publish, declared on May 17, 2007: "Estonia hit by 'Moscow cyber war.'"[7] On the same day, the British newspaper *The Guardian* wrote: "Russia accused to unleashing cyberwar to disable Estonia."[8]

While the *BBC* had used the term before, this was the first time that it had been used to describe real-life state-to-state attacks. In addition to declaring the events a "war," there was a great deal of description about how "E-stonia" essentially functioned off of its Internet applications. While it is true that Estonia has a high level of connectivity, Internet banking, online voting and all the rest, the tone of many articles illustrated a scene of near-meltdown and destruction. The *Times* reported that the attacks "came close to shutting down the country's digital infrastructure." *The Washington Post* wrote that the attacks "disrupted government e-mail and led financial institutions to shut down online banking." Jaak Aaviksoo, Estonia's defense minister, told *Wired* that Estonia's national security was threatened. However, the attacks, while annoying, did not do any permanent damage, nor was the society in immediate peril.[9] While there was little technical difference between the attacks against Estonia and Georgia, the first political "cyberwar," Estonia's technological landscape made the rhetoric used that much more dramatic.

4. Lay Off the Hyperbole – It's the Worst Thing Ever

If there is anything to be learned from the first "cyberwar," or the first "Web War One," (as *Wired* called it) is that hyperbole is a great weapon that can be used effectively to draw the attention of the world. I'll admit that I myself fell for it – my *Slate* piece in the aftermath of the attacks on Estonia was dubbed by my editors as "Cyberwar I."[10] In retrospect, the term "cyberattack" would have been more descriptive, as a war implies a congruous, more or less armed conflict between two clear entities. In this case, the metaphor of "war" is not very accurate, as it was not possible for Estonia (or any other cyberattacked country) to retaliate even if it wanted to. In a cyberattack, the only strategy is defense – there is no way to counter-attack, or to take out the online firing turret. Furthermore, it's impossible to have a war against an enemy even more faceless and intangible than international terrorist organizations. If ordinary, un-technically sophisticated people like Morozov can become "cyberwarriors" within an hour, does that mean, then, that they are protected under the Geneva Conventions? Using the language of war quickly breaks down.

In addition to using the term "cyberwar," everyone has been easily seduced by the armageddon-style rhetoric that Estonian government officials and associated figures have used to describe what had happened. Ene Ergma, the speaker of the Estonian parliament, in an interview with *Wired* magazine, compared the cyberattacks to a nuclear explosion, calling them "the same thing."[11] Linnar Viik, the Estonian Internet guru, told *The Washington Post*: "These attacks were an attempt to take one country back to the cave, back to the Stone Age."[12] Not only are these statements ludicrous on their face, but they're blatantly untrue. If the Kremlin or the Russian "hacktivists" had wanted to pummel Estonia, then the attacks wouldn't have ceased two weeks after they had begun. The attacks clearly were meant as a message, not as a war. With all due respect, it was wrong of Ergma and Viik to make such hyperbolic statements, and it was equally wrong of anglophone journalists to lap it up as easily as they did. Journalists have a responsibility to not take such ridiculous statements at face value, particularly ones who have a history of reporting on technology.

Journalists and Estonians alike would do well do remember the example set by President Bill Clinton in February 2000. This was just after major American tech companies including Yahoo, Buy.com and CNN were hit with denial-of-service attacks. In a press conference, the president was asked if this attack was the "electronic Pearl Harbor." Clinton replied: "Well, I hope not. (Laughter.) I think it was an alarm. I don't think it was Pearl Harbor. We lost our Pacific fleet at Pearl Harbor – I don't think the analogous loss was that great."[13]

5. Cyberattacks and Civilians

As a technology journalist, or as a cybersecurity professional, it's easy to have tunnel vision. It's easy to see botnets on every network and miscreants in every Internet forum. This is not to say that these threats are not real. Rather, it is important to step back from our bandwidth-fueled lifestyle and begin to examine how cyberattacks do or don't affect people in the real world. It is a luxury to have high levels of Internet services, and it is equally a luxury to be able to worry about whether or not these sites are affected by online "warfare."

While trying to report on the cyberattacks against Georgia in August 2008, I was embarrassed when calling the Georgian Ministry of Foreign Affairs in Tbilisi, and a spokesperson rebuked me for wanting to know about cyberattacks, when in fact the Ministry was far more concerned with protecting territorial integrity and Georgian citizens, rather than where the ministry's web site was going to be hosted. While it may be of great concern and worry to many cybersecurity professionals who have warned for years of coming cyberattacks – these types of attacks, at least in their current form, take a back seat to actual, physical warfare. After all, it is worth repeating that no one has died as a result from a cyber attack. Further, while the Estonian Internet security community was going into overdrive during the cyberattacks of 2007, the Estonian public did not seem to be touched by the attacks. In a survey by the Estonian newspaper *Postimees*, nearly half (over 49 percent) of the 1,243 Estonian surveyed said that they were not affected by cyberattacks.

6. Difficulty of Catching the Cyberattackers

If there's one point that should be made to journalists and policymakers alike, it's that after nearly a decade of major denial of service attacks, that there is neither a perfect way to secure against them, nor is there a good way to track the perpetrators. After the attacks against CNN back in 2000, Richard Power, an official of the Computer Security Institute, told the news network at the time that such attacks "will be one of the most difficult things to address."[14] Indeed, it seems that while the attacks may have gotten more sophisticated and larger, that the basic procedure and execution of such an attack has not changed hardly at all since an attack that unleashed an estimated 800 megabits per second of data on web servers. Estonia was only able to defend against the attack by severing, temporarily, its international data connection to the outside world. Smaller countries with a limited number of international pipes can employ this tactic, whereas a much larger online presence like the United States, are unable to.

Further, it should be underscored that it's very difficult to catch anyone who engages in a cyberattack. Even the attacks against Estonia, which were publicized and had a high-level of international involvement, have only resulted in the arrest and successful prosecution of one Estonian citizen, Dmitri Galushkevich.

The 19-year-old quickly confessed to attacking government computer networks, which is punishable – according to the Estonian Penal Code Section 206, subsection 2 – up to three years in prison.[15] But Galushkevich said that he acted alone, based on instructions that he read online, which were probably not unlike the ones that Morozov discovered. He didn't have any knowledge about who the masterminds or perpetrators in other countries might be.

It is important to remember that in the immediate months after the 2007 cyberattacks, the Estonian government attempted to request further information from Russian authorities. Officials had a list of IP addresses that appeared to originate from within Russia, and needed the help of their neighbor to conduct further investigations, and perhaps find new suspects. But the Russian Embassy in Tallinn and the Kremlin gave their Estonian counterparts the run-around, arguing that technicalities of the treaty between the two countries prevented Russia from providing this information. Further, the Russian constitution forbids the extradition of its own citizens, so there was no way for Estonian authorities to question or even depose any Russians. Partly because of the evidence that he's seen, and Moscow's reluctance to be cooperative leads made Estonian Chief Prosecutor Margus Kurm say that he is confident that the leaders of the attacks are in Russia, despite saying: "We have no evidence and no information that this was the Russian government."

Still, Kurm is pretty hopeless of ever gaining any further information that could be legally useful for prosecuting anyone for cybercrimes against the Republic of Estonia. In an interview in July 2007, he admitted to me: "The status is that we haven't got any information from Russia and I'm quite sure that we will not get any information."[16]

On January 25, 2008, Dmitri Galushkevich pled guilty to attacking Estonian websites. He had to pay a fine of 17,500 Estonian kroons, or around $1,700 and received only probation – no jail time. The case was closed, and no further legal action was taken against anyone, largely because, in the words of Kurm, "Russia refused to co-operate."[17]

What this means, is that for the foreseeable future, cyberattacks will remain an effective tactic countries between nations that are not exactly always friendly with one another, as is the case with Russia and many of its former Soviet satellites.

7. Suggestions for Researchers and Policymakers to Improve Media Coverage

In summation, there are three main points that I would like researchers, policymakers and journalists to come away with.

First, tone down the rhetoric, hyperbole, and watch your language. If you talk about "cyberwar," – the use of the word war has a very specific meaning and very specific consequences. A war usually implies two, more-or-less equal sides, with a clear objective. Cyberattacks generally are not always necessarily couched in the applications of political conflict – in fact, many attacks have more to do with organized crime or online mischief than they do actual warfare. As such, journalists should be wary of sources that compare cyberattacks to nuclear warfare and make similarly absurd comparisons. Further, researchers and policymakers need to be aware of the words that they use themselves.

Second, researchers and policymakers need to be more open (as much as possible) with the information that they do have. Journalists need to be able to verify data, and understand the data that they're looking at. When everything is construed as a "cyberwar," it's tough to determine how various "cyberwars" compare to one another. Was Estonia's attack the same as the one against Georgia? What about the 2009 attack against the United States and South Korea?

Third, and most importantly, policymakers and researchers need to understand how they can work together. Whether they like it or not, media can have a significant influence on public policy. It is the job of the media to inform the public and act as a watchdog on government's activities. The more information that public officials, corporations and researchers can provide to journalists, the better the journalists can do in presenting the case. However, one of the problems is that there simply aren't very many journalists that fully understand neither how cyberattacks work nor what they are. It would be helpful for journalists to participate in a workshop on cyberwarfare from their local governments, or perhaps from the CCDCOE to better understand how these attacks work from a technical standpoint.

References

[1] Rhoads, Christopher, "Kyrgyzstan Knocked Offline," *The Wall Street* Journal. , January 28 2009. http://online.wsj.com/article/SB123310906904622741.html?pagewanted=print
[2] Mackey, Robert, "Are 'Cyber-Militas Attacking Kyrgyzstan?', *The New York Times*
[3] *Led* Blog, February 5 2009. http://thelede.blogs.nytimes.com/2009/02/05/are-cyber-militias-attacking-kyrgyzstan/?pagewanted=print
[4] "Severed cable disrupts net access," *BBC* News, December 19, 2008 http://news.bbc.co.uk/2/hi/technology/7792688.stm

[5] "Georgia States Computers Hit By Cyberattack," *The Wall Street* Journal, August 12, 2008. http://online.wsj.com/article/SB121850756472932159.html

[6] Morozov, Evgeny, "An Army of Ones and Zeroes," *Slate*, August 14, 2008. http://www.slate.com/id/2197514/pagenum/all/#p2

[7] Landler, Mark, and Markoff, John. "Digital Fears Emerge After Data Siege in Estonia," *The New York Times,* May 29, 2007. http://www.nytimes.com/2007/05/29/technology/29estonia.html?pagewanted=print

[8] "Estonia hit by 'Moscow cyber war,' *BBC* News, May 17, 2007. http://news.bbc.co.uk/2/hi/europe/6665145.stm

[9] Traynor, Ian, "Russia accused of unleashing cyberwar to disable Estonia," *The* Guardian, May 17, 2007. http://www.guardian.co.uk/world/2007/may/17/topstories3.russia

[10] Davis, Joshua, "Hackers Take Down the Most Wired Country in Europe," *Wired,* August 21, 2007. http://www.wired.com/politics/security/magazine/15-09/ff_estonia?currentPage=all

[11] Farivar, Cyrus, "Cyberwar I," Slate, May 22, 2007. http://www.slate.com/id/2166749/fr/flyout

[12] Davis, Joshua, "Hackers Take Down the Most Wired Country in Europe," *Wired,* August 21, 2007. http://www.wired.com/politics/security/magazine/15-09/ff_estonia?currentPage=all

[13] Finn, Peter, "Cyber Assaults on Estonia Typify a New Battle Tactic," *The Washington* Post, May 19, 2007. http://www.washingtonpost.com/wp-dyn/content/article/2007/05/18/AR2007051802122.html

[14] "Remarks by the President in Photo Opportunity with Leaders of High-Tech Industry and Experts on Computer Security," *The White House Office of the Press* Secretary, February 15, 2000. http://www.fas.org/irp/news/2000/02/000215-secure-wh1.htm

[15] "Cyber-attacks batter Web heavyweights," CNN, February 9, 2000. http://archives.cnn.com/2000/TECH/computing/02/09/cyber.attacks.01/index.html

[16] "Tulemused – Teksid," http://www.legaltext.ee/et/andmebaas/tekst.asp?loc=text&dok=X30068K7&keel=en&pg=1&ptyyp=RT&tyyp=X&query=karistusseadustik

[17] Margus Kurm, in discussion with the author, July 19 2007. Email from Margus Kurm to the author, January 29 2008.

Part II

Technical Challenges and Solutions

The Virtual Battlefield: Perspectives on Cyber Warfare
C. Czosseck and K. Geers (Eds.)
IOS Press, 2009
© 2009 The authors and IOS Press. All rights reserved.
doi:10.3233/978-1-60750-060-5-191

Behavioral Analysis of Zombie Armies

Olivier THONNARD [a,1] , Wim MEES [a] and Marc DACIER [b]
[a] *Royal Military Academy, Polytechnic Faculty, Brussels*
[b] *Symantec Research Labs, Sophia Antipolis, France*

Abstract. *Zombie armies* - or botnets, i.e., large groups of compromised machines controlled remotely by a same entity - pose today a significant threat to national security. Recent cyber-conficts have indeed demonstrated that botnets can be easily turned into digital weapons, which can be used by cybercriminals to attack the network resources of a country by performing simple Distributed Denial-of Service (DDoS) attacks against critical web services. A deep understanding of the long-term behavior of botnet armies, and their strategic evolution, is thus a vital requirement to combat effectively those latent threats. In this paper, we show how to enable such a long-term, strategic analysis, and how to study the dynamic behaviors and the global characteristics of these complex, large-scale phenomena by applying different techniques from the area of knowledge discovery on attack traces collected on the Internet. We illustrate our method with some experimental results obtained from a set of worldwide distributed server honeypots, which have monitored attack activity in 18 different IP subnets for more than 640 days. Our preliminary results highlight several interesting findings, such as *i)* the strong resilience of zombie armies on the Internet, with survival times going up to several months; *ii)* the high degree of coordination among zombies; *iii)* the highly uneven spatial distribution of bots in a limited number of "unclean networks", and *iv)* the large proportion of home users' machines with high-speed Internet connexions among the bot population.

Keywords. Intelligence monitoring, Threat analysis, Zombie armies.

Introduction

In the recent years, many security experts have drawn attention to the increasingly important security problem related to *zombie armies* - also called botnets, which are groups of malware-infected machines that are remotely controlled and coordinated by a same entity. Still today, zombie armies and botnets constitute, admittedly, one of the main threats on the Internet, as they are used for different kinds of illegal activities (e.g., bulk spam sending, online fraud, denial of service attack, etc) [2,19]. More importantly, the analysis of recent "cyber conflicts", such as the presumed cases related to Estonia and Georgia [17,6,7], have lead experts to the conclusion that botnets can be easily turned into digital weapons, which can be used by cybercriminals (or dissidents) to attack the network resources of a country by performing very simple Distributed Denial-of Service (DDoS) attacks against critical web services (e.g., DNS servers, network routers, gov-

[1]Corresponding Author: Olivier Thonnard, Royal Military Academy, Avenue de la Renaissance 30, 1000 Brussels, Belgium; E-mail: olivier.thonnard@rma.ac.be.

ernment or financial websites, etc), which can lead to substantial economical or financial loss. Although no clear evidence of the implication of any governmental organization in those attacks could be underlined, one important lesson learned from these events is that botnets are primarily used by dissidents or activists to perform this type of attacks in periods of political disturbances. A deep understanding of the long-term behavior of botnet armies, and their evolution, is thus a vital requirement to be able to combat effectively those latent threats.

While most previous studies related to botnets have focused on understanding their inner working [24,5,1], or on techniques for detecting individual bots at the network-level [8,9], in this work we are more interested in studying the global behaviors of those armies from a strategic viewpoint. That is, we are not interested in studying a particular botnet from the inside, or in the analysis of the various protocols used by bots to communicate with their C&C server. But instead, we want to perform a **long-term**, **strategic analysis** of those armies from a behavioral point of view, i.e.: how long do they stay alive on the Internet, what is their average size and their spatial distribution, and more importantly, how do they evolve over time with respect to different criteria such as their origins, or the type of activities (or scanning) they perform.

The first contribution of this paper consists in introducing a systematic method that enables us to perform such a strategic analysis of zombie armies, based on the botnet scanning traffic observed in a global honeynet. Our approach is based on an appropriate combination of different knowledge discovery and data mining techniques, which consists of the following components:

1. detection and characterization of coordinated attack events;
2. unsupervised clique-based clustering, so as to discover correlations among attack events;
3. dimensionality reduction techniques, which allow us to visualize and to assess the cliques correlations;
4. a fuzzy, multi-criteria decision-making process that leverages the results obtained in the previous steps, in order to identify sequences of attack events that are very likely attributed to the same zombie army.

As second contribution, we present some preliminary results obtained from a proof-of-concept framework in which we implemented the techniques mentioned here above. The experiments have been performed on attack traces collected with a worldwide distributed honeynet, which has observed global attack activity in over 18 different IP subnets from Sep 2006 until July 2008 (i.e., about 640 days). Our experimental results highlight several interesting facets of the botnet phenomenon:

- with a mean lifetime of about 98 days, zombie armies seem to be quite resilient. In some extreme cases, we observed certain armies surviving for more than 18 months, which indicates that *taking down botnets still constitutes a real challenge*. On average, zombie armies had at least 8,500 distinct, observable sources during their lifetime.
- regarding the origins, malicious sources involved in zombie armies seem to be highly unevenly distributed in the IPv4 address space; they clearly form a relatively small number of tight clusters within a number of "unclean networks", which are thus responsible for a large deal of malicious activities related to server-side attacks (e.g., network scanning, bot propagation).

- over all zombie armies observed so far, at least 43% of the botnet population is made of home users' machines with high-speed Internet connexions (cable, DSL). Windows 2000 and WinXP Pro were the primarily operating systems among zombie machines (i.e., more than 90% of the bots).
- similarly to real-world armies, certain groups of zombie machines seem to be able to coordinate their efforts, e.g., by coordinating different tasks such as network reconnaissance and subsequent targeted attacks.
- finally, most of the identified zombie armies had a significant attack capability, not only in terms of the available bandwidth that can possibly be offered by all zombies together, but also the number of ports they are able to probe or to exploit.

The rest of the paper is structured as follows: in Section 1, we give a brief overview of the honeynet used in our experiments, and we define the notion of coordinated *attack events* as observed by the honeypots. In Section 2, we describe the components of our knowledge discovery framework that we use to identify global attack phenomena, whereof most are related to some activities of zombie armies. In Section 3, we present our experimental results and the kind of findings we can obtain by applying this method to a set of attack events collected on the Internet. Finally, we conclude in Section 4.

Note that this research builds on prior work in malicious traffic analysis. More particularly, we have presented in [28] a more formal and complete discussion of our framework, especially regarding the aspect fuzzy, multi-criteria decision-making. To make this paper as self-contained as possible, we have summarized as much as possible our previous contributions in Section 2. This paper will mostly focus on the practical results obtained in each step of our analysis framework, rather than the formal aspects of the different techniques.

1. Collecting Attack Traces with a Global Honeynet

1.1. Leurre.com Honeynet - Dataset Overview

Our data set is made of network attack traces collected with a distributed set of sensors (called *server honeypots*), which are deployed in the context of the *Leurre.com Project* [14,22]. Because honeypots are systems deployed for the sole purpose of being probed or compromised, any network connection that they establish with a remote IP can be considered as malicious, or at least suspicious.
Launched in 2003 by Eurecom, a research Institute based in Sophia Antipolis (France), this project maintains a worldwide distributed system of honeypots running in more than 30 different countries covering the five continents. The main objective of the project is to get a realistic picture of certain classes of global attack phenomena happening on the Internet, by collecting unbiased quantitative data in a long-term perspective. In the first phase of the project, the data collection infrastructure relied solely on low-interaction sensors based on *Honeyd* [23] to collect unsolicited traffic (also sometimes termed "Internet background radiation" [18]). In early 2008, a second phase of the project was started with the deployment of medium-interaction honeypots based on the *ScriptGen* [15] technology, in order to enrich the network conversations with the attackers. Scriptgen sensors are able to automatically learn about new protocol interactions, such that they can handle *0-day* exploits, and eventually capture shellcode samples and malware

Table 1. Overview of some prevalent types of activities observed in the honeynet, grouped by port sequence. The network traffic has been collected from Sep'06 until June'08.

Observed Port Sequence	Targeted Service	Volume of Sources (%)	Main Origins (countries)
|I	ICMP (Echo request/reply)	755,227 (28%)	US(20%),KR(11%),CN(10%),BR(6%), others(53%)
|1026U|1027U|1028U	Windows Messenger	373,361 (14%)	CA(100%)
|1026U	Windows Messenger	216,040 (8%)	US(50%),null(17%),CA(6%), others(27%)
|445T	Microsoft-DS	208,060 (8%)	CS(32%),RS(19%),US(6%), others(43%)
|I|139T, |I|139T|445T	ICMP (Allaple), MS-Netbios-ssn, Microsoft-DS	130,392 (5%)	KR(20%), others(80%)
|135T	Microsoft DCE/RPC	112,764 (4%)	JP(16%),US(13%),CS(7%),RS(7%), PL(6%),DE(6%), others(45%)
|5900T	VNC	104,238 (4%)	US(17%),CN(6%),FR(6%),KR(6%), others(51%)
|2967T	Symantec AntiVirus (ssc-agent)	101,062 (4%)	US(23%),CN(8%),JP(6%), DE(5%),PK(5%), others(53%)
|1433T	MS-SQL	87,332 (3%)	CN(32%),US(15%),others(53%)
|139T	MS-Netbios-ssn	50,781 (2%)	US(17%),CA(8%),TW(5%),FR(5%), others(65%)
|I|80T	ICMP, Web	48,649 (2%)	US(54%),KR(11%),CN(8%), CA(7%), others(20%)
|1434U	MS-SQL-Monitor (Slammer)	36,627 (1%)	CN(44%),US(14%),JP(6%), others(36%)
|22T	SSH	36,094 (1%)	CN(24%),US(13%),KR(8%), TW(5%), others(50%)
|80T	Web	28,005 (1%)	US(27%),CN(7%),FR(7%), DE(7%),null(5%), others(47%)
|137U	MS-Netbios-ns	25,630 (<1%)	US(16%),BR(9%),AR(6%), FR(5%),ES(5%), others(59%)
|I|445T	ICMP, Microsoft-DS	18,273 (<1%)	US(14%),CN(13%),TW(8%),FR(7%), JP(7%),null(6%),DE(5%), others(41%)
|4899T	Remote Admin	15,935 (<1%)	CN(15%),US(15%),KR(10%), RU(5%), others(54%)

binaries when they are targeted by code injection attacks. All network traces captured on the platforms are automatically uploaded into a centralized database. The collected traffic is also enriched with a diverse set of contextual information, such as: the geographical location and the ISP's of malicious sources (via Maxmind), reverse DNS lookups, VirusTotal[2] and Anubis[3] reports for each sample of downloaded malware, passive OS fingerprinting (with P0f), Snort IDS alerts, and more recently, we also added the correlation of the observed IP sources with different IP blacklisting services (e.g., Spamhaus[4], Emergingthreats[5] blocking lists, and a fast-flux bot tracker[6]).

For the purpose of this study, we have used a 640-day attack trace collected by 36 platforms located in 20 different countries and belonging to 18 different class A-subnets. Note that, in the scope of this paper, we only considered the traffic collected by low-interaction sensors; but we are actively looking into extending our analysis techniques to integrate the attack traffic gathered by the medium-interaction (ScriptGen) platforms. Table 1 gives an overview of the most prevalent types of activities grouped by targeted port sequences, and their origins, as observed in the honeynet.

From this traffic, we have then selected only the most prevalent types of activities observed on the sensors, i.e., about 130 distinct attack profiles for which an activity

[2]http://www.virustotal.com
[3]http://anubis.iseclab.org
[4]http://www.spamhaus.org
[5]http://www.emergingthreats.net
[6]http://dnsbl.abuse.ch

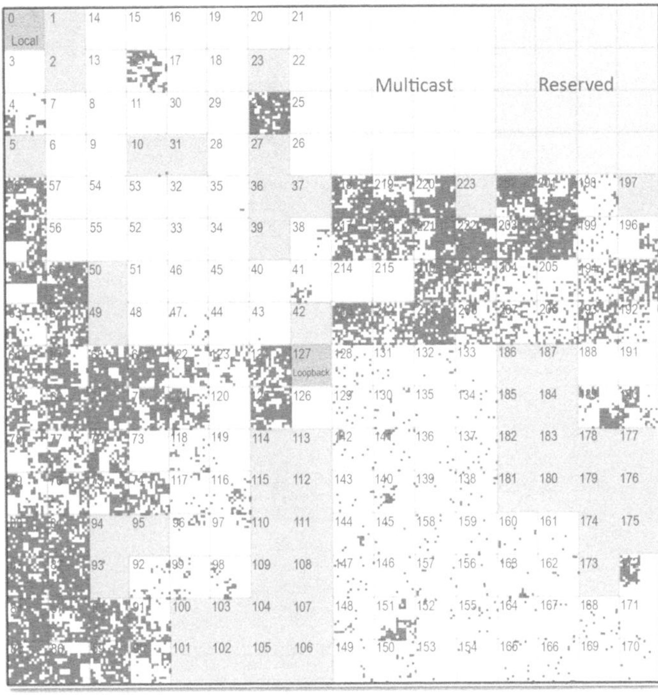

Figure 1. Distribution of malicious sources in the IPv4 address space using a fractal mapping (Hilbert curve).

involving a sufficient number of IP sources had been observed at least once on a given day. This data set comprises totally 1,195,254 distinct sources, which have sent about 3,423,577 packets to the sensors. Fig. 1 illustrates the distribution of malicious sources for these activities using a fractal mapping (e.g., a Hilbert curve). Note that spoofed IP addresses have already been filtered from this data set. As such, Fig. 1 and Table 1 give already some interesting viewpoints, as it clearly shows that most malicious sources seem to be clustered in a limited number of IP blocks (or AS'es). Nevertheless, this type of global analysis does not help us to get insights into the individual attack phenomena that occurred at a large scale (such as zombie armies). Moreover, such global trends do not allow us to learn about the *modus operandi* of the attackers, which is why we need to develop a more detailed analysis.

1.2. Coordinated Attack Events

We use a classical clustering algorithm to perform a first low-level classification of the raw network traffic. Hence, each IP source observed on a honeypot sensor is attributed to a so-called *attack cluster* [21] according to its network characteristics, such as the number of IP addresses targeted on the sensor, the number of packets and bytes sent to each IP, the attack duration, the average inter-arrival time between packets, the associated port sequence being probed (e.g., if a source sends first some ICMP packets followed by an exploit on port 445/TCP, then it is associated to the port sequence ⟨I-445T⟩), and the packet payload. Therefore, all IP sources belonging to a given attack cluster have left

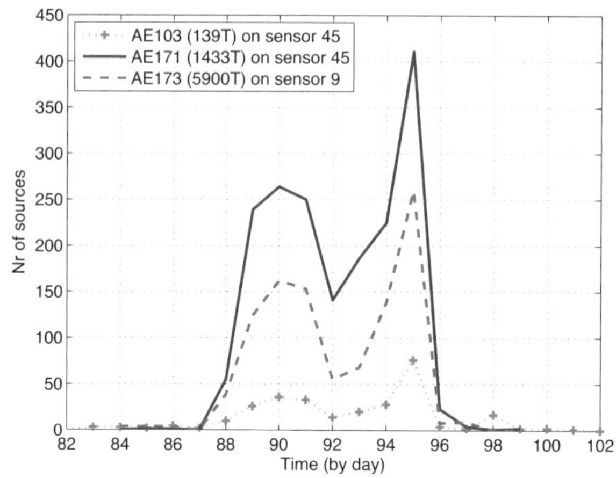

Figure 2. Illustration of 3 attack events observed on 2 different sensors, and targeting 3 different ports.

very similar network traces on a given sensor and consequently, they can be considered as having the same *attack profile*. This leads us then to the concept of attack event, which is defined as follows:

An *attack event* refers to a subset of IP sources having the same attack profile on a given sensor, and whose coordinated activity has been observed within a specific time window.

Fig. 2 illustrates this notion by representing the time series (i.e., the number of sources per day) of three coordinated attack events observed on two different sensors in the same time interval, and targeting three different ports. The identification of those events can be easily automated by using the method presented in [20]. By doing so, we are able to extract interesting events from the spurious, nonproductive traffic collected by our sensors, and we can focus on the most important events that might originate from coordinated phenomena, such as attack activities resulting from botnet reconnaissance scans, and bot propagation. As previous botnet studies have already showed [13], it seems that the botnet scanning behavior is ingrained to the botnets because this is an effective (and low-cost) way for them to recruit new bots. Therefore, botmasters will probably not give up scanning in the near future.

By using the technique described in [20], we have extracted from the whole data set about 351 attack events that were coordinated on at least two different sensors. In the rest of this paper, we will focus on the analysis of this set of attack events, which still accounts for 282,363 unique sources (23.6 % of the original data set), or 741,349 packets (21.5%), and we will show how to take advantage of different external attack characteristics to discover knowledge, and to identify individual phenomena related to zombie armies.

2. A Framework to Identify Global Attack Phenomena

2.1. Overview

Once we have identified a set of attack events occurring at different moments, how could we know in a reliable way which events can be attributed to the same root phenomenon? That is, how can we identify which *sequences of attack events* are very likely the consequence of the same zombie army scanning or probing one or several subnets, eventually during non-contiguous intervals of time?

In the realm of threat monitoring, this problem is sometimes referred to as "attack attribution", which is the process of effectively attributing new attack events to (un)-known phenomena, based on some evidence or traces left on one or several monitoring platforms. To address this problem in a systematic way, we have developed a framework that analyzes attack events with appropriate knowledge discovery (KDD) techniques. The main components of this framework are sketched in Fig. 3. Based on a set of attack events (as defined here above), the first KDD component extracts cliques of attackers in an unsupervised way, so as to identify meaningful correlations between events. That is, we want to know whether some groups of events are strongly correlated with respect to some given characteristics. For example, we could discover which groups of attack events share the very same spatial distributions (in terms of geographical or IP subnet distributions), or which other groups of attack events are targeting the same set of sensors in the same time interval, or which groups of attacks are similar in terms of activities (e.g., the port sequences targeted by malicious sources), and so on. We motivate our choice of attack characteristics used to discover knowledge in the next subsection. Then, we evaluate the consistency of the extracted cliques (or clusters) by using dimensionality reduction techniques, which enable us to visualize on a map the cliques results for each attack dimension. We refer to this step as "semantic mapping", since the distance between each pair of events on a given mapping has a certain meaning. Indeed, the distances are related to the degree of similarity between the underlying feature vectors of the attack events (i.e., the distributions of countries, subnets, etc).

In the next component of the framework, we have implemented a multi-criteria decision-making algorithm that is based on fuzzy inference systems (FIS). The objective consists in combining intelligently the previously extracted knowledge (i.e., the cliques and the semantic mappings), so as to build sequences of attack events that can be attributed to the same global phenomena with a high degree of confidence, thanks to the combination of different statistical measurements. Interestingly, a FIS does not need any training prior making inferences. Instead, it takes only advantage of the previously extracted knowledge to make sound inferences, so as to attribute incoming attack events to a given phenomenon. Each identified attack phenomenon is then modeled with a fuzzy inference system.

2.2. Defining Attack Characteristics

In most knowledge discovery applications, we must first define salient features that may provide some meaningful *patterns* [11]. So, we start by defining different attack char-

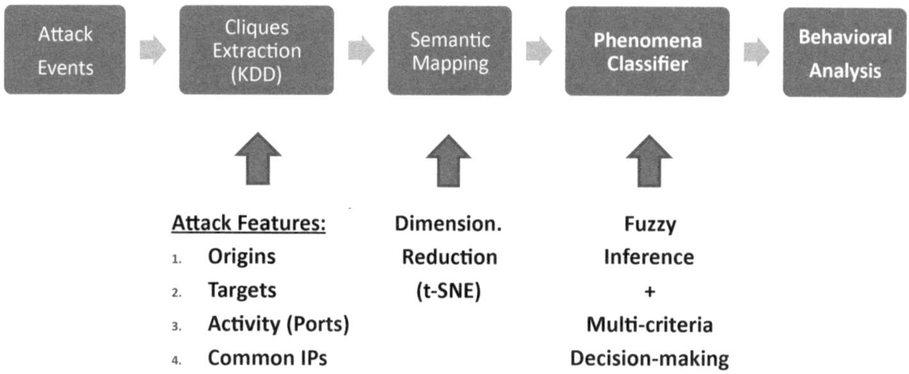

Figure 3. Components of a Knowledge Discovery Framework for Identifying Global Phenomena.

acteristics that we have used to extract knowledge from our set of attack events. In this specific case, we consider them as useful to analyze the root causes of global phenomena observed on our sensors, and as a result, to identify different zombie armies. However, we do not pretend they are the only ones that could be used in threat monitoring. Since other characteristics might prove relevant in the future, our framework is built such that additional features could be easily included when necessary (e.g., to include characteristics related to code injection attacks, shellcodes, or malware samples).

The two first characteristics retained are related to the *origins* of the attackers, i.e. their spatial distributions. First, the geographical distribution of malicious sources can be used to identify botnets that are located in a limited number of countries. Similarly, the IP network blocks provide also an interesting viewpoint on the attack phenomena, since it gives a good indication of the spatial "uncleanliness" of certain networks, i.e., the tendency for compromised hosts (e.g., zombie machines) to stay clustered within unclean networks [4]. So, for each attack event, we can create a feature vector representing either the distribution of originating countries, or of IP addresses grouped by Class A-subnet (i.e., by /8 prefix).

The next characteristic deals with the *targets* of the attackers, namely the distribution of sensors that have been targeted by the sources. Botmasters may indeed send commands at a given time to all zombies to instruct them to start scanning (or attacking) one or several IP subnets, which of course will create coordinated attack events on specific sensors. Therefore, it seems important to look at relationships that may exist between attack events and the sensors they have been observed on.

Besides the origins and the targets, the type of activity performed by the attackers seems also relevant to us. In fact, bot software is often crafted with a certain number of available exploits targeting a reduced set of TCP or UDP ports. In other words, we might think of each botnet having its own *attack capability*, which means that a botmaster will normally issue scan or attack commands only for vulnerabilities that he might exploit to expand his botnet. So, it seems to make sense to take advantage of this feature to look for similarities between the sequences of ports that have been targeted by the sources of the attack events.

Table 2. Some experimental clique results obtained from a honeynet dataset collected from Sep 06 until June 08. (1) the given patterns represent the average distributions for the most prevalent cliques, i.e. the ones lying in the upper quartile in terms of number of sources. For the IP subnets (resp. targeted platforms), the numbers refer to the distributions of originating (resp. targeted) class A-subnets.

Attack Dimension	**Nr of Cliques**	**Max.size (nr events)**	**Min.size (nr events)**	**Volume of sources (%)**	**Most prevalent patterns found in the cliques**(1)
Geolocation	31	40	3	84.4	⟨CN,CA,US,FR,TW⟩, ⟨IT,ES,FR,SE,DE,IL⟩, ⟨KR,US,BR,PL,CN,CA⟩ ⟨US,JP,GB,DE,CA,FR,CN,KR⟩, ⟨US,FR,JP,CN,DE,ES,TW⟩, ⟨CA,CN⟩ ⟨PL,DE,ES,HU,FR⟩
IP Subnets (Class A)	25	51	3	91.2	⟨87,82,151,83,84,81,85,213⟩, ⟨222,221,60,218,58,24,124,121,219,82,220⟩ ⟨201,83,200,24,211,218,89,124,61,82,84⟩, ⟨24,60⟩ ⟨83,84,85,80,88⟩, ⟨193,195,201,202,203,216,200,61,24,84,59⟩
Targeted platforms	17	86	2	70.1	⟨202⟩, ⟨88, 192⟩, ⟨195⟩, ⟨193⟩, ⟨194⟩ ⟨129, 134, 139, 150⟩, ⟨24, 213⟩
Port sequences	22	66	4	93.2	⟨I⟩, ⟨1433T⟩, ⟨I-445T⟩, ⟨5900T⟩, ⟨1026U⟩, ⟨135T⟩, ⟨50286T⟩ ⟨I-445T-139T-445T-139T-445T⟩, ⟨6769T⟩, ⟨1028U-1027U-1026U⟩

Finally, we have also decided to compute, for each pair of events, the ratio of common IP addresses. We are aware of the fact that, as time passes, some zombie machines of a given botnet might be cured while others may get infected and join the botnet. Additionally, certain ISPs apply a quite dynamic policy of IP address allocation to residential users, which means that bot-infected machines can have different IP addresses when we observe them at different moments (i.e., DHCP churn effect). Nevertheless, and according to our domain experience, it is reasonable to expect that if two distinct attack events have a high percentage of IP addresses in common, then the probability that those two events are somehow related to the same global phenomenon is increased (assuming that the time difference between the two events is not too large).

2.3. Clique-based Knowledge Discovery

For each attack characteristic considered here above, we have applied a clique-based clustering on our set of attack events. That is, we use a graph-based approach to formulate the problem: the vertices of the graph represent the feature vectors of each attack event (e.g., the distribution of countries, subnets, targeted sensors, etc), and the edges express the similarity relationships between those vertices. Clearly, the choice of a similarity metric is very important, as it has an impact on the properties of the final clusters, such as their size, quality, and consistency. To reliably compare the kind of empirical distributions mentioned here above, we have chosen to rely on strong statistical distances, such as Pearson's χ^2, or the Jensen-Shannon divergence (JSD) [16], which derives itself from the Kullback-Leibler divergence [12]. Finally, the clustering is performed by extracting so-called *maximal weighted cliques* (MWC) from the graph, where a maximal *clique* is defined as an induced sub-graph in which the vertices are fully connected and it is not contained within any other clique. We refer the interested reader to [27,26] for a more detailed description of this clique-based clustering technique applied to honeynet traces.

Table 2 presents a high-level overview of the cliques obtained for each attack dimension separately. As we can see, a relatively high volume of sources could be classified into cliques for each dimension. The last colon with the most prevalent patterns gives an indication of which countries or class A-subnets (e.g., originating or targeted IP subnets) are most commonly observed in the cliques that lie in the upper quartile with respect to the number of sources. Interestingly, it seems that many coordinated attack events are coming from a given IP sub-space. Regarding the targeted platforms, several cliques involve a single class A-subnet. About the type of activities, we can observe some commonly targeted ports (e.g., Windows ports used for SMB or RPC, or SQL and VNC ports), but also a large number of uncommon high TCP ports that are normally unused on standard (and clean) machines (such as 6769T, 50286T, 9661T, ...). A non-negligeable volume of sources is also due to UDP spammers targeting Windows Messenger popup service (ports 1026 to 1028/UDP).

2.4. Visualizing Cliques - Knowledge Consolidation

In order to assess the consistency of the resulting cliques of attack events, it can be useful to see them charted on a two-dimensional map so as to *i)* verify the proximities among clique members (*intra-clique* consistency), and *ii)* understand potential relationships between *different* cliques that are somehow related (i.e. *inter-clique* relationships). Moreover, the statistical distances used to compute those cliques make them intrinsically coherent, which means also that certain cliques of events may be somehow related to each other, although they were separated by the clique algorithm.

Since most of the feature vectors we are dealing with have a high number of variables (e.g., a geographical vector has more than 200 country variables), the structure of such high-dimensional data set cannot be displayed directly on a 2D map. Multidimensional scaling (MDS) is a set of methods that can help to address this problem. MDS is based on dimensionality reduction techniques, which aim at converting a high-dimensional dataset into a two or three-dimensional representation that can be displayed, for example, in a scatter plot. The aim of dimensionality reduction is to preserve as much of the significant structure of the high-dimensional data as possible in the low-dimensional map. As a consequence, MDS allows an analyst to visualize how far observations are from each other for different kinds of similarity measures, which in turn can deliver insights into the underlying structure of the high-dimensional dataset.

Because of the intrinsic non-linearity of real-world data sets, we have applied a recent MDS technique called *t-SNE* to visualize each dimension of the data set, and to assess the consistency of the cliques results. t-SNE [29] is a variation of *Stochastic Neighbour Embedding*; it produces significantly better visualizations than other MDS techniques by reducing the tendency to crowd points together in the centre of the map. Moreover, this technique has proven to perform better in retaining both the local and global structure of real, high-dimensional datasets in a single map, in comparison to other non-linear dimensionality reduction techniques such as Sammon mapping, Isomaps or Laplacian Eigenmaps [10].

Figure 4 shows the resulting two-dimensional plot obtained by mapping the geographical vectors on a 2D map using t-SNE. Each datapoint on this map represents the geographical distribution of a given attack event. The coloring refers to the clique membership of each event, and the dotted circles indicate the clique sizes. We could easily verify that two adjacent events on the map have highly similar geographical distributions (even from a statistical viewpoint), while two distant events have clearly nothing in common in terms of originating countries. Quite surprisingly, the resulting mapping is far from being "chaotic"; it presents a relatively sparse structure with clear datapoint group-

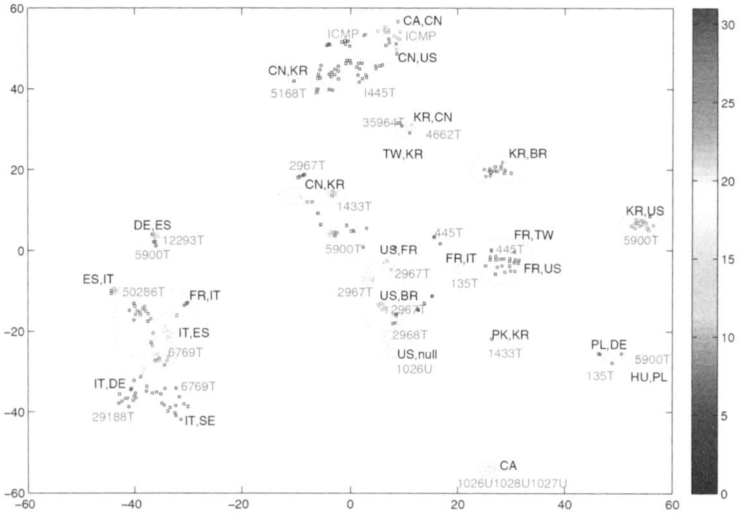

Figure 4. Visualization of geographical cliques of attackers. The coloring refers to the different cliques and the dotted circles indicate their sizes on the low-D map. The superposed text labels indicate the two first attacking countries of the distribution of certain attack events, as well as some of the targeted port sequences (in red).

ings, which means also that most of those attack events present very tight relationships regarding their origins. Due to the strict statistical distances used to calculate cliques, this kind of correlation can hardly be obtained by chance only.

Similar "semantic mapping" can naturally be obtained for the other dimensions (e.g., subnets, platforms, etc), so as to help assessing the quality of other cliques of attackers. As described in the next Section, those different mappings will be used by the multi criteria decision-making component of our framework to identify global phenomena, i.e. by combining efficiently different sets of cliques.

2.5. *Identification of Zombie Armies using Fuzzy Inferences*

The final objective consists in re-constructing *sequences of attack events* that can be attributed with a high confidence to the same root phenomenon in function of multiple criteria. In other words, we want to build an inference engine that takes as input the extracted knowledge (cliques and mappings) to classify incoming attack events into either "known phenomena", or otherwise to identify a new phenomenon when needed (e.g., when we observe a new zombie army). To do this, we have implemented a multi-criteria decision-making algorithm that relies on fuzzy inferences. Our motivation is that: i) we have *a priori* zero-knowledge of the expected output, which means that we can not provide training samples showing the characteristics of the output we are looking for; and ii) we want to include some domain knowledge to specify which type of combinations we expect to be promising in the root cause identification. Also, we favor the "white-box" approach (or a transparent reasoning process), which allows an expert to understand why the system has grouped a given set of events into the same root phenomenon.

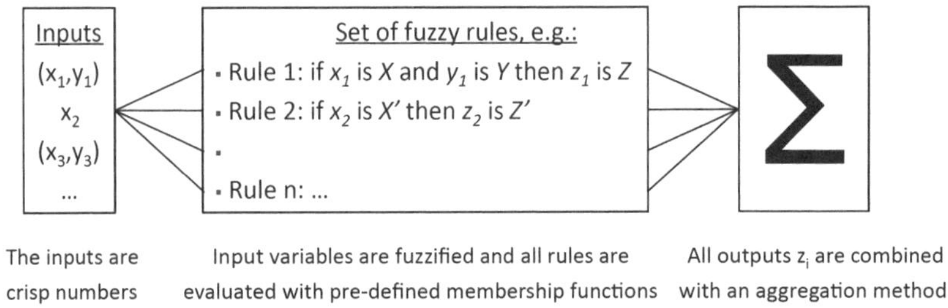

Figure 5. Main components of a Fuzzy System.

Although large-scale phenomena on the Internet are complex and dynamic, our intuition is that two **consecutive** attack events should be linked to the same root phenomenon if and only if they share at least two different attack characteristics. That is, our decision-making process will attribute two attack events to the same phenomenon when the events characteristics are "close enough" (from a statistical viewpoint) for any combination of **at least two** attack dimensions out of the complete set of criteria: $\{origins, targets, activity, common_{IP}\}$. In other words, we hypothesize that real-world phenomena may perfectly evolve over time, which means that two consecutive attack events of the same zombie army must not necessarily have all their attributes in common. For example, the bots' composition of a zombie army may evolve over time because of the cleaning of infected machines and the recruitment of new bots. From our observation viewpoint, this will translate into a certain shift in the IP subnet distribution of the zombie machines for subsequent attack events of this army (and thus, most probably different cliques w.r.t. the origins). Or, a zombie army may be instructed to scan several consecutive IP subnets in a rather short interval of time, which will lead to the observation of different events having highly similar distributions of originating countries and subnets, but those events will target completely different sensors, and may eventually use different exploits (hence, targeting different port sequences).

On the other hand, we consider that only one correlated attack dimension is not sufficient to link two attack events to the same root cause, since the result might then be due to chance only (e.g., a large proportion of attacks originate from some large or popular countries, certain Windows ports are commonly targeted, etc). However, by combining intelligently several attack viewpoints, we can reduce considerably the probability that two attack events would be attributed to the same root cause whereas they are in fact unrelated.

We still need to formally define what is the "relatedness degree" between two attack events, certainly when they do not belong to a same clique but are somehow "close" to each other. Intuitively, attack events characteristics in the real world have unsharp boundaries, and the membership to a given phenomenon can be a matter of degree. For this reason, we have developed a decision-making process that is based on a fuzzy inference system (FIS). Fuzzy Inference is a convenient way to map an input space to an output space with a flexible and extensible system, and using the codification of common sense and expert knowledge. The mapping then provides a basis from which decisions can be made. The main components of an inference system are sketched in Fig. 5. To map the input space to the output space, the primary mechanism is a list of if-then statements called rules, which are evaluated in parallel, so the order of the rules is unimportant.

Instead of using crisp variables, all inputs are *fuzzified* using membership functions in order to determine the degree to which the input variables belong to each of the appropriate fuzzy sets. If the antecedent of a given rule has more than one part (i.e., multiple 'if' statements), a fuzzy logical operator is applied to obtain one number that represents the result of the antecedent for that rule.

Concretely, we use the knowledge obtained from the extraction of cliques to build the fuzzy rules that describe the behavior of a given phenomenon. The characteristics of new incoming attack events are then used as input to the fuzzy systems that model the phenomena identified so far. In each of those fuzzy systems, the features of the *most recent* attack event shall define the current parameters of the membership function used to evaluate the following simple rules: if x_i is $close$ AND if y_i is $close$ then z_i is $related$, $\forall i \in \{geo, subnets, targets, portsequence\}$. The membership functions referred to as "is close" in the fuzzy rules are thus defined by the characteristics of the cliques to which the attack events belong. The calculation of the rule output $z_i \in [0, 1]$ is just the intersection between two curves, which quantifies the inter-relationship between the cliques (and hence, between the attack events).

The results of all rules are then combined and distilled into a single, crisp value using an appropriate multi-criteria aggregation function. In this case, we use an Ordered Weighted Average (OWA) operator, which allows to model more complex requirements such as "most of", or "at least two" criteria to be satisfied in the overall decision function [30]. We refer the interested reader to [28] for a more detailed discussion of our multi-criteria decision-making algorithm.

3. Behavioral Analysis of Zombie Armies

3.1. Global Characteristics

In this Section, we provide some experimental results obtained by applying our multi-criteria inference method to our set of attack events introduced in Section 2 (clique analysis). Over the whole collection period (640 days), we found only 32 global phenomena. In total, 348 attack events (99%) could be attributed to a large-scale phenomenon. An in-depth analysis has revealed that most of those phenomena (apart from the noisy network worm W32.Rahack.H [25], also known as W32/Allaple) are quite likely related to *zombie armies*, i.e. groups of compromised machines belonging to the same botnet(s). We conjecture this for the following main reasons: *i)* the apparent coordination of the sources, both in time (i.e., coordinated events on several sensors) and in the distribution of tasks (e.g., scanners versus attackers); *ii)* the short durations of the attack events, typically a few days only, whereas "classical" worms tend to spread over longer, continuous periods of time; *iii)* the absence of known classical network worm spreading on many of the observed port sequences; and *iv)* the source growing rate, which has a sort of exponential shape for worms and is somehow different for botnets [13].

To illustrate the results, Table 3 presents an overview of some global phenomena found in our dataset. Thanks to our method, we are able to characterize precisely the behaviors of the identified phenomena or zombie armies. Hence, we found that the largest army had in total 57 attack events comprising 69,884 sources, and could survive for about 112 days. The longest lifetime of a zombie army observed so far was still 586

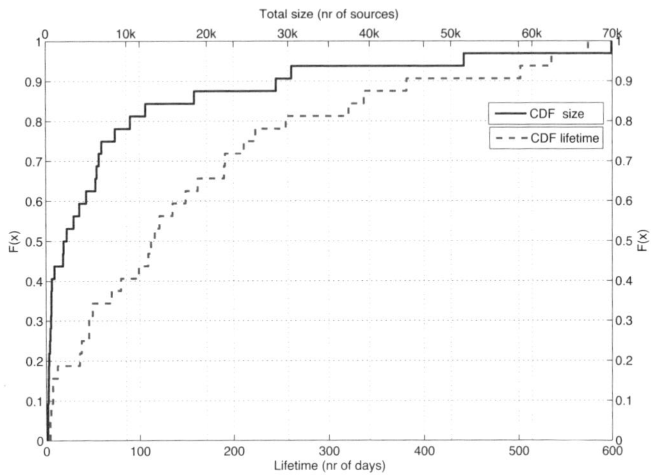

Figure 6. Empirical CDF of the size and lifetime of zombie armies.

days. Fig. 6 shows the cumulative distributions (CDF) of the lifetime and size of the identified armies. Those figures reveal some interesting aspects of their global behaviors: according to our observations, at least 20% of the zombie armies had in total more than ten thousand observable[7] sources during their lifetime, and the same proportion of armies could survive on the Internet for at least 250 days. On average, zombie armies have a total size of about 8,500 observed sources, a mean number of 658 sources per event, and their mean survival time is 98 days.

Regarding the origins, we observe some very persistent groups of IP subnets and countries of origin across many different armies. On Fig. 7, we can see the CDF of the sources involved in the zombie armies of Table 3, where the x-axis represents the first byte of the IPv4 address space. It appears clearly that malicious sources involved in those phenomena are highly unevenly distributed and form a relatively small number of tight clusters, which account for a significant number of sources and are thus responsible for a large deal of the observed malicious activities. This is consistent with other prior work on monitoring global malicious activities, in particular with previous studies related to measurements of Internet background radiation [3,18,31]. However, we are now able to show that there are still some notable differences in the spatial distributions of those zombie armies with respect to the average distribution over all sources (represented with the blue dashed line). In other words, certain armies of compromised machines can have very different spatial distributions, even though there is a large overlap between "zombie-friendly" IP subnets. Moreover, because of the dynamics of this kind of phenomena, we can even observe very different spatial distributions within a *same army* at different moments of its lifetime. This is a strong advantage of our analysis method that is more precise and enables us to distinguish *individual* phenomena, instead of global trends, and to follow their dynamic behavior over time.

[7] It is important to note that the sizes of the zombie armies given here only reflect the number of sources we could *observe* on our sensors; the actual sizes of those armies are most probably much larger.

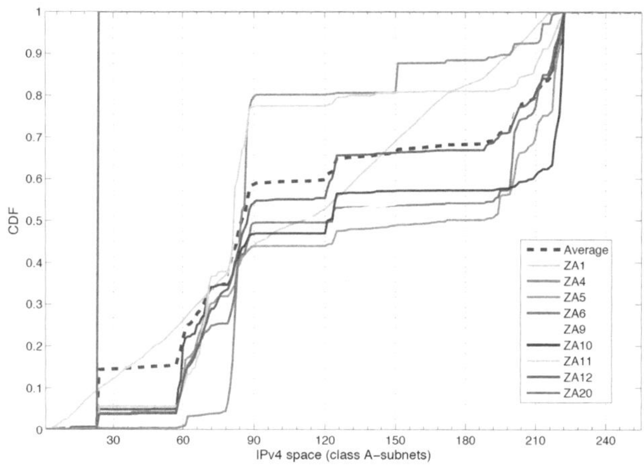

Figure 7. Empirical CDF of sources in IPv4 address space for the 9 zombie armies illustrated in Table 3.

Another interesting observation on Fig. 7 is related to the subnet CDF of ZA1 (uniformly distributed in the IPv4 space, which means randomly chosen source addresses) and ZA20 (a constant distribution coming exclusively from the subnet 24.0.0.0/8). A very likely explanation is that those zombie armies have used spoofed addresses to send UDP spam messages to the Windows Messenger service. So, this indicates that IP spoofing is still possible under the current state of filtering policies implemented by certain ISP's on the Internet.

Then, in terms of *attack capability*, we observe that about 50% of the armies could target at least two completely different ports (thus, probably two different exploits, at least), and one army had even an attack capability greater than 10. Table 4 provides additional details on the characteristics of malicious sources involved in those zombie armies. Regarding the operating systems (detected through passive OS fingerprinting with P0f), we can see that a large majority of the sources are running either Windows 2000 SP or Windows XP Pro. Finally, by analyzing the hostnames of the sources (obtained via reverse DNS lookups), we infer the ratio of home users's machines by looking for typical strings such as '%DSL%', '%PPP%', '%CABLE%'. Over all zombie armies observed so far, we found that at least 43% of the botnet population is made of residential users with high-speed Internet connections. If we take 256kbps as a lower-bound estimate of the average upstream bandwidth for this kind of connection, then we observe that most of those zombie armies could have an aggregate network capacity of several gigabits per seconds, which can easily be used to exhaust almost any type of network resources on the Internet by launching Distributed Denial of Service attacks.

3.2. Some Detailed Examples

In this Section, we further detail two zombie armies to illustrate some typical behaviors we could observe among the identified phenomena, e.g.:

i) a move (or drift) in the origins of certain armies (both geographical and IP blocks) during their lifetime;

Table 3. Overview of some large-scale phenomena found in a honeynet dataset (Sep'06 until Jun'08.

Id	Nr of events	Total size (nr sources)	Lifetime (nr days)	Targeted sensors (Class A- subnets)	Attack capability	Main origins (countries / subnets)
1	10	18,468	535	24.*,193.*,195.*,213.*	1026U	US,JP,GB,DE,CA,FR,CN,KR,NL,IT 69,128,195,60,81,214,211,132,87,63
4	82	26,962	321	202.*	12293T,15264T,18462T,25083T, 25618T,28238T,29188T, 32878T,33018T,38009T,4152T, 46030T,4662T,50286T,...	IT,ES,DE,FR,IL,SE,PL 87,82,83,84,151,85,81,88,80
5	13	9,644	131	195.*	135T,139T,1433T,2968T,5900T	CN,US,PL,IN,KR,JP,FR,MX,CA 218,61,222,83,195,221,202,24,219
6	15	51,598	>1 year	> 7 subnets	ICMP (W32.Rahack.H / Allaple)	KR,US,BR,PL,CN,CA,FR,MX,TW 201,83,200,24,211,218,89,124
9	23	11,198	218	192.*,193.*,194.*	2967T,2968T,5900T	US,CN,TW,FR,DE,CA,BR,IT,RU 193,200,24,71,70,213,216,66
10	57	69,884	112	128.*,129.*,134.*,139.*,150.*	I-I445T	CN,CA,US,FR,TW,IT,JP,DE 222,221,60,218,58,24,70,124
11	14	2,636	110	129.*,134.*,139.*,150.*	I-445T-139T-445T-139T-445T	US,FR,CA,TW,IT 82,71,24,70,68,88,87
12	14	27,442	183	192.*,193.*,194.*,195.*	1025T,1433T,2967T	US,JP,CN,FR,TR,DE,KR,GB 218,125,88,222,24,60,220,85,82
20	10	30,435	337	24.*, 129.*, 195.*	1026U,1026U1028U1027U,1027U	CA,CN 24,60

Table 4. Some detailed characteristics related to the composition of different zombie armies.

Zombie Army Id	Home Users (DSL, Cable, PPP)	Operating Systems (P0f)
1	spoofed IP's	-
4	69%	Windows 2000 SP (68%), Windows XP Pro (5%)
5	27%	Windows 2000 SP (50%), Windows XP Pro (21%)
6	38%	Windows 2000 SP (2%), unknown (98%)
9	29%	Windows 2000 SP (63%), Windows XP Pro (16%)
10	34%	Windows 2000 SP (10%), unknown (87%)
11	61%	Windows 2000 SP (56%), unknown (35%)
12	26%	Windows 2000 SP (61%), Windows XP Pro (17%)
20	spoofed IP's	-

ii) a large scan sweep by the same army targeting several consecutive class A-subnets;

iii) within a same army, multiple changes in the port sequences (or exploits) used by zombies to scan or to attack;

iv) a coordination between different armies.

Zombie army 12 (ZA12) is an interesting case in which we can observe the behaviors *ii)* and *iii)*. Fig. 8 represents the output of the fuzzy system modeling this phenomenon. Each bar graph represents the fuzzy output z_i for a given attack dimension, whereas the last plot shows the final aggregated output from which the decision to group those events together was made (i.e., $F(z_i)$). We can clearly see that the targets and the activities of this army have evolved between certain attack events (e.g., when the value of z_i is low). That is, this army has been scanning (at least) four consecutive class A-subnets during its lifetime (still 183 days), while probing at the same time three different ports on these subnetworks.

Then, the largest zombie army observed by the sensors (ZA10) has showed the behaviors *i)* and *iv)*. On Fig. 9, we can see that this army had four waves of activity during which it was randomly scanning 5 different subnets (note the almost perfect coordination among those attack events) on Windows ports (445T, 139T), preceded by ICMP. When inspecting the subnet distributions of those different attack waves, we could clearly ob-

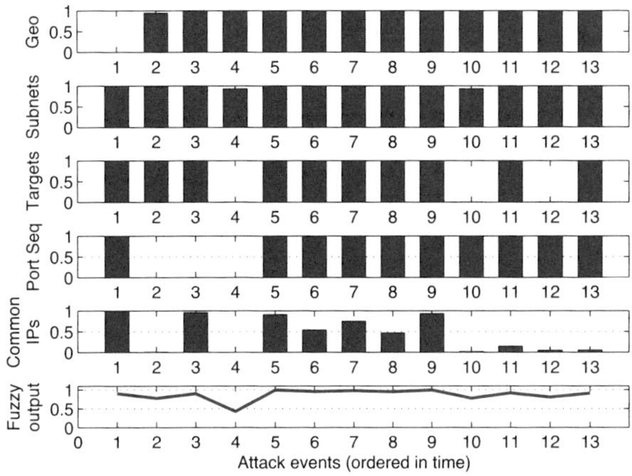

Figure 8. Output of the fuzzy inference system (z_i and $F(z_i)$) modeling the zombie army nr 12.

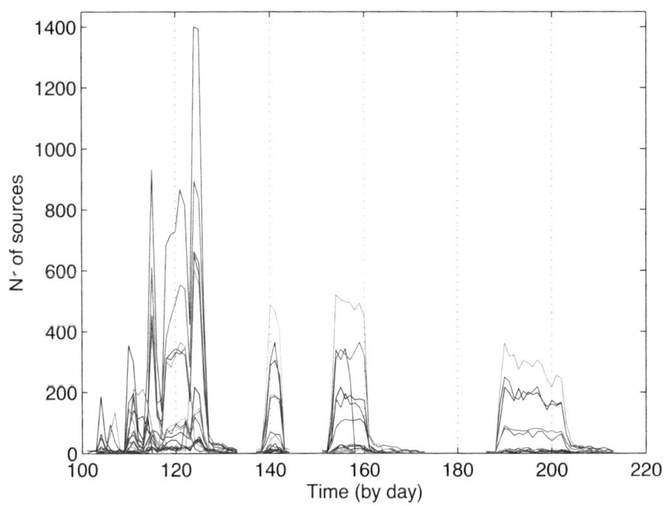

Figure 9. Time series of coordinated attack events for zombie army ZA10 (Nr of sources / day).

serve a drift in the origins of those sources, quite likely as certain machines were infected by (resp. cleaned from) the bot software. Finally, we found another smaller army (ZA11) that is clearly related to ZA10 (e.g., same temporal behavior, similar activity, same targets); but in this case, a different group of zombie machines, resulting in very different subnet CDF's on Fig. 7), was used to attack only specific IP addresses on our sensors, probably by taking advantage of the results given by the army of scanners (ZA10). The scanners were probably using some OS fingerprinting techniques to detect Windows operating systems, since only those ones were targeted by the attackers on ports 445 and

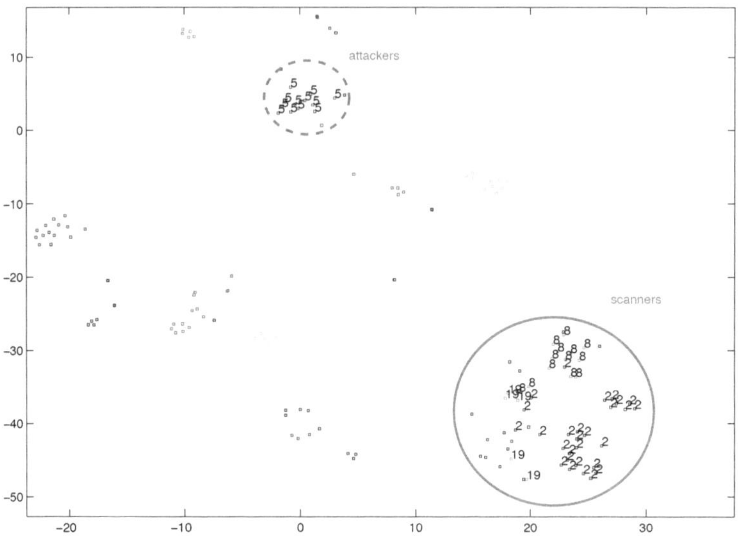

Figure 10. Visualization of the distributions of subnets of origins for Zombie armies 10 and 11, which involve two distinct communities of machines (scanners and attackers). The labels indicate the cliques' memberships of the attack events represented by the data points.

139 (and not the Linux honeypots). The distinction between scanners and attackers is even more visible on the 2D mapping (illustrated on Fig 10) obtained from the subnets distributions of these two zombie armies.

4. Conclusions

In this paper, we have introduced an analysis framework to identify, observe and characterize zombie armies on the Internet, based on the attack traces they have left on distributed sensors. Recent cyber-conflicts have showed that zombie armies and botnets can be easily turned into digital weapons and used to perform DDoS attacks against the network infrastructure of a Nation. It is thus very important to understand the long-term behavior of botnet armies, and their strategic evolution, in order to deploy effective countermeasures against those latent threats. Our analysis is based on the application of appropriate knowledge discovery techniques and a multi-criteria decision-making process. A key aspect of the proposed method is the exploitation of external characteristics of malicious sources, such as their spatial distributions in terms of countries and IP subnets. Our experiments on a set of real-world attack traces have also highlighted some interesting aspects of the global characteristics of such zombie armies, such as their high resilience and the high attack capacity that zombie machines can potentially offer. As future work, we envisage to extend our method to other data sets, such as high-interaction (client) honeypot data, or malware data sets, and to include even more relevant attack features so as to improve further the inference capabilities of the system, and thus also our insights into malicious behaviors observed on the Internet.

Acknowledgements

This work has been partially supported by the European Commission through project FP7-ICT-216026-WOMBAT funded by the 7th framework program. The opinions expressed in this paper are those of the authors and do not necessarily reflect the views of the European Commission.

References

[1] Paul Barford and Vinod Yegneswaran. *An Inside Look at Botnets*. Advances in Information Security. Springer, 2006.

[2] David Barroso. Botnets - the silent threat. In *European Network and Information Security Agency (ENISA)*, November 2007.

[3] Zesheng Chen, Chuanyi Ji, and Paul Barford. Spatial-temporal characteristics of internet malicious sources. In *Proceedings of INFOCOM*, pages 2306–2314, 2008.

[4] M. P. Collins, T. J. Shimeall, S. Faber, J. Janies, R. Weaver, M. De Shon, and J. Kadane. Using uncleanliness to predict future botnet addresses. In *IMC '07: Proceedings of the 7th ACM SIGCOMM conference on Internet measurement*, pages 93–104, New York, NY, USA, 2007. ACM.

[5] Evan Cooke, Farnam Jahanian, and Danny McPherson. The Zombie roundup: Understanding, detecting, and disrupting botnets. In *Proceedings of the Steps to Reducing Unwanted Traffic on the Internet (SRUTI 2005 Workshop)*, Cambridge, MA, July 2005.

[6] Crime-Research. Cyberwar: Russia vs estonia, http://www.crime-research.org/articles/cyberwar-russia-vs-estonia/, [may 09].

[7] Darkreading. Botnets behind georgian attacks offer clues, http://www.darkreading.com/security/app-security/showarticle.jhtml?articleid=211201216, [may 09].

[8] G. Gu, R. Perdisci, J. Zhang, and W. Lee. BotMiner: Clustering analysis of network traffic for protocol- and structure-independent botnet detection. In *Proceedings of the 17th USENIX Security Symposium*, 2008.

[9] Guofei Gu, Junjie Zhang, and Wenke Lee. BotSniffer: Detecting botnet command and control channels in network traffic. In *Proceedings of the 15th Annual Network and Distributed System Security Symposium (NDSS'08)*, February 2008.

[10] Geoffrey Hinton and Sam Rowels. Stochastic neighbor embedding. In *Advances in Neural Information Processing Systems 15*, volume 15, pages 833–840, 2003.

[11] A.K. Jain and R.C. Dubes. *Algorithms for Clustering Data*. Prentice-Hall advanced reference series, 1988.

[12] S. Kullback and R. A. Leibler. On information and sufficiency. *Annals of Mathematical Statistics 22: 79-86.*, 1951.

[13] Wenke Lee, Cliff Wang, and David Dagon, editors. *Botnet Detection: Countering the Largest Security Threat*, volume 36 of *Advances in Information Security*. Springer, 2008.

[14] C. Leita, V.H. Pham, O. Thonnard, E. Ramirez-Silva, F. Pouget, E. Kirda, and Dacier M. The Leurre.com Project: Collecting Internet Threats Information Using a Worldwide Distributed Honeynet. In *Proceedings of the WOMBAT Workshop on Information Security Threats Data Collection and Sharing, WISTDCS 2008*. IEEE Computer Society press, April 2008.

[15] Corrado Leita, Ken Mermoud, and Marc Dacier. Scriptgen: an automated script generation tool for honeyd. In *Proceedings of the 21st Annual Computer Security Applications Conference*, December 2005.

[16] J. Lin. Divergence measures based on the shannon entropy. *Information Theory, IEEE Transactions on*, 37(1):145–151, Jan 1991.

[17] Arbor Networks. http://asert.arbornetworks.com/2007/05/estonian-ddos-attacks-a-summary-to-date, [may 09].

[18] Ruoming Pang, Vinod Yegneswaran, Paul Barford, Vern Paxson, and Larry Peterson. Characteristics of internet background radiation. In *IMC '04: Proceedings of the 4th ACM SIGCOMM conference on Internet measurement*, pages 27–40, New York, NY, USA, 2004. ACM.

[19] Markus Kötter Georg Wicherski Paul Bächer, Thorsten Holz. Know your enemy: Tracking botnets. In *http://www.honeynet.org/papers/bots/*.

[20] V. Pham, M. Dacier, G. Urvoy Keller, and T. En Najjary. The quest for multi-headed worms. In *DIMVA 2008, 5th Conference on Detection of Intrusions and Malware & Vulnerability Assessment, Paris, France*, Jul 2008.

[21] F. Pouget and M. Dacier. Honeypot-based forensics. In *AusCERT2004, AusCERT Asia Pacific Information technology Security Conference 2004, 23rd - 27th May 2004, Brisbane, Australia*, 2004.

[22] The Leurre.com Project. http://www.leurrecom.org.

[23] Niels Provos. A virtual honeypot framework. In *Proceedings of the 12th USENIX Security Symposium*, pages 1–14, August 2004.

[24] M. Abu Rajab, J. Zarfoss, F. Monrose, and A. Terzis. A multifaceted approach to understanding the botnet phenomenon. In *IMC '06: Proceedings of the 6th ACM SIGCOMM conference on Internet measurement*, pages 41–52, New York, NY, USA, 2006. ACM.

[25] Symantec Security Response. W32.rahack.h, [april 2009].

[26] Olivier Thonnard and Marc Dacier. A framework for attack patterns' discovery in honeynet data. *DFRWS 2008, 8th Digital Forensics Research Conference, August 11- 13, 2008, Baltimore, USA*, 2008.

[27] Olivier Thonnard and Marc Dacier. Actionable knowledge discovery for threats intelligence support using a multi-dimensional data mining methodology. In *ICDM'08, 8th IEEE International Conference on Data Mining series, December 15-19, 2008, Pisa, Italy*, Dec 2008.

[28] Olivier Thonnard, Wim Mees, and Marc Dacier. Addressing the attack attribution problem using knowledge discovery and multi-criteria fuzzy decision-making. In *KDD'09, 15th ACM SIGKDD Conference on Knowledge Discovery and Data Mining, Workshop on CyberSecurity and Intelligence Informatics, Paris, France*, Jun 2009.

[29] Laurens van der Maaten and Geoffrey Hinton. Visualizing data using t-sne. *Journal of Machine Learning Research*, 9:2579–2605, November 2008.

[30] Ronald R. Yager. On ordered weighted averaging aggregation operators in multicriteria decisionmaking. *IEEE Trans. Syst. Man Cybern.*, 18(1):183–190, 1988.

[31] Vinod Yegneswaran, Paul Barford, and Johannes Ullrich. Internet intrusions: global characteristics and prevalence. In *SIGMETRICS*, pages 138–147, 2003.

THE INTERNATIONAL INSTITUTE
FOR STRATEGIC STUDIES
ARUNDEL HOUSE
13-15 ARUNDEL STREET
LONDON WC2R 3DX

The Virtual Battlefield: Perspectives on Cyber Warfare
C. Czosseck and K. Geers (Eds.)
IOS Press, 2009

© 2009 The authors and IOS Press. All rights reserved.
doi:10.3233/978-1-60750-060-5-211

Proactive Botnet Countermeasures An Offensive Approach

Felix LEDER, Tillmann WERNER, and Peter MARTINI
Institute of Computer Science IV, University of Bonn, Germany

Abstract. Botnets, consisting of thousands of interconnected, remote-controlled computers, pose a big threat against the Internet. We have witnessed the involvement of such malicious infrastructures in politically motivated attacks more than once in recent years. Classical countermeasures are mostly reactive and conducted as part of incident response actions. This is often not sufficient. We argue that proactive measures are necessary to mitigate the botnet threat and demonstrate techniques based on a formalized view of botnet infrastructures. However, while being technically feasible, such actions raise legal and ethical questions.

Keywords. Botnets, cyberwar, DDoS, defense strategies, countermeasures

Introduction

A botnet is an alliance of interconnected computers infected with malicious software (a bot). Bots are commanded by an operator and can typically be advised to send Spam mails, harvest information such as license keys or banking data on compromised machines, or launch distributed denial-of-service (DDoS) attacks against arbitrary targets. What's more, they often interfere with regular operation rendering infected machines unstable or unusable. Thousands of such botnets exists, with each containing thousands to millions of infected systems. The result are major direct and indirect consequences for economy as well as for the political life [2].

In the past, the economic damage caused by botnets has been related to bandwidth and CPU resources bound by Spam, DDoS attacks, and propagation of the malware. More recent reports show that the damage is largely increasing due to the number of stolen credit card information and banking credentials [17]. As more and more botnets are incorporating functionality to collect this data, the damage will likely increase over the next years. Besides this, distributed denial-of-service attacks originating from botnets disrupt business at attacked sites. The measures for handling these attacks, like forensic analysis, moving sites into different networks, data recovery etc., cost up to several million US dollars per incident, let alone the collateral damage, which is hard to measure [23].

Recent developments show that botnets are not only harmful to companies and consumers but are also involved in politically motivated activity. Largely organized DDoS attacks conducted by botnets in 2007 and 2008 cut off major Government sites in Eastern Europe from the rest of the Internet. This drastically shows how the vast number of remotely controlled machines has the potential to be used as a powerful weapon in a cyberwar rather than just being an annoying phenomenon affecting only some individuals. The spreading of botnets is conducted actively: Remote systems are automatically attacked and exploited, mails are sent that trick the reader into opening malicious programs or web pages which actively exploit the visiting computer. On the contrary side, measures against botnets are very often passive and defensive. To date, active measures are often taken in the context of responses to an ongoing incident only. We have developed methodologies and prototypes for infiltrating botnets that can be used to tackle them from the inside. Such offensive countermeasures can be used to mitigate or extinguish existing botnets. We present our different approaches and demonstrate how they can be applied to existing botnets in case studies.

The remainder of this paper is organized as follows: The next section presents a brief overview of common botnet topologies. Section 2 reviews classical countermeasures. Proactive approaches will be explained in section 3. In section 4 some case studies will be presented. We will discuss legal and ethical aspects in section 5. Section 6 concludes the paper.

1. Botnet Topologies

The two things needed to set up a botnet are an addressing mechanism to identify and reach a command-and-control instance, and a communication protocol to distribute commands to the bots. The latter is often referred to as an *overlay network* that forms the botnet's communication channel. Different botnets are using different strategies here which is reflected in the topology used: We differentiate between *centralized*, *decentralized* and *locomotive* botnets. The kind of topology is extremely important for the selection of containment strategies.

Centralized topologies as depicted in figure 1 are the classical botnet structures. The box in the middle denotes the central C&C server with seven connected bots and a commander (the star symbol). Examples are the IRC-based *Agobot*, *Rbot*, and *Sdbot* families [1]. A static command-and control server is contacted by bots via its IP address (which generally requires resolving a DNS name first). Centralized botnet infrastructures often rely on existing network protocols on top of IP that implement standard client-server architectures, like IRC or HTTP. For this reason, they are obviously completely extinguishable by taking down their C&C server.

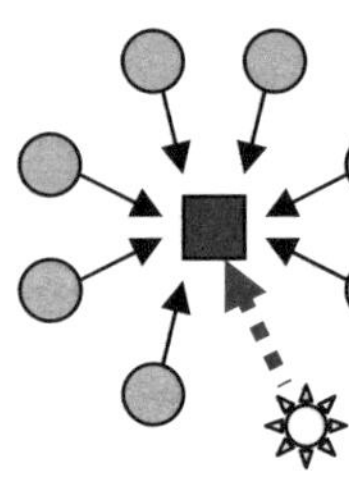

Figure 1. A centralized botnet with seven bots and a commander

The communication in a centralized botnet can either follow a *push strategy* (as in IRC-based communication) where each bot stays connected to a server which then distributes commands simultaneously to all hosts in a broadcast-like manner. Or the server has to be *polled* by the clients on a regular basis (as in HTTP-based botnets). In the latter scenario, the general method is to set up and update a central resource like a web page which can be browsed by the bots. Both approaches have their advantages, e.g., IRC botnets can be built upon an existing IRC infrastructure with multiple self-synchronizing servers, providing load-balancing and reliability. HTTP, on the other hand, is more stealthy and better suited for bypassing security gateways and hiding amongst regular traffic patterns.

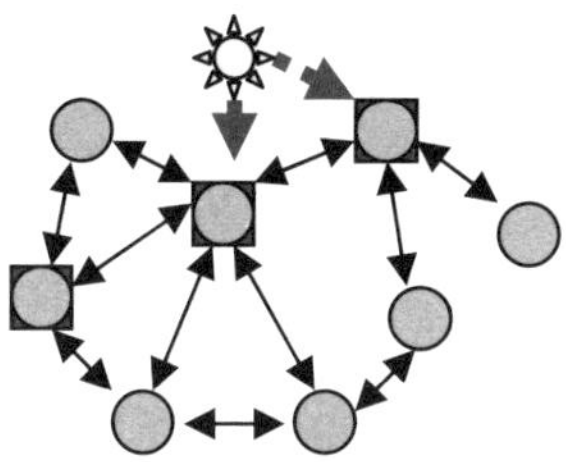

Figure 2. A decentralized botnet with three bots acting as C&C servers

In a d*ecentralized topology*, no single command-and-control component exists. Instead, each bot seeks for a commander using some upstream query mechanism. A schematic structure is depicted in figure 2: Each bot knows some neighbors and receives and forwards commands. Three bots act as C&C servers and are advised to distribute commands in the network. Well-known representatives are the *Storm Worm* [3], or *Conficker* [5]. The two-tiered approach allows the botnet owner to easily change the C&C backbone, making it much harder to take it down. As in centralized botnets, commands can be pushed to bots, which requires that they can be reached instantly, or infected machines pull commands from their individual C&C server (the latter being the most common case). Bots can be implemented to automatically re-establish a C&C session on disconnects. Most decentralized botnets seen so far were based on peer-to-peer (P2P) technology that allows for both information queries as well as host addressing, the two features needed for the communication between a bot and a command server. In a common P2P botnet some peers are controlled by the botnet owner and used to issue and propagate information (i.e. commands) to other peers. Taking advantage of the flexible self-organizing network infrastructure, these nodes are easily replaceable with other hosts.

The decentralization can be taken even further by designing fluxy registration of C&C servers at the query layer (i.e., a pool of command servers returned to queries which is kept highly dynamic through automated subscriptions). This situation is visualized in figure 3 on the next page: The shaded structures are past C&C servers that have been replaced by other ones automatically. Bots recognize the change and contact the new server instead. In most cases these C&C servers are also infected hosts, temporarily playing the role of a commander.

Another way would be to change the query interface, e.g., by choosing time-dependent domain names. We call such botnets *locomotive* because of their constantly moving structure. One example is the HTTP-driven *Torpig* botnet [4]. *Conficker*, in addition to its P2P structure, also makes use of constantly changing DNS names [5-7]. There is no standard implementation of such botnets. In fact, the overall structure is often even more complex than outlined here.

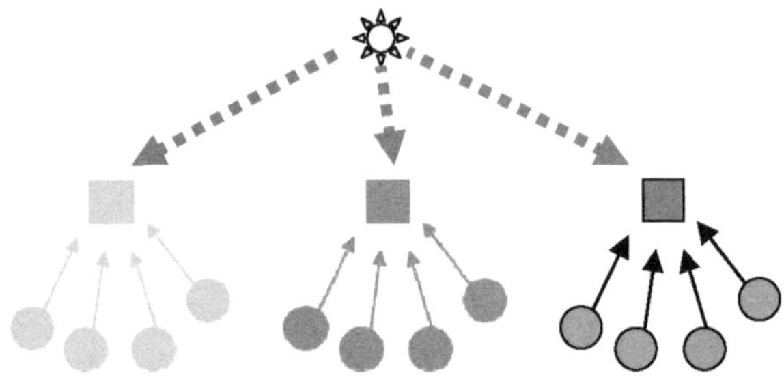

Figure 3. A locomotive botnet with C&C servers that move over time

In reality the boundaries between centralized, decentralized, and locomotive botnets are blurred: A similar strategy was already commonly implemented in classical botnet infrastructures where a DNS entry was used to transparently switch between servers. However, this does not really provide more security as it only displaces the single point against which takeover attempts could be mounted.

2. Classical Countermeasures

Traditional ways of counter-measuring botnets is generally restricted to spotting a central weak point in their infrastructure that can be manipulated, disrupted or blocked. The most common way is to cooperate with an Internet service provider to gain access and shut down the central component, resulting in a loss of control for the botnet owner: The botnet cannot be commanded anymore. Such actions are often performed during emergency response to an ongoing incident like a DDoS attack. While this course of action has proven effective (e.g., shutting down an IRC-based C&C server prevents bots from receiving commands, and machines already involved in an attack are rebooted sooner or later), it requires access to the machine, and, most notably, the willingness to cooperate at the responsible institution. Classical countermeasures against botnets have three different points to attack:

1. The command and control (C&C) server
2. The botnet traffic
3. The infected computers

We will explain the countermeasures with their chances and difficulties in the following. Our goal is to show their differences and why we need more discussion about offensives approaches against botnets.

2.1. Taking Down the C&C Server

The most promising approach is to remove the base of a botnet, which is the C&C server. Pulling the plug of the command-and-control host allows to extinguish the whole botnet in one go. Unfortunately this is only possible if all of the following conditions are met:

1. The botnet uses a centralized structure.
2. The location of the C&C server is known.
3. The provider cooperates.

If one of those conditions is not met, removing the C&C server is impossible. More and more botnets are not solely relying on a centralized structure anymore. Instead they use peer-to-peer (P2P) functionality or multi-proxy structures to hide their central origin. It is often hardly possible to find the location of the C&C server of such botnets. If multiple, fixed servers are used, all of them must be removed. When the location is known, the provider hosting the C&C server must be cooperative. Very often, botnets are controlled from locations hosted by so-called *bullet-proof hosters*, that are not responsive to abuse requests or, even worse, move the server to affiliated partners as soon as the pressure to take the host down rises. Those providers are found in almost all countries, Germany and the U.S. being amongst the most prominent ones according to our observations. Law enforcement is often one step behind when the hosted services are suddenly switched to another provider. In the lucky situation where the location of a centralized botnet to be within a cooperative provider's network, the provider must still be notified and has to agree to all actions. But different organizations that track attacks in the Internet receive so many hints about possible C&C servers that they cannot handle, follow, and verify actions against each C&C individually. The number of conditions is part of the problem that such a large number of botnets exist, and it is still increasing.

Taking the C&C servers down is not always similar to removing the root of the botnet. Infected machines can also contain functionality to spread autonomously, as well as other fall-back logic that gets executed in case the C&C cannot be reached anymore. This creates additional traffic and can lead to more infected machines.

Some cases are known where a botnet takeover was performed with the goal to issue commands that make the bots stop an attack or deinstall themselves. While this approach is more delicate with respect to responsibility for effects caused on infected machines, it is extremely successful at the same time. Attacks are stopped immediately and the botnet is eventually shut down conclusively without the chance to be brought back up by the owner. However, the success rate depends on whether cooperation with the responsible infrastructure providers is possible or not.

2.2. Sinkholing Malicious Traffic

Sinkholing is the term for redirecting network traffic or connection attempts to a special purpose server. If the C&C server cannot be taken down, another option is to redirect malicious traffic to sinkholes, a strategy that found its way into recent mitigation techniques, either locally [24] or globally [14]. The sinkholes record malicious traffic, analyze it and drop it afterwards such that it cannot reach the original target it is meant for. One example of sinkholing is *DDoS null-routing*. In case traffic belongs to an ongoing DDoS attempt is observed it is dropped and sometimes counted for later analysis. DDoS null-routing at border-routers is a promising approach to mitigate DDoS attacks but comes with the challenges of reliable identification of attack related traffic and clean dissection of high-bandwidth data streams at an early stage. This is generally only possible at ISP level. A collaborative worldwide initiative between providers would be another option, but is obviously beyond all question.

2.3. Cleaning Infected Systems

The most sustainable countermeasure against botnets is probably to clean all infected systems and remove the bots installed. While this removes the full power of a botnet, it is also the most complex and most difficult to manage countermeasure. To date, the owners or administrators are responsible for keeping their systems clean from infections. Only recommendations and technical advice can be given to them. As most users are not even aware of their machine being infected, let alone the ability to remove a malware, a global cleaning is impossible. The huge media campaigns about Conficker and the number of still infected systems show that even with intense warnings a large-scale client-side cleanup performed by the individual owners is not feasible.

The standard recommendation to keep systems safe from botnets is to use firewalls and up-to-date anti-virus (AV) software. Firewalls are a preventive feature that in many cases only block attacks from the outside. The increasing number of drive-by-exploits, using bugs in the user's browser to infect a system, and the mobility of malicious data on laptops or USB-sticks opens up a range of new infection vectors that bypass firewalls. This development has been very obvious with Conficker infections [5,14]. Anti-virus software is a reactive feature. Before it is able to detect anything, signatures must be available and the malicious data has to be on the targeted computer. If signatures do not yet exist, the systems cannot be defended. Tests of different AV engines have shown that some detection rates are as low as 80% [18]. Once a system is infected, the bot can spread and perform malicious actions until AV signatures are available. AV engines are often outdated and not updated on a regular basis. Furthermore, different bots disable AV scanners or hide in ways that cannot be detected by regular scanners [19].

All in all, a global cleanup, as it would be required in order to effectively take power from botnets, seems to be infeasible when approached at an organizational level only.

2.4. Conclusion on Classical Strategies

The observations discussed in this section demonstrate that to date the success rate of botnet countermeasures depends mainly on organizational and political general conditions. Given that setting up cooperation or diplomatic agreements takes time we come to the conclusion that establishing an appropriate relationship that legitimates cooperation for collaborated actions is not suited as an ad-hoc scheme for fighting ongoing attacks.

The situation gets worse considering that modern botnet infrastructures do not fall under responsibility of one entity. Instead, distributed peer-to-peer networks operate globally, thus shutting down local parts (often no more than single machines) would be no effective solution. All in all, countermeasures that require close cooperation are today generally infeasible for both technical and political reasons.

There have been discussions where experts stated that shutting down C&C servers has become useless as they would be replaced with new, better protected systems almost immediately. This accelerated arm's race would eventually lead to sophisticated botnet technology sooner than without mitigation. We think that this view is by no means bearable: It ignores the fact that botnets cause harm against other organizations. A hands-off approach leaves potential target sites alone with the existing threat. In the end, restricting mitigation techniques to eluding from or block ongoing attacks is an admission of powerlessness. We propose a combination of classical techniques with additional proactive strategies which we discuss in the following section.

3. Proactive Measures

The classical countermeasures described in the previous section are very good steps to mitigate the power of botnets, but recent developments show that they are only applicable to a certain extent: Newer botnets use more sophisticated obfuscation techniques that deny the use of classical approaches due to the difficulties explained above. While the newer structures introduced by recent botnets complicate the applicability of some strategies, they are open to more offensive tactics. This section explains general principles that can be exploited to create offensive countermeasures against botnets. We focus on the technical possibilities.

Exploring the structure of a botnet is often the first step for finding starting points for possible countermeasures. An immanent property of all botnets is that they have to allow new machines, which run on untrusted platforms, to join the network [25]. This is an important aspect for countermeasure approaches: We are not restricted to acting from the outside – we can join the network, perform investigations while being part of the infrastructure ourselves and might even be able to contain the botnet or take it down from the inside. Furthermore, bots are spreading to infect more systems and make the network grow. Malware samples, which are not hard to obtain, can be analyzed (i.e., reverse-engineered) to learn about their internals. With the knowledge about a bot's functionality, it is often possible to create a *fake bot* and link it into the botnet to monitor or perturb the internal communication. This procedure is always possible, as all information about the initial bootstrapping has to be included in the malware binary and can thus be cloned. Many approaches presented in this section rely on the infiltration of botnets, a technique that was discussed in different flavors in case studies before [3-5,9].

Offensive strategies can be split into three different categories: *Mitigation*, *manipulation*, and *exploitation*. The extent to which corresponding actions are possible depends largely on the topology used by botnet. Especially, decentralized and locomotive topologies offer multiple chances for countermeasures.

Strategies for *mitigation* are offensive, technical means that slow botnets down, by consuming resources for instance. Examples can be temporary DoS attempts against C&C servers, trapping and holding connections from infected machines, or blocking of malicious domains. *Manipulation* strategies make use of the command layer. The knowledge about command protocols is essential to manipulate and inject commands. The required knowledge about the protocols does also include cryptography used. Even though cryptography may completely deny the inspection of botnet data exchange, our Waledac case study shows how this can be achieved even when cryptographic methods like RSA and AES are used [10,11]. Possible manipulation can be the alteration or removal of DDoS or Spam commands as well as commands to download and execute programs, which allows a remote cleanup of infected machine. Less invasive options include dropping collected personal data, like credit card or banking details, replacing them by fake information, or issuing commands to make bots stop the collection. Lastly, *exploitation* is a special strategy that makes use of bugs found in bots. Like bugs in other products, these can be used to perform actions on the infected machines.

Even though, this category is the most powerful, it is the one with the highest risk involved because exploits can easily crash and damage systems if not designed carefully.

Not every strategy can be applied to every botnet. Some of them depend largely on the botnet's topology. Especially non-centralized botnets offer a range of possibilities. We will explain different technical possibilities in the following.

3.1. The Addressing Layer

In this paragraph we discuss strategies targeting the routing and the addressing layer of a botnet infrastructure. It is important to understand that the routing mechanism used in a botnet is needed for addressing hosts, or C&C servers respectively. The command layer, in contrary, works on top of the addressing scheme to provide a communication channel to the interconnected machines.

The most common way for a bot to address a central C&C server is a DNS name that resolves to an IP address – the addressing takes place in two phases. Each phase makes a potential starting point for intervention. For instance, DNS requests are generally handled by a local DNS resolver which, in turn, forwards the request to an authoritative DNS server. This local resolver is controlled by the site administrator and can easily be instructed to return a specially crafted response to specific queries. The same holds for IP routing: Local routers can be equipped with routing table entries to sinkhole certain addresses or redirect them to different hosts. As a consequence, both steps result in bots in the local network being unable to contact the original C&C server and might even be controlled by a pseudo-server. An intervention as described above always requires a man-in-the-middle position. However, it is not always necessary to change the configuration of inline devices. Approaches exist that demonstrate the live modification of relevant network traffic [11].

Modern botnets use more complex addressing schemes which are also run as an overlay network on top of the IP-based Internet. Examples are peer-to-peer networks like *Storm* or *Waledac*. They provide their own addressing scheme with the goal to increase flexibility and decentralization. Both examples will be discussed in more detail in section 4. Again, a strategic position is necessary to infiltrate the addressing layer of these botnets. A general approach is to inject a carefully monitored and controlled node, e.g., a clone of an original peer.

Even when C&C servers cannot be physically accessed, they must be reachable over the Internet because bots have to contact them to receive commands. This fact can be used to mitigate the botnet by creating a DoS situation at the server. A controlled allied DDoS from would make the server unreachable. Additionally, botnets often rely on technology that is prone to specific attacks by design, like the Transmission Control Protocol (TCP). For example, a C&C server's TCP backlog queue can be filled up with connection attempts to trigger denial-of-service conditions, turning the botnet's weapons against itself. This is especially useful for most HTTP based bots where new connections are established for every command request. We have evaluated different service and operating system combinations and found a temporary TCP DoS attack to be easily conductible with only very few resources. During our research we have been able to reliably decrease the probability to establish connections to TCP servers to less than 5% with only one offensive machine. A single host can keep the victim service's backlog queue filled, blocking all further connection attempts and thus hindering bots from requesting commands. Such an action can be crafted in a way that it is not possible to tell apart the connection attempts from the ones issued by bots. As a result any counter-action intending to block the requests would also block all "legitimate" bots. Our tests showed that one single machine can keep a TCP service permanently unresponsive just by initiating and completing 3-way handshakes and keeping connections up as long as possible. Such an attack results in less bots being able to contact the C&C server and participate in malicious activity.

Flooding the link or network where the C&C server resides with packets that consume all available bandwidth is another similar attack. It obviously requires more resources, though, as more packets must be sent. A reflection attack can be used to amplify the amount of traffic sent. However, that would incorporate third-party resources and probably permission by the affected site owners which is apparently not granted.

3.2. The Command Layer

Attacking the command layer of a botnet requires knowledge of the protocol used. An easy example would be an IRC-based network where a *remove command* instructs bots to uninstall themselves from infected systems. Many classical bots implement such an instruction [1], and it was shown before that it can be used to disintegrate a botnet [24]. The injection of a command requires either control over the C&C server, or bots have to be redirected to a different server by performing an attack against the addressing layer (as described in the previous paragraph), which then distributes the removal instruction. Other bots do not have an uninstall option but offer an update functionality that can be used to replace the malware with a innocuous binary or a program that scans for a bot and eventually removes it (similar to a virus scanner).

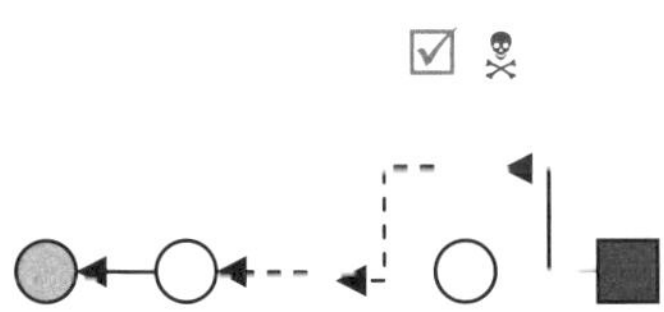

Figure 4. Interception and modification of commands from a C&C server to a bot

In combination with infiltrating the addressing layer other approaches become feasible: Original commands can be monitored, intercepted and modified. This situation is depicted in figure 4 where a fake bot positioned in between the path to other bots and the C&C server intercepts the communication and replaces commands with other information. A protocol could implement checks to render such manipulations impossible. However, such measures were not seen in botnets so far.

In general, to actually conduct a botnet infiltration attack, a combination of actions on both the addressing and the command layer is necessary. Redirecting bots to a controlled server either for sinkholing or to command them to perform a self-removal is probably one of the most effective countermeasures on the infrastructure level.

3.3. Exploitation

Exploit based strategies make use of the fact, that even botnets contain bugs and programming flaws that result in vulnerabilities which can be exploited to gain control either over a central component (like a C&C server) or over bot-infected machines. Such vulnerabilities may range from misconfiguration, like e.g., an insecure IRC server setup that allows other users to control a channel, to security holes in software, like remotely-exploitable buffer overflows.

Mitigation and manipulation strategies are mostly not invasive for the infected machines themselves. An exception are commands that download and execute programs. The exploitation of bugs is even more invasive than executing regular programs because exploit code is often required to be specifically tailored to the targeted host operating system and language. Frameworks like metasploit [21] help in developing generic exploit code. All in all, there still is a higher risk that remote systems are crashed this way. This has to be taken into consideration especially in scenarios where infected systems control critical infrastructures.

Before exploiting the bugs, infected systems have to be found. For decentralized topologies they can be enumerated by counting connection attempts to injected bots. In locomotive topologies this information can be extracted from sinkhole data. Other options are the use of honeypots, IDS signatures or scanners that scan network ranges for infected machines. In very rare cases lists of other bots are available from central IRC C&C servers.

Exploitable vulnerabilities in bots have been found before [13]. Many Rbot and Sdbot variants share the same code base that contains vulnerable functions like this. A potential way to take down botnets would be to identify infected machines, exploit a vulnerability in the bot, and inject and execute code that shuts the malware down. Vulnerable code can still be found in recent malware. Conficker.B uses the MD6 cryptographic hash function for its digital signatures. The MD6 algorithm was found to contain a buffer overrun vulnerability and fixed in an update release that was immediately incorporated in Conficker.C [7]. While this particular vulnerability in Conficker.B was not exploitable, it demonstrates that even sophisticated malware is not immune to critical security holes. Actively attacking bot-infected machines raises lots of ethical and legal questions. We provide a summary of the most important of these aspects in section 5.

3.4. Conclusion on Offensive Strategies

The number of technically feasible strategies shows that there are plenty of possibilities to pro-actively act against botnets before they cause any harm. The use cases presented in the following section show that the offensive strategies are not purely theoretical but based on our practical research. While technically possible, the ethical and legal problems those strategies bring up have to be taken in consideration in practice. Before starting to use (especially the invasive opportunities), an extensive discussion about those topics and authorities is required. The last sections of this paper are to be seen as a step towards this.

A general challenge about many offensive approaches is that they have to be performed stealthy. Otherwise mitigation attempts can be countered by the botnet commanders. Manipulation possibilities can quickly be outdated with small protocol changes or the use of digital signatures. Furthermore, exploitable bugs can usually be fixed in a short time. In case a botnet is to be shut down, this must be performed globally and quickly to not leave any time to the botnet commanders for countermeasures themselves.

Experts consider prosecution of botnet constructors unlikely to have a strong impact on the global threat [20]. Instead, botnets must be fought on a technical level. Proactive measures must be taken as a joint effort of international security teams with local authority. This approach has proven successful and should be followed more consequently in the future [22].

4. Case Studies

This section contains some brief case studies where we present our research on the feasibility and effectiveness of proactive countermeasures on real botnets. We focused on more sophisticated bots, rather than standard IRC or HTTP based networks, as these are far more challenging and our methods must work for them as well. However, we cannot discuss all the technical details for lack of space and refer to the references for more information.

4.1. Kraken

If a botnet's communication protocol is known and messages can be forged, it is possible to inject commands that will be reacted upon by the bots. In case of the Kraken botnet commands are requested from a server after selecting and resolving an entry from a list of domain names. By registering some of those domains and accepting connections from Kraken machines it is possible to send arbitrary commands to the bots. In [15], we have described the encryption used in the protocol. [16] have demonstrated how a remote cleanup can be conducted by issuing an update command that instructs bots to start a removal tool.

4.2. Storm Worm

The *Storm Worm* (also known as just *Storm*) is probably one of the most known bots worldwide [3,9]. While other specimens that use P2P technology were seen before, Storm was the first malware that used it in a way that the botnet could exist for more than three years. Storm is interesting for different reasons: First, spreading was almost only based on social engineering through sending Spam – people even started talking about *Spam campaigns* as the topics were linked to current news or dates like the Iran War or Christmas.

Storm uses an encrypted version of the *Edonkey* peer-to-peer protocol. We have been able to extract the 40-bytes XOR key through reverse-engineering of a storm sample and have built our own Storm P2P client to be able to infiltrate the network [9]. In P2P botnets like Storm, all nodes take part in the infrastructure and perform routing or searches for other bots. Being able to communicate with other nodes, the Storm network routing infrastructure can be infiltrated and disrupted [3,8,9]. However, we have found a less complex yet more powerful approach [9]: We were able to extract Storm's algorithm responsible for the privilege calculation and to displace the original commanders. This makes it possible to issue own commands to all bots in the network.

Storm's command set has been reverse-engineered by us. Consequently, we would have been able to instruct Storm nodes to download and run an arbitrary binary, e.g., to remove the bot from the system. All in all, a complete take down was possible by combining attacks on the infrastructure and the command layer while exploiting a design flaw in the P2P protocol. Today, only an insignificant number of Storm machines is still existing.

4.3. Waledac

Waledac is another P2P bot that tunnels all communication through HTTP. Additionally, each message is encrypted using a hybrid encryption scheme that applies the AES and RSA implementations of OpenSSL. To be able to spy on the traffic, we conducted a man-in-the-middle attack and intercepted the RSA key exchange. One important observation was that the AES key used for further encryption was static rather than dynamically chosen, a design flaw that enables us to also decrypt Waledac messages offline, without the need for a man-in-the-middle proxy.

Being able to snoop on the traffic, we were able to manipulate the communication between two nodes and even developed a tool to construct and inject valid messages ourselves. A takeover strategy based on these findings would be to announce oneself to other Waledac hosts as a proxy node to achieve a prominent position and then drop important commands like DDoS instructions. We could also modify update commands to make bots download and execute our own binary instead of the one provided by the commander.

4.4. Conficker

The first variants of the Conficker worm implement a C&C query algorithm similar to the one used by Kraken. Every day, a list of domain names are generated. Some randomly selected names are then resolved and the corresponding hosts are contacted.

The Conficker Working Group [15] has organized a collaborative effort for pre-registering and sinkholing these domains to make them unavailable to the Conficker constructors. Furthermore, vulnerabilities exist in Conficker's code that would theoretically allow for exploitation and execution of arbitrary commands on infected machines. We have developed a network scanner for reliable identification of Conficker hosts [5]. These techniques can be combined in a proactive defense strategy to take down the botnet.

5. Legal and Ethical Aspects

The technical feasibility of the presented countermeasures does not justify their use in practice [26, 30]. The conduction of these countermeasures may interfere with law or current ethical beliefs depending on their invasiveness and impact on third-parties. On the one hand, many people fear the debates and political consequences, and therefore the general tendency is to stick to conservative approaches [29]. On the other hand, the enormous damage caused by botnets cannot be simply overlooked [23]. Since classical means have not proven to keep up with the increasing threat, discussions have to be initiated about more active strategies. The recently published "Cyberspace Policy

Reveiw" [26] by the White House discusses a strategy to make the Internet more secure from an general viewpoint and identifies privacy and the question of responsibility as important challenges. In fact, this is not really surprising as both of these topics have a strong judicial impact.

This section aims at bringing up the most important consequences of traffic manipulation and countermeasures against control servers. Further, we briefly discuss some ethical, legal, and liability aspects of remote bot disinfections.

5.1. Targeting Control Servers

Most countermeasures that target C&C servers only can be regarded as non-critical. We assume that the commanders of botnets are cyber criminals. Taking down their C&C server literally disarms them. The same holds for regular DoS attacks on those servers. However, DDoS attacks that use lots of bandwidth and processing power, yield to a trade-off between the large amount of resources consumed by the botnet and resources for DDoS countermeasures.

5.2. Targeting Traffic

Traffic manipulation is generally considered to be ethically and legally feasible as long as affected parties agree to it. Such alterations might be offered as a service to prevent DDoS attack, for example. Many users don't know about the threat and therefore don't take steps towards such agreements. Inspecting their traffic and modifying it is a legal problem in many countries, even though more and more countries, like the U.K [31]., pass laws that allow traffic inspection from certain official organizations. Traffic inspection and traffic modification at ISP level would allow to remove Spam and DDoS commands as they are passed in known botnet traffic.

Such actions can also be seen as an indirect protection of the ISP's infrastructure. However, on the one hand, such courses of action raise ethical problems as users may interpret them as a kind of censorship [32]. On the other hand, many users don't know about their infections and would really appreciate if their systems are not misused. A default policy included in contracts to allow ISP to perform those actions in conjunction with a possibility to withdraw would be a solution that is actually already evaluated at several sites.

5.3. Targeting Infected Systems

The most controversial discussion takes place about more invasive strategies, like a remote removal of bots from infected computers. This raises different issues:

1. Ethical: This bypasses the responsibility of users to keep their systems clean.
2. Legal: In most countries it is illegal to run software without the system owner's permission.
3. Liability: Who takes the consequences if the cleanup actions fail or cause problems?

A remote cleanup, in most cases, requires running a removal tool on the infected computer, which has been shown to be technically feasible for a range of botnets. This kind of cleanup has to be performed fast because otherwise new commands to kill the removal tools can be issued by the botnet commanders. Thus, asking all users is not feasible.

Up to now, users are responsible for their own systems. Remote cleanup with automated removal tools bypasses the user, his autonomy, and his responsibility. While some users interpret this as an intrusion into their privacy, a wide range of users would be very grateful for this kind of support to keep their systems clean. All in all, it keeps them a little safer from getting their banking or credit card details stolen. The typical use of AV software as an install-and-forget means supports this view.

Downloading and running software on a remote computer without the owner's permission is illegal in many countries because it is seen as an act of hacking into the system [27, 28]. However, some countries, like the Netherlands, require criminally motivated deeds for the applications of those laws. Since it is a general belief that botnets are run by criminals and cyber terrorists, the disinfection of hosts clearly states good will [28].

This may lead to the conclusion to run proactive strategies only for systems in specific countries or organizations that agreed on such actions. This selective approach is not very effective and yields at maximum in a mitigation but not removal of the considered botnet. Technically it is not always feasible to identify hosts from specific countries or organizations in the overlay network of the botnet. The cleanup of only selected systems rises the problem that the left-over partition may react, adjust to the new situation, and conduct a counter attack. Similarly, concurrent criminal organizations may observe those actions and may use the information to issue their own “updates”, which simply replaces the bot.

Cyber criminals act globally. Thus, countermeasure can only be effective when performed on a global level or at least in large parts of the Internet. A global take down would be the ideal situation. However, a global disinfection rises political questions because most countries would not agree on another country's forces to remotely run software on their systems. This holds especially for infected governmental systems. The foundation of a global organization with legitimation to perform those actions might be a solution. The discussions and consents about such an initiative have to take place on a political level.

Even the best software contains bugs. Invasive countermeasures, like removal tools, can lead to instability of the disinfected system, even with a low probability. This risk increases when bugs in a malware are exploited. In the unlikely case that this happens, the liability is an important question. Who takes responsibility for this happening?

Closely linked to the liability question is the ethical question on the consequences of such actions on medical devices or critical infrastructures, for instance. However, the probability of malicious software causing harm is much more likely. During Conficker outbreaks in hospitals, medical devices were infected and stopped working properly. In the end, it is also a question of responsibility to leave no stone unturned – and that might even include a proactive botnet takedown to prevent further harm.

6. Conclusion

While technically possible, we argue that pro-actively fighting botnets requires immediate political and international consensus. It is a matter of the impact whether people would agree to offensive approaches or not. The affected systems' criticality have to be balanced against potential damage caused by countermeasures. This is, however, also the case for classical mitigation techniques. The portfolio of measures demonstrated in this paper range from more passive ones, like sinkholing, to offensive ones, like exploiting bot hosts to take them over and clean them. We believe that a framework for a staged approach that combines both defensive and offensive techniques should be prepared as part of an emergency response toolkit.

We have seen that cooperation is one of the most important aspects when it comes to successful and sustainable botnet mitigation. This holds for the technical and the political level likewise. Trusted forums must be strengthened and extended to be capable of reacting to botnet incidents effective and immediately. A laissez-faire policy does not lead anywhere.

References

[1] E. Stinson, and J.C. Mitchell, *Characterizing Bots' Remote Control Behavior*, Springer Verlag, Proc. of the 4th intl. conference on Detection of Intrusions and Malware, and Vulnerability Assessment, 2007

[2] S.W. Korns, J.E. Kastenberg, *Georgia's Cyber Left Hook,* 2009

[3] T. Holz et al., *Measurements and Mitigation of Peer-to-Peer-based Botnets: A Case Study on Storm Worm*, Proc. of the 1st Usenix LEET'08

[4] B. Stone-Gross et al., *Your Botnet is My Botnet: Analysis of a Botnet Takeover*, UCSB Tech. Rep., 2009

[5] F. Leder, T. Werner, *Know Your Enemy: Containing Conficker*, Honeynet Project, 2009

[6] P. Porras, H. Saidi, and V. Yegneswaran, A*n analysis of Conficker's Logic and Rendezvous Points*, SRI International Technical Report, 2009

[7] P. Porras, H. Saidi, and V. Yegneswaran, *Conficker C Analysis*, SRI International Technical Report, 2009

[8] C. Kreibich et al., *On the Spam Campaign Trail*, First USENIX LEET'08

[9] G. Wicherski et al., *Stormfucker: Owning the Storm Botnet*, 25th Chaos Communication Congress, 2008

[10] F. Leder, *Waledac is wishing merry christmas*, http://www.honeynet.org/node/325, 2009

[11] F. Leder, *Speaking Waledac*, https://www.honeynet.org/node/348, 2009

[12] Team Cymru, *A Taste of HTTP Botnets*, 2008

[13] Sasser Ftpd Exploit, http://www.securiteam.com/exploits/5AP0J0ACUM.html

[14] The Conficker Working Group, http://confickerworkinggroup.com

[15] F. Leder, P. Martini, *NGBPA Next Generation BotNet Protocol Analysis,* IFIP SEC 2009

[16] P. Amini, C. Pierce, *Kraken Botnet Infiltration*, http://dvlabs.tippingpoint.com (2009)

[17] Symantec, *Symantec Global Internet Security Threat Report 2008*, 2009

[18] Malware Research Group, http://malwareresearchgroup.com (2009)

[19] J. Rutkowska, *Subverting the Vista Kernel for Fun and Profit*, Blackhat Briefings 2006

[20] R. Lemos, "*Arrests unlikely to impact bot net threat, say experts*", http://www.securityfocus.com/news/11344 (2009)

[21] The Metasploit Project, http://metasploit.com

[22] J. Stewart, Interview: "*Researcher argues for CERTs with teeth*", http://www.securityfocus.com/brief/950

[23] Ponemon Institute, *2008 Annual Study: Cost of a Data Breach*, 2009

[24] V, Thomas, N. Jyoti, "*Bot Countermeasures*", Journal in Computer Virology, 2007

[25] R Vogt, J Aycock, M Jacobson, "*Army of botnets*", ISOC Symposium on Network and Distributed Systems Security, 2007

[26] Panel: Ethics in Botnet Research, LEET 09, Boston, April 21, 2009

[27] T. O'Connor, Cybercrime, Cyberlaw, and Cybercriminals, http://www.apsu.edu/oconnort/3100/3100lect02b.htm , Dec. 2007

[28] B. Koops, "Cybercrime Legislation in the Netherlands", Cybercrime and Security, Vol. 2005/4, D.Ferry

[29] D. Fisher, Botnet disruption raises ethical concerns among researchers, http://searchsecurity.techtarget.com/news/article/0,289142,sid14_gci1311711,00.html, Apr. 2008

[30] vnunet, Security experts blast BBC over botnet stunt, http://www.vnunet.com/vnunet/news/2238682/experts-blast-bbc-botnet-stunt, Mar. 2009

[31] EPIC, European Commission Seeks to Protect Internet Privacy, Apr. 2009

[32] CCC, Internet censorship: blocking objectionable content only protects felons, http://www.ccc.de/press/releases/2009/20090212/?language=en Feb. 2009

The Virtual Battlefield: Perspectives on Cyber Warfare
C. Czosseck and K. Geers (Eds.)
IOS Press, 2009
© 2009 The authors and IOS Press. All rights reserved.
doi:10.3233/978-1-60750-060-5-226

When Not to Pull the Plug – The Need for Network Counter-Surveillance Operations

Scott KNIGHT[1,a], Sylvain LEBLANC[a]
[a]*Royal Military College of Canada*

Abstract. The classic response to attack in computer networks has been to disconnect the effected system from the network, preserve the information on the system (including evidence of the attack for a forensic investigation), and restore the system. However, it can be argued that this type of response is not appropriate in many situations. This paper argues that understanding the adversary is essential to effective defence. Instead it may be appropriate to respond with a Network Counter-Surveillance Operation to observe the activity of the attacker. The aim of this research is to enable this new kind of operation through the identification and development of the new tools and techniques required to carry it out. This paper is an omnibus presentation of a group of research projects associated with satisfying this aim, namely tools to help observe the attacker's actions on the compromised system, tools to provide a realistic environment on the compromised system, and tools to mitigate the risks associated with the attacker's use of the compromised system. The argument for the tools and techniques described is presented in the context of an illustrative Network Counter-Surveillance Operation.

Keywords. Network Counter-Surveillance Operation, Computer Network Security, Network Defence

Introduction

The classic response to the compromise of a computer system has been to remove the system from service (perhaps preserving memory images of the system for forensic analysis), clean/reimage the system, and restore the system to service (perhaps after patching the suspected vulnerability) [1]. However, we likely do not know who the attacker really was, what his mission was in attacking this computer system, what are the attacker's capabilities, to what extent has the attacker compromised our systems, what are the attacker's strategic goals, etc. When under targeted attack from a dangerous adversary it may not be appropriate to pull-the-plug as a response; maintaining contact with the attacker and managing the risk of the attacker's presence on the network may be an opportunity too valuable to ignore. There is a need for new tools and techniques to adequately observe the attacker and to manage the risk of such operations.

---

[1] Corresponding Author: Scott Knight, Department of Electrical and Computer Engineering Royal Military, College of Canada; Email: knight-s@rmc.ca

The world-wide computer network environment is recognized as being a new battlespace. As such it would seem logical to apply lessons from more traditional battlespaces where they seem suitable. For example, in no traditional battlespace would a defender apply the defensive tactics, as described in the last paragraph, as a primary response. That is, building firewalls, hard perimeter defences, defence in depth, and then break contact with the enemy as quickly as possible as soon as he shows up. The continued use of this kind of a response may accomplish short term tactical aims such as restoring network services, but abandons any thought to identifying or achieving strategic goals geared towards discovering the attacker's identity, capabilities or objectives. As in other battlespaces, it is the achievement of strategic goals that will win the conflict.

A basic tenet of area defence as prescribed by the U.S. Army Field Manual is that gaining and maintaining contact with the enemy is vital to the success of defensive operations [2]. "As the enemy's attack begins, the defending unit's first concerns are to identify committed enemy units' positions and capabilities, determine the enemy's intent and direction of attack, and gain time to react." [2] In the sphere of naval operations where conflict at sea can be a cat and mouse game of detecting, stalking and engaging, commanders have been censured for failing to maintain contact with the enemy [3].

The recurring directive to maintain contact with the enemy arises from the need to know who the enemy is, what his capabilities are, and what his intention is. This is especially important when the enemy is hard to detect in the first place and there is an incomplete understanding of his capabilities and objectives.

However, maintaining contact with an attacker is a hard thing to do in a modern computer network environment. The attacker will be attempting to conceal himself using encryption, hidden processes and rootkits [4]. If the attacker becomes aware that he has been detected he is likely to change his tactics, techniques, and procedures (TTPs) resulting in defenders loosing contact or being fed misinformation. Of course there are also risks inherent with maintaining contact with the attacker. In the scenarios of interest to this research the attacker has been detected on our system(s) by the defenders. It may be difficult to contain the attacker on the compromised system(s) and mitigate the hazards such a presence poses to the rest of the network. Indeed it may not be possible to maintain contact without accepting some residual risk. However, the risk posed by maintaining this contact may be acceptable when considering the alternative risk of breaking contact, and what that implies.

The aim of this paper is to identify the need to make Network Counter-surveillance Operations (NCSOs) an accepted and standard response in certain cases of computer network compromise. The paper also aims to present the research and development of new tools to enable such Network Counter-Surveillance Operations.

Section 1 of the paper makes the argument for why NCSOs are needed and why they are an appropriate response to some instances of network attack. The section will also briefly identify what capabilities and tools are implied by the need for NCSOs. Section 2 will present three current projects that address the need for these new capabilities. These projects present techniques for covertly observing an attacker, maintaining a cover-story or outward appearance of normal activity while an NCSO is in progress, and a tool for containing the attacker and mitigating the risk associated with the presence of the attacker on the network. The last section provides a conclusion to the paper.

1. The Need for NCSOs

The remove-clean-restore (RCR) response to network attack may be a useful tactic in mitigating the risk associated with a broad non-targeted attack, such as a rapidly propagating virus or worm, or a script-based attack that is exploiting a published software vulnerability. But such a response is not likely to be effective in mitigating the risk associated with targeted attack. Targeted attacks by criminal organizations, non-state actors, or foreign governments are the most serious threat to government/military systems in terms of loss or damage to information assets. The RCR approach leads to some feeling of security in winning the short-term battle, but frustrates the strategic objective of winning the cyber war with the enemy mounting the targeted attack.

By limiting themselves to the RCR responses the defenders may win every battle (i.e. remove the attacker from the compromised systems), but still not prevent the attacker from achieving his strategic goals. To win the cyber war at the strategic level will ultimately require identifying and understanding the enemy. The immediate application of an RCR response denies the defender understanding of who the attacker is, what capabilities the attacker has, and what his objectives are (both his immediate tactical goal, and ultimately his strategic objectives). Controlled surveillance of the attacker's activities, TTPs and unfolding his communications links can provide the defenders with intelligence on the attacker and understanding of his objectives.

Modern government/military computer systems and networks are extremely large and complex systems. The technology and topology of the systems mean that they inherently have large and poorly defined perimeters. Weaknesses are routinely exploited by attackers in every layer of a system's architecture from the network switching equipment to the desktop applications. Current protection technologies make it impossible to prevent successful attack on such large network perimeters. This is an asymmetric conflict environment where a relatively small, covert attacker can effectively engage a strong, well-resourced defender. The RCR response is actually counterproductive in this situation because it will move the attacker away from an attack-lane that is observable (and perhaps controllable) thus breaking contact with the enemy. Moving the attacker from the attack-lane where he has been discovered does not effectively deny access to the system. In a targeted attack scenario the attacker will very likely be back, using another attack-lane. In this battlespace the enemy is hard to find; therefore the defender may not detect the new attack and thus loose the opportunity to observe or control the attacker. Additionally, the RCR response is likely to alert the attacker that he has been detected and will quite likely force him to change his TTPs as a result.

In many cases the RCR response is not available to the defenders of the network because the system(s) compromised cannot be removed from service. This may be because the system is providing some critical service that cannot be disrupted. In this case both the attacker and the defenders are sharing a common infrastructure to support their missions, which are the resources and services of the compromised system. The defenders will have to contain the attack and battle for control of the live compromised system. Preparation for that battle will require proper surveillance and understanding of the attacker, either on the compromised system itself or further back along the attacker's communications chain.

1.1. Communality with Modern Warfare Doctrine

Consider that the scenario we are investigating has a number of common elements with the urban warfare battlespace. Characteristics from the urban warfare battlespace [5][6] that are common with the computer network battlespace are listed below:

- Complex battlespace terrain (i.e. many complex layers of intersupporting technologies, communications mechanisms and applications).
- This complex terrain is inhabited by non-combatants (i.e. legitimate users of the system, their processes and data).
- An infrastructure upon which both the attacker and the non-combatants depend to accomplish their goals, missions.
- Many internal vital points that cannot be completely defended (i.e. complex ill-defined perimeter to the battlespace).
- Asymmetric threat agents.
- An enemy that is hard to locate and identify.
- An enemy that is hard to separate from non-combatants.

All NATO nations train their forces in general to operate adopting the manoeuvrist approach in their plans to defeat the enemy. This approach has been adapted for urban area operations [6]. The Understand, Shape, Engage, Consolidate and Transition (*USECT*) framework is used to conduct such operations [5][6]. The manoeuvrist approach moves the focus from the traditionally predominant *Engagement* element to the *Understand* element (*usEct* to *Usect*). This fits well with our argument that immediate engagement using an RCR response may not be appropriate, and that there are cases where we want to remain in contact to conduct a surveillance operation in order to develop understanding of the attacker, and to control the actions (i.e. shape the battlespace).

Tables 1 and 2 present elements from the NATO doctrinal recommendations for *understanding* and *shaping* that seem especially applicable (edited to reflect the computer network battlespace) [6].

Table 1. Understand Capabilities

NUMBER	CAPABILITY REQUIREMENT
U5	Establish a psycho-sociological profile of the potential enemy
U6	Determine intent, aim, location, movement, status, capabilities, support structure of the potential enemy
U7	Acquire an accurate understanding of the infrastructure, the systems and the dynamics of the computer network environment and their impact on operations (identify the key components/technologies and their vulnerabilities)

Table 2. Shaping Capabilities

NUMBER	CAPABILITY REQUIREMENT
S2	Selective control of infrastructure, utilities and communications
S4	Restrict enemy movement/intentions
S6	Provide own users/assets with adequate protection against the entire threat
S8	Isolate the computer network battlespace
S14	Deny the enemy from operating effective C4ISTAR systems
S15	Deceive enemy as to own force intentions and actions

The concept of NCSOs as a response to network attack is motivated by achieving these capabilities though operations that emphasize maintaining contact with the attacker. As with other operations, surveillance combined with stealth is often sufficient to maintain contact, and is the preferred method for doing so [2]. The NCSO is designed to provide an understanding of the attacker and shaping of the battlespace. Shaping the battlespace through isolation is aimed at denying the attacker any advantages of occupying the compromised computer system. Isolation will also protect friendly users and assets within the limits of an acceptable risk envelope for the operation.

1.2. Requirements for NCSO Toolset

Application of a manoeuvrist approach to computer network defence using the USERT framework implies that NCSOs must be conducted with a view to enable understanding of the attacker and shaping of the network battlespace. This in turn implies the need to satisfy the capabilities identified in the paragraphs above. There are currently no technologies or supporting tools to satisfy these required capabilities. An initial set of required capabilities might be broken down into the following areas for further research and development:

- A toolset for covertly monitoring an attacker's processes and communications activity on a compromised computer system (i.e. the attacker cannot be aware of the surveillance),
- A toolset for maintaining an adequate cover-story on the compromised computer system (i.e. synthetic user activity that maintains the appearance that a system is still being used in a normal way), and
- An internal network firewall to isolate the attacker's activity in order to contain the attack and the risk to other friendly assets while maintaining the covert nature of the surveillance (i.e. through blocking, spoofing, modifying the attackers interaction with friendly systems).

2. NCSO Capability Development

The toolset requirements that we have discussed above represent new areas of research that have not been addressed in the research literature. The Computer Security Laboratory (CSL) of the Royal Military College of Canada has begun a research thrust entitled *Network Intelligence Surveillance Toolset (NIST)* which begins exploratory research into such tools. The following sub-sections will describe an operating scenario that provides context for the research and three of the CSL's NIST research projects that would support NCSOs.

2.1. Representative Scenario

To help motivate the circumstances in which NIST tools are expected to operate consider a scenario similar to the *GhostNet* cyber spying operation discovered in March of 2009 [7]. For our purposes, let us imagine that a sophisticated attacker, representing a hostile government, has exploited the computer system of a senior embassy staff officer located in an enclave inside the hostile country that is being protected by the defender.

The attacker has many options for effecting the initial compromise, from a social engineering attack to many different compromises. These are of little interest to us here, because the short duration of the initial attack makes it unlikely that the defender will discover the compromise. However, once the system has been compromised, the attacker will ensure that he will be able to regain access through the network by installing back doors. The attacker will install malicious tools such as rootkits [4] allowing him to hide the processes that he is running on the compromised system, thus decreasing the likelihood that his compromise will be detected. To further disguise his actions, the attacker will encrypt all communications with the compromised system. Finally, having gained a foothold in the network, the attacker will consider both the value of the information of the system he has compromised, and the potential of using it as a launch point for further attacks. The compromised system could be used against other systems inside the protected enclave or against other networks, thus providing the attacker a level of deniability.

2.2. Covertly Monitoring an Attacker

The first of the capabilities required above was to the development of tools that allow a defender to covertly observe an attacker's actions on a compromised computer, and to remain unobserved by the attacker. For the sake of minimizing detection in cases such as our representative scenario (Section 2.1), we must assume that the attacker has managed to gain administrator (root) access, therefore making any actions made by us on the compromised machine visible to him. This lack of effective isolation motivates research into protecting observation tools by raising the compromised operating system (OS) into a virtual machine and moving the observation tools into an underlying virtual machine monitor (VMM), as depicted in Figure 1. The interface between the OS's virtual machine and the VMM provides strong security guarantees, but this comes at the expense of operating with out the abstractions provided by the OS and so severely degrades the quality of information that can be collected about the attacker. This tension between isolation and abstraction visibility can be partially resolved with virtual machine introspection — the process of reconstructing the OSs abstractions in the VMM. Previous techniques for virtual machine introspection have had shortcomings that render them unsuitable for the observation of attackers: they either rely on software agents in the virtual machine under observation, defeating the purpose of migrating to the virtual machine monitor, or rely on prior extensive knowledge of the OS kernel's internal layout and implementation, making them unsuitable for observing closed-source OSs. Recently developed techniques [8] significantly narrow the gap between the semantic visibility possible from tools behind the virtual machine interface and those naturally available in the compromised guest OSs. Implementations of these techniques are shown to be effective in observing guest system calls and retrieving information exposed by guest OS system call interfaces.

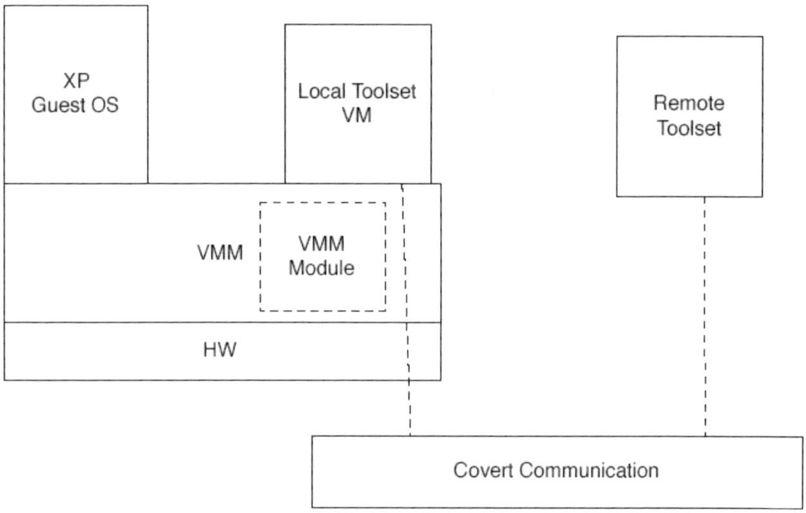

Figure. 1. The architecture for the intrusion surveillance toolset suggested by the scenario and problem-space.

In order to maintain the appearance to the attacker that the machine is still being used by a normal user (cover story in Section 2.2), we cannot physically work on the infected machine; we must work remotely and establish a session with a local surveillance toolset. This session must remain invisible to the attacker, so obviously it cannot run on the compromised machine either. At the same time, this toolset must provide us the ability to inspect the state of the machine and to monitor the attacker's actions. This scenario introduces several implicit constraints to the problem. Based on this, we have identified that our surveillance system must contain four major components [9]: a remote toolset, a local toolset, a VMM module, and a covert communication channel between the remote and local toolsets. This high-level candidate architecture is depicted in Figure 1.

To start an investigation in the context of the representative scenario (Section 2.1) used in this paper, we will conduct surveillance in an attempt to identify the origin of the attacker's network traffic. Effective surveillance will require the ability to perform a number of tasks. For our purposes we define the surveillance system as the components which allow the analyst to perform introspection on the compromised machine from a remote location. In this case, introspection refers to the ability to intercept and reconstruct system calls, and to inject system calls and gather their results. The surveillance system must provide the ability to list all running processes, despite that attacker's attempts to hide them. It must do this using multiple techniques, in case the attacker has managed to hide himself from one method. The surveillance system must provide the ability to inspect the network activity, including open connections and the process-to-port binding map. Since we can see the encrypted traffic, we should know which port our attacker is using, so this may point us to a process id (pid) for the attacker's process. The surveillance system must provide the ability to browse the infected machine's file-system in order to look for files belonging to the attacker. The ability to do file inspection "on-the-fly" is also required. It is also a requirement to be able to copy files back to our trusted domain in order to perform a thorough inspection. The surveillance system must provide the ability to inspect the registry of the compromised machine. After discovering where the attacker "lives" on the system, the analyst needs to monitor him. In our scenario, the attacker is controlling the machine via encrypted communications with a backdoor process. To monitor his actions will therefore require observing the communications between encryption process and the backdoor process. The surveillance system should provide the analyst the ability to watch the attacker's actions in "real-time." The surveillance system must provide this same observation capability in a 'background' mode, writing data to a file while providing the analyst the ability to perform other tasks in the foreground. The identification and development of the surveillance system is the subject of current research [9].

The current research project has developed and validated useful techniques for virtual machine introspection and is now developing a more complete suite of counter-surveillance tools for the intelligence analyst monitoring the attacker.

2.3. Maintaining a Cover-story

A targeted attack such as the one we described in the representative scenario is likely carried out by a sophisticated attacker. Such an attacker is likely to go to great lengths to ensure that the system that he has compromised is indeed a high value target. In fact, the literature has many examples of techniques used by attackers to detect traps like *honeypots*: such as looking for evidence of tampering with the system call table [10], or finding differences between the memory reported by a kernel and the memory it actually uses [11].

We argue that an attacker would also be expecting to see activity at the human interface devices if the compromised system is a user workstation. Such activity at the human interface devices can also be used to characterize the compromised system and we use the term *Vitality Detection* to describe such characterization efforts by the attacker [12]. We posit that the attacker can gather statistics on the mouse and keyboard events being generated to derive user activity on the compromised system. The attacker can then compare this derived activity to models of user behaviour to decide if it appears anomalous.

The targeted attack described in the scenario would be a significant undertaking for the attacker, and he would likely wish to maximize the benefits that he can derive from access to the compromised system. Just as the defenders wish to be able to covertly monitor the attacker on the compromised system (Section 2.1), the attacker also wishes to remain unobserved from the user of the compromised system. This limits the attacker's vitality detection options. For example, the attacker will not risk the bandwidth required to stream the entire human interface event stream and he will limit the processing carried out on the compromised system.

To fool the attacker's vitality detection capability, we suggest a *Synthetic User Environment (SUE)* that would generate human interface device events in a manner that is consistent with a human user. Given a target document, SUE would be able to create a document production model that would cause the release of the stream of human interface device events that would make it appear as if that target document had been input on the compromised system by a human user. Because humans do not input a document starting at the first capitalized letter of the title and ending the final period, SUE has to be able to generate errors and text editing choices that are consistent with a human user. This process is depicted at Figure 2.

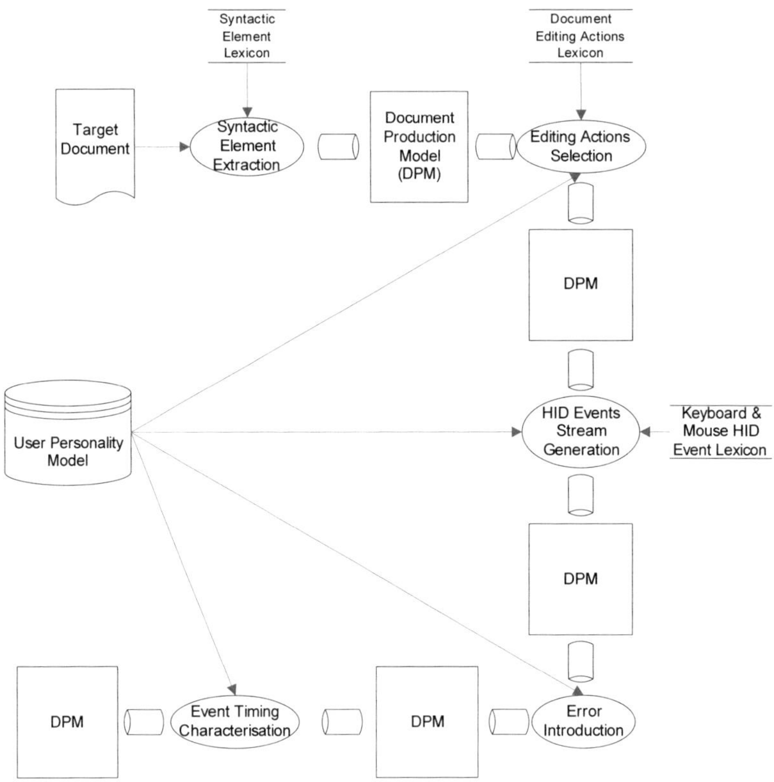

Figure 2. Generation of Human Interface Device Events by a Synthetic User Environment

In addition to helping maintain a cover story on the compromised system, this research is expected to provide additional benefits. It will contribute better models of user activity at the level of human interface devices, which are poorly documented in the open literature. This research would also make it easier to efficiently monitor the attacker's activity on the compromised system. This is because the activity generated by SUE is known to the defender, therefore making it easier to attribute observed activity on the compromised system to the attacker.

2.4. Isolating and Containing the Attacker

The CSL research project dealing with attacker containment and isolation is called ApateX. ApateX is an intelligent transparent network bridge which controls communications traversing it. Its key capabilities are to allow, block, spoof and/or modify communications traversing it [13]. The tool is designed to isolate the attacker's activity in order to contain an attack and manage the risk to other friendly assets while maintaining the covert nature of the surveillance.

An essential concept for this tool is that of a *risk domain*. Risk can be defined as a function of an asset's value, the agents threatening the asset and its vulnerability. For the purpose of this tool, a risk domain is defined as a subset of networked components (ex computers, storage devices, network infrastructure, etc) that share similar asset value, threats and vulnerabilities. A risk policy for an operation can be enunciated by specifying the kinds of traffic that can be allowed between the risk domain that contains the system compromised by the attacker and all other risk domains. Note that the internet external to the protected enclave (i.e. the outside world) may be defined as a risk domain.

Similarly to a firewall ApateX operates by examining the packets that traverse it and making decisions based on characteristics of those packets. The criteria upon which ApateX makes its decisions can be one or more of: the IP addresses, port numbers, and payload strings of the packets. However unlike a classic firewall, ApateX can modify and redirect packets in addition to simply passing or blocking them.

We can examine the capability of the ApateX tool to isolate and control an attacker on a compromised computer system, while maintaining the covert nature of the surveillance. This capability allows the defenders to mitigate the risk associated with maintaining contact with the attacker. Consider the following cases in the context of Figure 3.

a) We may want to allow packets associated with the attacker's communications links to the outside world to pass. In this case we would allow packets to the attacker's IP address to pass to Risk Domain Foxtrot. To cut off the communications links would isolate the compromised machine, thereby breaking contact with the attacker and exposing our knowledge of the compromise.
b) Risk Domain Charlie may contain very sensitive information assets and we may not want the attacker to gain any access to that sub-network. In this case we would block any attacker packets to or from Risk Domain Charlie. This will cause that network to effectively disappear from the perspective of the compromised machine. Alternatively, ApateX can redirect traffic for Risk Domain Charlie to a dummy system/network and provide cover for the operation.
c) We may want to allow domain name services (DNS) for the attacker. In this case we would pass packets to and from Risk Domain Echo. DNS is common infrastructure and is needed by the attacker and the defender. To disallow DNS traffic would be unusual and alert the attacker that he may have been discovered. We can limit the attacker's access to the DNS port and restrict connection attempts to other services/ports to contain his ability to attack the DNS server.
d) Access to the file server and mail server in Risk Domain Delta may be considered too dangerous to allow the attacker to have access. The attacker may have captured or cracked the passwords on the machine he has compromised. These passwords may be valid on the file and/or mail server accounts. We can inspect packet on the fly using ApateX to look for login attempts. The

account/password information can be modified on-the-fly as it passes through ApateX to Risk Domain Delta. This will result in invalid login attempt messages being returned to the attacker. This preserves the covert nature of the surveillance operation because, from the attacker's perspective, it is not possible to tell that the login information has been modified. From his perspective it may just appear that the login information he has gathered is not valid on the machines he is trying to use it on.

e) The attacker may try to attack another external computer somewhere on the internet from our network, e.g. a DDoS attack. We can use ApateX to limit the speed or number of packets allowed to Risk Domain Foxtrot. The attacker's DDoS software seems to work normally from his perspective but never gets to the target. The attacker cannot readily tell at what point in the communications path any filtering of the attack traffic is being done.

ApateX has been fully implemented and is a working system with the capability to allow, block or modify packets traversing it based on criteria passed to it through a user defined policy[13]. Work is continuing with ApateX to develop more protocol awareness through deep-packet inspection.

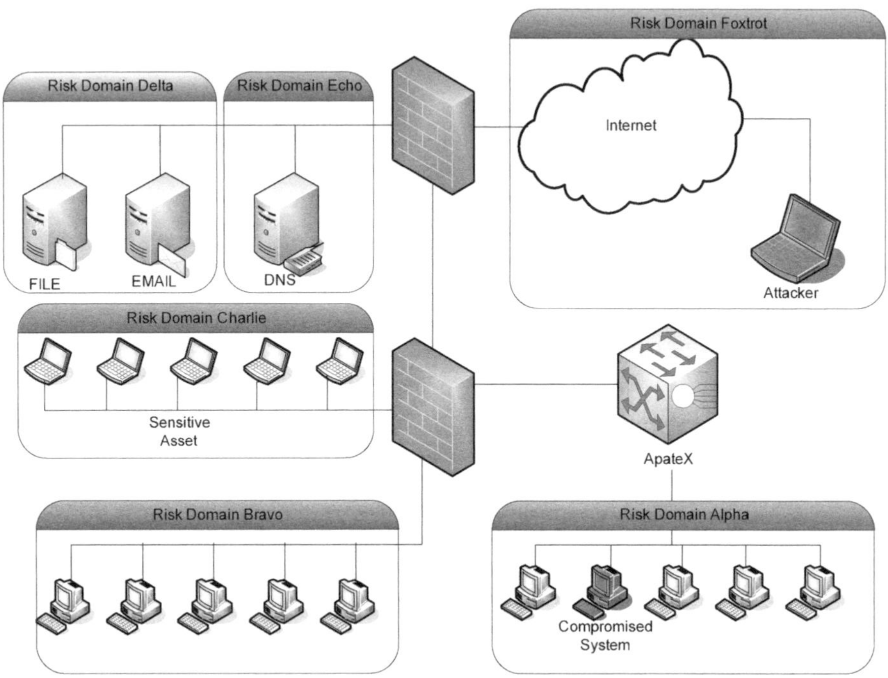

Figure 3. Representative ApateX Deployment

Conclusion

There are many parallels between the computer network battlespace and other modern warfare environments, and we should draw lessons from those other aspects of military operations, including urban area warfare and manoeuvrist doctrine. We have demonstrated that there are situations where it is not appropriate to follow the remove-clean-restore response to a network compromise because it breaks contact with the attacker. In such situations, it may be appropriate to mount a Network Counter-Surveillance Operation to gather intelligence on the attacker. An NCSO can help the defenders gather intelligence on the attacker, including his identity, his immediate tactical goal and his strategic objective. This maintenance of contact can also allow the defenders to follow the attacker's communication paths, which may ultimately lead to the discovery of other compromised systems.

The RMC CSL research projects that we have introduced in this paper begin the exploration necessary to mount successful NCSOs. While we have made inroads in the areas of covert attacker observation, maintenance of a realistic environment for the attacker, and attacker isolation and containment, much work remains to be done. In order to properly gather intelligence on the attacker we will need to reorganize our computer network defence organizations, to earmark resources and develop proper operational templates. NCSOs will have training implications for operators and require the development of new TTPs and doctrine. NCSOs will be expensive and hazardous, but we cannot afford to ignore this requirement; doing would prove more expensive and hazardous in the long run.

References

[1] Computer Emergency Readiness Team (CERT), *Steps for Recovering from a UNIX or NT System Compromise*, Online Available: http://www.cert.org/tech_tips/win-UNIX-system_compromise.html, 20 May 2009.

[2] The United States Army, *Army Field Manual FM 3-90*, Department of the Army, Washington, DC , 4 July 2001.

[3] Sweetman, Jack, *The great admirals: command at sea, 1587-1945*, Naval Institute Press, 1997.

[4] Hoglund, Greg, and Butler, James, Rootkits*: Subverting the Windows Kernel*, Addison-Wesley, 2006.

[5] U.S. Department of Defence, *Joint Publication 3-06 Doctrine for Joint Urban Operations*, DoD, 16 Sep 2002.

[6] North Atlantic Treaty Organisation Research and Technology Organisation, *RTO Technical Report 71Urban Operations in the Year 2020 - RTO-TR-071*, NATO, April 2003.

[7] JR02-2009, *Tracking GhostNet: Investigating a Cyber Espionage Network*, Information Warfare Monitor, 29 March 2009 [Online] Available: http://www.f-secure.com/weblog/archives/ghostnet.pdf

[8] Major, Daniel, *Exploiting System Call Interfaces to Observe Attackers in Virtual Machines*, Master's Thesis, Royal Military College of Canada, 2008.

[9] Heywood, Harley, *Investigating Architectural Design Pressures on a Virtual Machine Based Intrusion Surveillance Toolset*, ECE Dept., Royal Military College of Canada, 2009.

[10] M. Dornseif, T. Holz, and C. N. Klein, "Nosebreak-attacking honeynets," *Arxiv preprint cs.CR/0406052*, 2004.

[11] C. K. Tan, \Detecting sebek win32 client," SIG 2 G-TEC -, Jun. 2004. [Online]. Available: http://www.security.org.sg/vuln/sebek215.html

[12] Leblanc, Sylvain Paul, *Toward the Creation of a Synthetic User Environment – An Active Network Defence Enabler*, Depth Research and Doctoral Proposal, ECE Dept., Royal Military College of Canada, January 2008.

[13] Vessey, David and Smith, Pat, *DID-08 Detailed Design Document - ApateX*, ECE Dept., Royal Military College of Canada, 2008.

The Virtual Battlefield: Perspectives on Cyber Warfare
C. Czosseck and K. Geers (Eds.)
IOS Press, 2009
© 2009 The authors and IOS Press. All rights reserved.
doi:10.3233/978-1-60750-060-5-238

Autonomic Computer Network Defence Using Risk State and Reinforcement Learning

Luc BEAUDOIN[a], Nathalie JAPKOWICZ[b] and Stan MATWIN[b]
[a]*Defense Research and Development Canada*
[b]*University of Ottawa*

Abstract. Computer Network Defence is concerned with the active protection of information technology infrastructure against malicious and accidental incidents. Given the growing complexity of IT systems and the speed at which automated attacks can be launched, implementing timely and efficient network incident mitigating actions, whether proactive or reactive, is a great challenge. We refer to the automation of action selection and implementation in this domain as Autonomic Computer Network Defence. In this work, we suggest that Autonomic Computer Network Defence can be achieved using Reinforcement Learning and dynamic risk assessment to learn the optimal action sequence, or policy, to recover from given computer network risk situations. Such a policy could then be used by commercial network management and security products to implement selected mitigating actions automatically, as risk states are sensed.

Keywords. Autonomic Computer Network Defence, Reinforcement Learning, risk, simulation.

Introduction

Autonomic Computer Network Defence (CND) aims to provide a self-protection capability of information technology (IT) networks in order to limit the risk caused by malicious and accidental events. This requires an automated controller with a policy, which selects the most appropriate action in any undesired network state. Due to the complexity and constant evolution of the CND environment, a-priori design for an automated controller is not effective. A solution for generating and continuously improving CND decision policies is needed.

A system capable of achieving autonomic CND must be able to iterate through the CND decision cycle in an automated manner. This cycle typically involves the following steps: sensing network changes, analyzing their impact, selecting an appropriate mitigation action, and implementing this action back onto the network. This process forms a control loop which employs available resources to continuously protect the IT infrastructure. Various commercial products and research prototypes support individual steps of this control loop. However, the search for an adaptive controller design capable of steering IT networks towards an acceptable and stable equilibrium in the face of security events is in its infancy. Related research areas include autonomic computing, autonomic networking and automated security policy management [1] [2].

In our work, we investigated the suitability of different Reinforcement Learning algorithms paired with dynamic risk assessment to form the basis of an autonomic CND controller. We developed an experimental framework, which includes Discrete Event Dynamics System (DEDS) simulation and graph models of the environment, to iterate through CND policies in various scenarios and attempt to minimize business risk. We show that Reinforcement Learning algorithms can learn efficient CND policies. We also show that the difference between policies becomes less significant as the resources available to implement CND actions are increased.

1. Risk in CND Decision Making: The Proactive-Reactive Dilemma

The North Atlantic Treaty Organization (NATO) has defined Computer Network Defence as: "Actions taken through the use of computer networks to protect, monitor, analyze, detect and respond to unauthorized activity within information systems and computer networks"[1]. Generally, CND actions can be implemented either proactively or reactively, and are triggered by metrics such as up/down asset status, security alerts, disclosure of new vulnerabilities, or capacity engineering. An important difference between proactive and reactive actions is the notion of risk. A proactive action aims to reduce the probability **p** of occurrence of a damaging event (**p<1**), whereas a reactive action typically aims to reduce the damage of an actual event (**p=1**). Often, there are situations of conflicting priorities, where resources must be rationed between proactive and reactive actions. An example of such proactive-reactive dilemma would be Network Operations Centre Staff (NOC Staff) needing to decide between first patching a vulnerable system, or fixing a simultaneous outage on another system. To solve this proactive-reactive dilemma, we need a single risk metric which can account for both potential (**p<1**) and actual (**p=1**) incidents. This risk metric would also need to account for the combined effects of time, new security event arrivals and the mitigation actions selected.

This leads to a combinatorial problem whereby risk for every network state, action and event timing is path-dependent. This means that the overall risk exposure of a given situation depends on the entire sequence of actions taken to recover from it.

To illustrate this important concept, a decision tree for a simple scenario with three assets, only one of which being exposed to the Internet, is shown in Figure 1. The first decision is triggered by a new vulnerability on asset 1, which is exposed to external exploit sources. From the resulting decision tree, we show a sample of seven potential risk outcomes, depending on the timing of the first exploit event, the duration of each action, and the mitigation strategy used. Deciding to immediately patch asset 1's vulnerability (node 1) may lead to an optimal path (green arrow starting at node 2, which means that low risk is assumed from this point on). However, the same decision may also lead to the worst path, in the case an exploit occurs before the patching action is completed, and spreads to the two other assets. For this reason, isolating the vulnerable host first, prior to patching it, may be an advantageous decision path. Although isolating asset 1 results in self-denial of service, it also prevents any exploits

[1] NATO publication 3000 TI-3/TT-1162

from using this host to spread to other hosts. Once exploited, an asset has to be "fixed", which typically means restoring a clean disk image on the host.

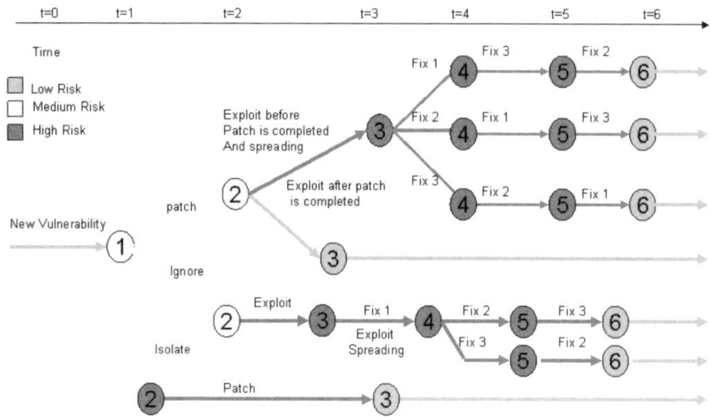

Figure 1. Example of a simple CND decision tree for a new vulnerability event

In more complex scenarios, new vulnerabilities, exploits and outages can occur at any point in time, resulting in new decision branches for each arrival, many of which can be followed concurrently, or serviced through a queue, depending on the number of available resources. If we also consider that assets may have varying levels of business importance, the number of potential outcomes quickly becomes unmanageable. This situation captures the problem we are interested in: given vulnerability, exploit and outage events, an IT network, stated business needs for IT services and limited resources, how can we decide which is the next action to take amongst options of fixing, patching, isolating risk-exposed assets, or simply waiting?

2. Reinforcement Learning for Autonomic Computer Network Defence

In a decision tree such as the one previously introduced, assessing the cumulative risk for every possible branch, in a greedy way, is not a practical strategy. We need to be able to sample this state-action space and steer exploration towards the most promising action selection strategy. This strategy is also known as a policy, and the optimization goal we seek is to minimize business risk over a given time horizon. The search for such an optimal policy is referred to as the Generalized Policy Iteration (GPI) problem, and different approaches have been used to solve it, including Reinforcement Learning [3]. Reinforcement Learning has been successfully used in complex control loop systems, namely in automated packet routing problems [4] and automated server resources allocation problems [1]. One of Reinforcement Learning's benefits is that it can be used online (adaptation to situations as they occur), offline (planning for anticipated situations), on-policy and off-policy. The two last terms refer to whether the learned policy is used to influence exploration (on-policy), or whether the policy iteration is controlled through another mechanism independently, such as random selection, human control, or heuristic rules (off-policy). Finally, Reinforcement Learning can make use of state generalization methods such as function approximators, to scale in continuous state space applications [5].

In Reinforcement Learning, an agent is rewarded after reaching a goal, and this reward is discounted to each preceding action through exploration and policy iterations (also known as *epochs*). Since our objective is to find a policy leading to the minimum risk exposure, we considered both immediate risk reduction, ΔR(t), and the integral of R(t) over the full simulation period, as potential reward functions. We then experimented with various Reinforcement Learning algorithms such as Q-learning, and parameters such as eligibility traces, learning rates, discount factors and random exploration thresholds, to attempt to converge to a desirable CND policy.

3. Autonomic CND Experimentation Framework Architecture

Our experimentation framework is presented in Figure 2. It is broken down into seven main modules, shown in dark blue, and which we describe in this Section.

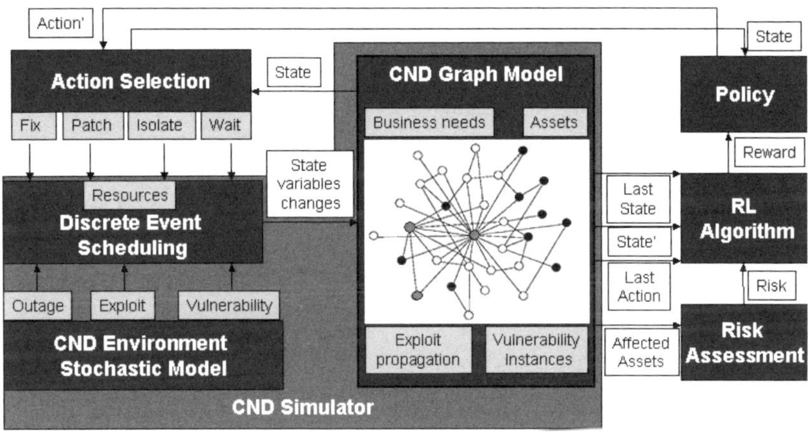

Figure 2. Autonomic CND Experimentation Framework Architecture

Action Selection: For every CND environment state change, the *Action Selection* module searches the *policy* for the best next action, *action'*, to implement given the current *state*. This is performed either through a greedy search or using a *softmax* Boltzman probability driven choice. The resulting *action'* (*fix, patch, isolate* an *asset*, or *wait*) is than passed to a *resource* and scheduled using *Discrete Event Scheduling*.

Discrete Event Scheduling: The scheduling module generates event duration and puts the *action* event in a queue, ordered by time. In some scenarios, it also receives exogenous events from the *CND Environment Stochastic Model*. The scheduling module then advances the simulation clock to the next event in the queue and communicates the associated *State variables changes* to the *CND Graph model*. We implemented this module using the DEDS Java library called ABCMod from University of Ottawa [6], which also supports random seed management and sample dataset collection.

CND Environment Stochastic Model: This module generates arrival times for *outages, vulnerabilities* and *exploits* according to predetermined probability distributions. Leaving the details of these distributions to other forums, we used

Poisson distributions, implemented using the CERN Colt Java library[2], with the following means parameter λ values: 0.0134 vulnerabilities per hour, 0.0093 exploits per hour per vulnerable/exposed host, and 0.00036 outages per hour per host (which includes maintenance related outages). These values were derived from a simple statistical and empirical analysis of incident reports[3] and public data sources[4]. Details can be found in [7] and [8].

CND Graph model: The *CND Graph model* keeps track of the status of each *assets*, their interdependencies, and their support to the *business processes* (needs). This module also enforces the rules for asset status changes (OK, vulnerable, outage or exploited), considering safeguards, actions and exogenous events. This model was implemented using the JGraphT[5] Java library.

Risk Assessment: The *Risk Assessment* module queries the *CND graph Model* for the list of *affected assets*, computes the instantaneous *risk* R(t) and its integral over the simulation run, then passes these scalars to the *RL algorithm*. The risk is updated periodically and considers cumulative effects of potential and actual damages. The dynamic risk assessment algorithm is shown in Eq. (1). Its details are kept to other forums [].

$$R(t) = \sum_{i=1}^{n} \sum_{j=1}^{m} d_i(t) * p_j(t) \tag{1}$$

Where:

- **n** is the number of affected assets;
- **m** is the number of events;
- $\mathbf{d_i(t)}$ is the damage function incurred by the business at time **t** for asset **i;**
- $\mathbf{p_j(t)}$ is the likelihood of occurrence of an exploit for a vulnerability event **j** at time **t**; 1 otherwise.

RL algorithm: Before each *action* implementation is completed, the *RL algorithm* queries the *CND Graph Model* for the current *state*. After the *action* implementation, the RL module queries the Graph Model for the new *state'* and updates the *policy* with these quantities and the associated *reward* ΔR(t), or the integral of R(t), depending on the training strategy. This module is implemented using the QConnectionist [9] Java package[6], and University of New South Whales Reinforcement Learning Java package by Time Eden, Anthony Knittel and Raphael van Uffelen[7] .

Policy: The *policy* updates its state-action map with *rewards* and *states* received from the RL algorithm. It also provides the *action selection* module with the preferred action for a given state. This was implemented using the neural network provided in the QConnectionist framework [9] for the state generalization policy, and standard Java vectors objects for the table policies.

---

[2] The Colt Java library is available at http://acs.lbl.gov/~hoschek/colt/
[3] Trouble Tickets recorded at the Canadian Forces Network Operations Centre
[4] Primarily from the National Institute of Standards and Technology (NIST)
[5] JGraphT is an open source project available at http://jgrapht.sourceforge.net/
[6] The QConnectionist source code is available at http://www.elsy.gdan.pl/
[7] Source available at: http://www.cse.unsw.edu.au/~cs9417ml/RL1/applet.html

4. Experiment

We implemented a simple CND environment model, as shown in Figure 3. It includes eleven nodes regrouped under four interconnected sites in order to create multiple service paths. One site hosts a DNS server, one provides access to the internet gateway (Router-4), and two other sites host each an email server and a client. The business needs include email communications between Email-2 and Email-5, as well as browsing the internet from Browser-3. Functional dependencies exist between the email clients and the email servers, as well as between the browser and the DNS. These business needs and functional dependencies form logical dependencies between assets, shown by orange arcs. Using this model, we ran five simulation experiments: fixing four concurrent outages, fixing eleven concurrent outages, patching eleven concurrent vulnerabilities, patching a vulnerability or fixing exploits, and patching-fixing-isolating-waiting in a continuous event arrival simulation. We used five different policies for each experiment: reinforcement learning with a table and a neural network, as well as heuristic policies including *fixing or patching assets randomly*, *fix or patch the asset with the highest value first*, and doing nothing.

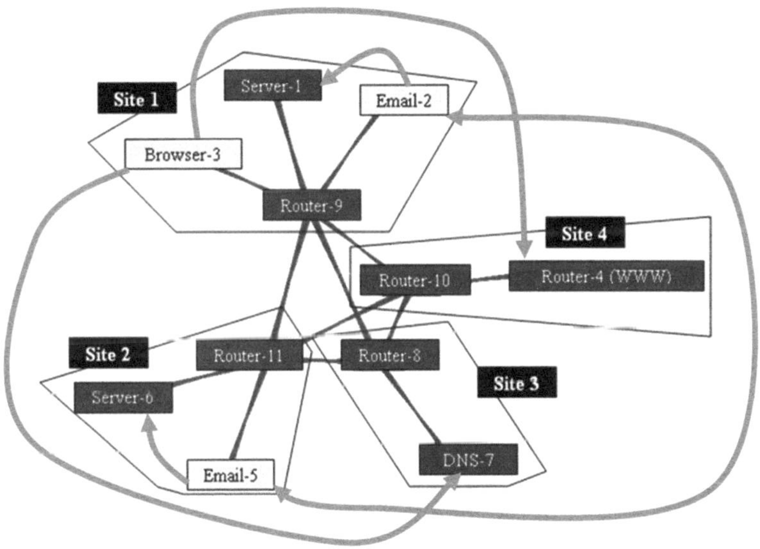

Figure 3. Simple CND environment with eleven inter-dependant assets

For each simulation, random number generator seeds for DEDS event timings were managed to assure independence of results, avoid over fitting local timing conditions and allow policy comparisons.

5. Results and Discussion

Prior to running simulations, we computed the contribution of each asset to the stated business needs. This was accomplished using an asset valuation algorithm based on the ratio of business-enabling network paths supported by each asset described in [8]. In our simple CND environment, we found sixty-six paths through greedy, depth-first, search. The resulting asset values, presented in Table 1, were later used as damage metrics by the dynamic risk assessment module.

Asset name	Asset Value	Stated business needs
Server-1	0.7	
Email-2	0.7	0.3
Browser-3	0.3	0.3
Router-4 (WWW)	0.3	
Email-5	0.7	0.4
Server-6	0.7	
DNS-7	0.3	
Router-8	0.72	
Router-9	1.0	
Router-10	0.72	
Router-11	0.88	

Table 1. CND environment asset value results

We then conducted simulation runs for each scenarios, which we repeated in the form of epochs to train our RL policies. In the first case, the learning task was to fix four concurrent asset outages in an optimal sequence to minimize risk (risk equal damages in this case, since **p=1** for outages). Both RL policies converged to the same risk performance, which was lower than the three other heuristic policies, as shown in Table 2.

Policy	Integral of R(t)
Let risk grow	41.4
Fix random	31.73438
Fix highest	17.5375
Q-Connectionist	16.3825
Q-Learning RL Table	16.3825

Table 2. Risk integral results of different policies for fixing four concurrent outages.

A sample of the exploration of the solution space by the RL agent can be seen in Figure 4. The agent initially explored, rather randomly, various actions leading sparsely distributed risk results, than converged to action sequences optimizing its reward (minimizing risk in this case). We notice a trend around 32, which is the result for the random action sequence used for exploration. Any results larger than 32 were caused by choosing "wait" actions. These choices were punished through the risk reward and became less frequent as training progressed. We can also observe a secondary periodic value after convergence at around 17. This value corresponds to *fix or patch the asset with the highest value first*, which is an expected equilibrium since asset values contribute largely to the risk rewards for this scenario. The upper bound of this graph is approximately 41 and corresponds to *waiting* for the entire simulation period.

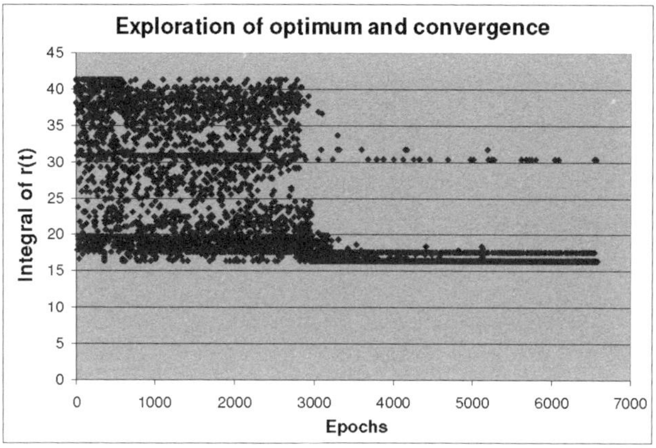

Figure 4. Exploration and convergence pattern for a neural network Reinforcement Learning agent

The other scenarios trialed had significantly larger solution spaces and statistical analysis was required to analyze the results. For this purpose we used Chi-Square metric with the Student distribution and we adjusted the number of runs to maintain our results within approximately 10%, as a quality measure. In all scenarios, the policy learned by the Reinforcement Learning agents achieved lower risk in a statistically significant way when compared to the random policy. These results are shown in Table 3.

Scenario	Random (avg risk)	Q-Learning Table policy (avg risk)	Improvement (avg risk)	QConnectionist Neural Network policy (avg risk)	Improvement (avg risk)
1. Fix 4 outages or wait.	31.73	16.38	**-15.35**	16.38	**-15.35**
2. Fix 11 outages	17.38	16.45	**-0.93**	13.42	**-3.96**
3. Patch 11 vulnerabilities	1.87	1.60	**-0.27**	1.77	**-0.10**
4. Patch 1 vulnerability or fix exploit	1.88	1.61	**-0.27**	1.81	**-0.07**
5. Patch, fix, isolate, or wait with continuous event arrivals	18933	18402	**-531**	18327	**-606**

Table 3. Reinforcement Learning policies results compared against the random policy.

In Figure 5, we show a sample simulation run for scenario 2, where 11 concurrent outages had to be fixed in an optimal sequence. The graph presents various decision points and the risk function R(t) for the five trialed policies. Because the random seed is the same for all actions timing in this case, we observed that some policies were more effective at finding root-cause outages early, hence lowering their overall risk score.

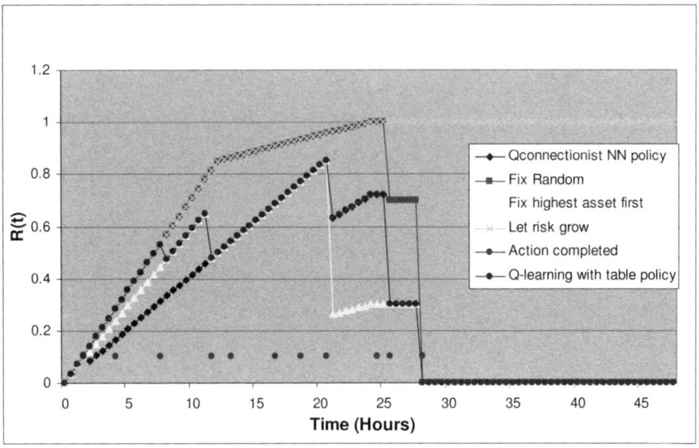

Figure 5. Policy comparison for fixing eleven outages in a single simulation run.

Over a 10-year continuous simulation, we observed that policies may be locally optimal for risk, but globally very poor. Figure 6 shows the last 15 days of such a simulation run. The turquoise fill represents the random policy area under R(t). Although, the random policy had the highest risk over the full simulation period, it achieved local optimums in regions marked by the red dashed ovals in the graph.

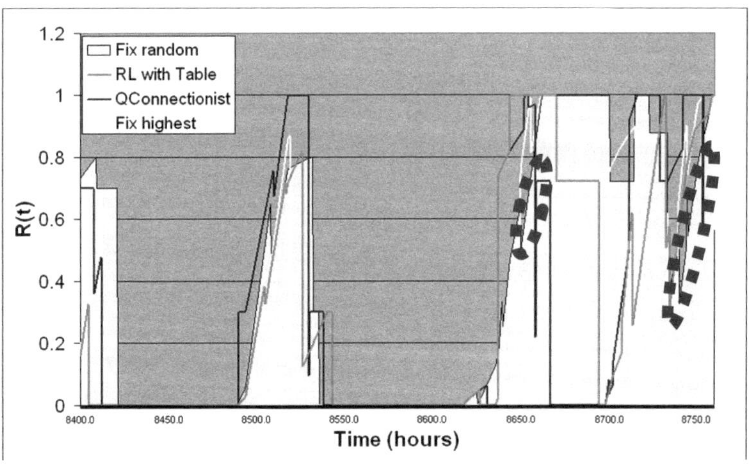

Figure 6. Comparison of policies over fifteen days of a 10-year continuous simulation.

In our experiments, the Reinforcement Learning agents seemed capable of learning generally good policies, but they could not account for all possible situations caused by the various event arrivals and action implementation delays. Even for our simple CND environment, in order to achieve globally optimal policies, more information may have been required (using a different CND state representation, as an example) as well as different learning strategies (more epochs and different parameters).

We were finally interested in evaluating the effect of resources on risk and policies. We conducted a number of 1-year simulation runs, with increasing numbers of NOC staff resources, for two different policies: *fixing or patching assets randomly* and *fixing or patching the highest asset value first*. The averaged results are shown in Figure 7.

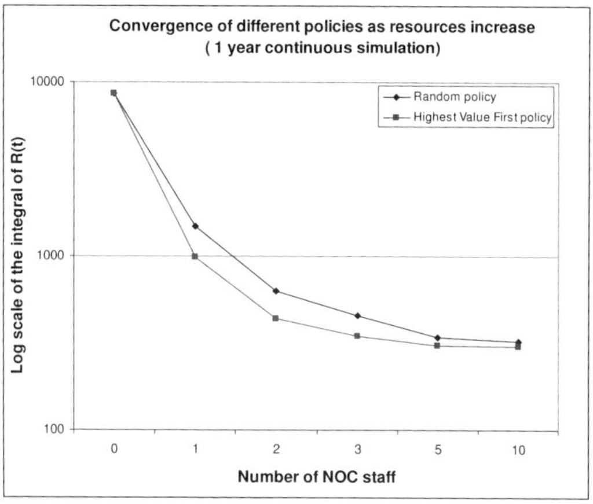

Figure 7. The effect of resource availability on overall risk.

We can observe that as the response capacity (number of NOC staff) is increased, not only is the integral of R(t) decreasing, but so is the difference between both policies[8]. Indeed, if there are no resource constraints, there is no need to prioritize responses since there is no event queue, and decision-making becomes essentially trivial. This observation clearly supports the value of automation in CND.

6. Conclusion and Future Work

In this research effort, we applied Reinforcement Learning to the problem of finding an optimal policy for Autonomic Computer Network Defence. We argued the need for a controller able to dynamically iterate through various policies and retain the best performing one. We have shown that risk maybe a good metric to steer such a controller, as long as it accounts for actual and potential events. We presented our experimentation framework and validated our concept using a simple CND environment and five scenarios. Our results show that Autonomic CND using risk states and Reinforcement Learning is possible, but that policies obtained, although generally good, did not represent global optimums.

As future work, we propose investigating further dynamic risk assessment algorithms and methodologies. We also suggest investigating different Computer Network Defence state representations, for use in conjunction with Reinforcement Learning agents herein tested, to see if better policies can be obtained. Namely, we propose investigating text mining techniques, including feature extraction, and consider modeling the CND environment as a "*bag-of-words*" to leverage these techniques. Finally, we suggest looking into scalability issues, namely investigating distributed

[8] Note that the upper bound of Figure 7 represents having no response capacity, which makes all policies equivalent by default.

policies and the use of Collaborative Reinforcement Learning to achieve superior risk results and stability.

References

[1] Tesauro, *Reinforcement Learning in Autonomic Computing*, IBM T.J. Watson Research Center, IEEE 2007.

[2] Benjamin, Pal, Webber, Atighetchi, Ruber, *Automating Cyber-Defense Management*, ACM Workshop on Recent Advances in Intrusion Tolerant Systems, 2008.

[3] Sutton, Barto, *Reinforcement Learning:An Introduction*, MIT press, 1998.

[4] Boyan, Littman, *Packet Routing in Dynamically Changing Networks: an RL approach*, Advances in Neural Information Processing Systems, Morgan Kaufmann, San Francisco CA (1993), volume 6, 671-678.

[5] Sutton, McAllester, Singh, Mansour, *Policy Gradient Methods for Reinforcement Learning with Function Approximation*, Advances in Neural Information Processing Systems 12, 2000.

[6] Birta, Arbez, *Foundation on Modeling and Simulation*, University of Ottawa, 2006.

[7] Alhazmi, Malaiya, *Quantitative Vulnerability Assessment of Systems Software*, Reliability and Maintainability Symposium, 2005.

[8] Beaudoin, Japkowizc, Matwin, *Autonomic Computer Network Defence Using Risk States and Reinforcement Learning*, Thesis manuscript to be submitted, University of Ottawa, 2009.

[9] Kuzmin, *Connectionist Q-Learning in Robot Control Task*, Riga Technical University, 2002.

[10] Cao, *From Perturbation Analysis to Markov Decision Processes and Reinforcement Learning*, DEDS: Theory and Application, 2003.

[11] Chairman of the Joint Chiefs of Staff Instruction, *Information Assurance and Computer Network Defense*, US DoD, 2004.

[12] Dobson, Denazis, Fenandez, Gaiti, Gelenbe, Massacci, Nixon, Saffre, Schmidt, Zabonelli, *A survey of Autonomic Communications*, ACM Autonomous and Adaptive Systems, Vol. 1, No. 2, 2006.

[13] Kotenko, *Multi-agent Modelling and Simulation of Cyber-Attacks and Cyber-Defense for Homeland Security*, IEEE International Workshop of Intelligent Data Acquisition and Advanced Computing Systems, 2007.

[14] Lefebvre, Grégoire, Froh, Beaudoin, *Computer Network Defence Situation Awareness Information Requirements*, MILCOM 2006.

[15] Moitra, Konda, *The Survivability of Network Systems: An Empirical Analysis*, CMU SEI, 2000.

[16] Ryan, *iWar: A new threat, its convenience – and our increasing vulnerability*, NATO review, 2007.

The Virtual Battlefield: Perspectives on Cyber Warfare
C. Czosseck and K. Geers (Eds.)
IOS Press, 2009

© 2009 The authors and IOS Press. All rights reserved.
doi:10.3233/978-1-60750-060-5-249

Enhancing Graph-based Automated DoS Attack Response

Gabriel KLEIN [a], Marko JAHNKE [a], Jens TÖLLE [a,1]
Peter MARTINI [b]
[a] *Research Institute for Communication, Information Processing and Ergonomics (FGAN-FKIE), Germany*
[b] *Insitute of Computer Science IV, University of Bonn, Germany*

Abstract. Timely and appropriate reactions to detected denial-of-service attacks against computer networks are crucial in both civilian and military settings. GrADAR is an intuitive graph-based approach for assessing the effects of DoS attacks against computer networks so that response measures can be automatically selected without human intervention. However, GrADAR has limitations insofar as implicit effects of countermeasures are only taken into account by propagation towards user nodes. Possible effects in the other direction are only considered if they are explicitly specified. For this, they need to be exactly known in advance which is often infeasible. This contribution presents an extension to GrADAR, in which we consider resource workload and processing capabilities and their effects on resource availability. We incorporate workload measurements into the GrADAR model which are done by passive analysis of network traffic. We further augment the active availability probes with passive measurements. This ensures more accurate availability values because additional measurement traffic that might falsify results only needs to be injected when resources are currently not accessed.

Keywords. denial-of-service attacks, automated response, response evaluation, passive availability measurement

Introduction

In recent years, the number of attacks against computer networks has steadily increased. They cannot only be observed in civilian scenarios (e. g. e-commerce or online banking) but also in military settings. Among these attacks, denial-of-service attacks are the most prevalent, often resulting in inaccessibility of services and/or entire networks. This can result in enormous financial losses, in case of commercial applications, and negatively impact battle readiness where military networks are concerned.

In typical wired networks, such attacks can be detected with a high degree of accuracy by incorporating intrusion detection systems into the networks' perimeter defence. Once detected, security personnel can then suitably react to these attacks. However, because of the increased speed and reliability, it is desirable to react to detected attacks in an automatic fashion. For the automatic selection of response measures, it is necessary

[1]Corresponding Author: Gabriel Klein, FGAN-FKIE, Neuenahrer Str. 20, 53343 Wachtberg, Germany; E-mail: `g.klein@fgan.de`

to quickly and accurately estimate the effects of the respective countermeasure on the network resources.

In previous work, we have proposed GrADAR [1,2], an intuitive approach to create and maintain a model of a computer network and the availability of its resources from the observations of deployed monitoring systems. The graph-based model is able to express both the effects of DoS attacks and the results of available response measures prior to their application in the real-world network. Thus, the approach provides a methodology for automatically selecting response measures to detected attacks. The most appropriate response is chosen based on metrics which are well-known from the pragmatic view of network security officers.

This contribution proposes an extension to our previous GrADAR approach that seeks to incorporate the effects of network and resource workload into the availability estimation. This will permit a more detailed modelling of the current network state. Further, it will allow the specification of the effects of more complex DoS countermeasures. To further improve the availability measurements and reduce the workload placed on the network and resources, passive measuring is employed instead of active probing.

The rest of this paper is structured as follows: section 1 introduces related work done in the area of passive measurements. Subsequently, section 2 gives an overview of the GrADAR approach as well as some of its limitations. Section 3 presents the proposed GrADAR extensions, describing workload and the measurement framework in more detail. Section 4 portrays first results and, following that, section 5 presents a summary and an outlook on future work.

1. Related Work

The area of passive network analysis has been a research area of interest for a number of years. In the context of this contribution, approaches concerned with the inference of server workload and the identification of network flows are of particular interest.

In [3], Barford and Crovella describe an architecture for actively and passively measuring the effects of Web server and network workload on the quality of client connections. They show that increasing network load causes a deterioration of connection quality. However, they observe a positive effect of server load on traffic characteristics and attribute this to a reduced burstiness of traffic.

Eriksson et al. [4] reiterate a methodology for determining the network structure by passive measuring of hop counts. They address the issues of missing hop count data and how to extract topological information from large hop count databases.

Passive measurement is also a popular method for determining the characteristics of network connections such as bandwidth, latencies or packet loss statistics. Seshan et al. [5] propose SPAND, a system in which sensors distributed throughout the network perform passive measurements and report to central performance servers where the data is collated. In [6], Lowekamp presents a report on the Wren project which deals with the development of network performance monitoring solutions. Here active and passive measurements of traffic statistics are combined to reduce the amount of artificial traffic wherever possible.

To passively determine availability and workload, observed traffic needs to be associated with the resources between which it flows. Distinct flows can, for example, be identified using NetFlow [7], but correlations between within the same or different flows also need to be ascertained subsequently. This is often done by simple payload inspection, although this potentially requires large amounts of memory. However, recent approaches for the identification of upper layer protocols include machine learning [8,9] and multi-scale gamma models [10].

2. GrADAR Overview

In the GrADAR approach, a simplified model of the real-world network is created in order to predict the effects of available response measures against denial-of-service attacks. Figure 1 shows an overview of the approach.

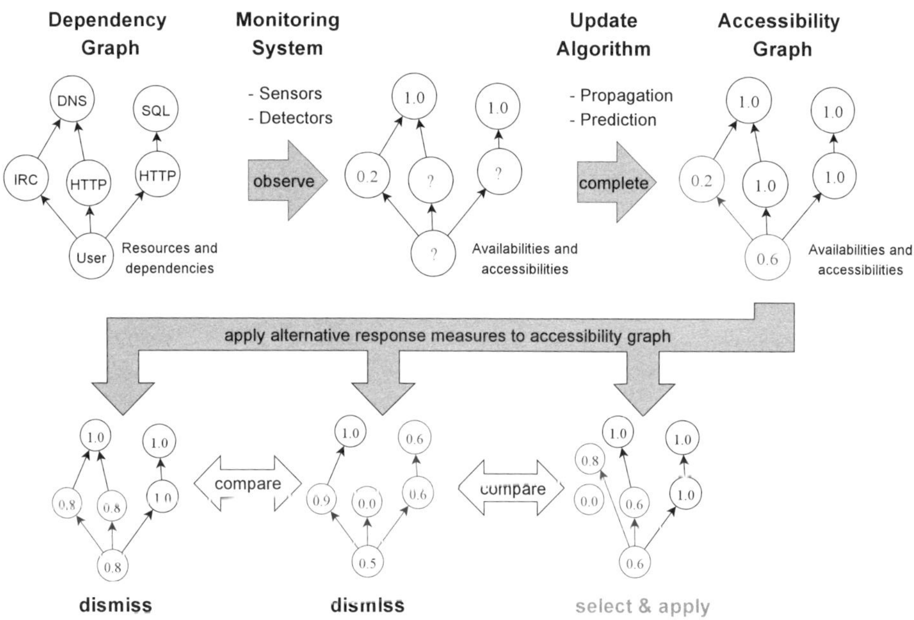

Figure 1. Schematic overview of the GrADAR approach.

2.1. Nomenclature

The core concept of GrADAR is based on the *availability* of *resources*. Resources (as suggested in [11]) can be either services provided by hardware or software components (denoted as $\mathcal{S}$), or users (denoted as $\mathcal{U}$). Therefore, the set of resources is $\mathcal{R} = \mathcal{S} \cup \mathcal{U}$.

Each resource r has an associated value $A(r) \in [0, 1]$, signifying the extent to which it is available to other resources. In [12] and [13], the concept of resource-typical transactions was proposed. We adopt this concept and define a resource's availability as the time needed for a transaction with the resource. Since a resource typically requires interaction with other resources to function correctly, we assume that its availability is the result of two independent factors, an internal state (the *intrinsic* availability) and the values of other associated resources (the *propagated* availability). Thus, a resource's availability is defined as

$$A(r) = A_I(r) \cdot A_P(r) \tag{1}$$

for each $r \in \mathcal{R}$.

A resource r may be dependent on other resources $s_1, \ldots, s_n$ (denoted as $r \rhd s_1, \ldots, r \rhd s_n$). In this case, the degree to which r depends on each of these may vary [14,15] and can be specified by weighting the respective dependency. This can be formalised as

$$A_P(r) = D_r\left(w_{r,s_1}\left(A\left(s_1\right)\right), w_{r,s_2}\left(A\left(s_2\right)\right), \ldots, w_{r,s_n}\left(A\left(s_n\right)\right)\right), \tag{2}$$

where $D_r : [0,1]^n \to [0,1]$ is a dependency function and $w_{r,s_i} : [0,1] \to [0,1]$ are corresponding dependency weighting functions. In optimal conditions,

$$D_r\left(w_{r,s_1}(1), \ldots, w_{r,s_n}(1)\right) = 1.$$

A more detailed discussion of this can be found in [2].

2.2. Dependency Graph

To represent the availability dependency relationships between the set of resources $\mathcal{R}$, the resources in the real-world network are modelled as a directed acyclic graph $\hat{G} = (\hat{V}, \hat{E})$ with $\hat{V} \subseteq \mathcal{R}$ and $\hat{E} \subseteq ((\mathcal{S} \cup \mathcal{U}) \times \mathcal{S})$. Its vertices correspond to the resources and the edges correspond to the dependency between the respective resources. These resource dependencies need to be determined beforehand, either analytically or experimentally. $\hat{G}$ contains an edge (r, s) iff $r \rhd s$. These edges are labelled with the corresponding weighting function $w_{r,s}$. This graph is called the *dependency graph* and reflects the ideal state of the network. Let $\hat{\mathcal{G}}$ be the set of all possible dependency graphs.

2.3. Accessibility Graph and Overall Availability

A DoS attack typically affects the availability of resources. Thus, there is the possibility that some resources might no longer be accessible to others. Therefore, a second graph is required, the so-called *accessibility graph*. Mutual accessibility of a set of resources $\mathcal{R}$ is expressed by a graph $G = (V, E)$ with $V \subseteq \mathcal{R}$ and $E = ((\mathcal{S} \cup \mathcal{U}) \times \mathcal{S})$, in which an edge (r, s) exists when a resource s is directly accessible from r. The vertices $r \in \mathcal{R}$ of the accessibility graph are labelled with the corresponding resource's availability $A(r)$.

The availability of user nodes is interpreted as the user-perceived availability of the network. Since the network supports one or more users or groups of users in conducting a common mission, we define the overall availability of the network as the weighted average of all user nodes' availability values:

$$A(G) := \sum_{u \in \mathcal{U}} m(u) \cdot A(u),$$

where $m(u)$ is the *relative importance* of user u to the common mission which needs to be determined beforehand or adaptively, and $\sum_{u \in \mathcal{U}} m(u) = 1$. Let $\mathcal{G}$ be the set accessibility graphs.

Usually, monitoring systems deployed in the network will only be able to observe availability values for some of the network's resources. This is especially true for users, for which an availability cannot be objectively measured. Thus, the availability of resources for which values cannot be observed need to be estimated. For a resource r with $r \rhd s_1, \ldots, r \rhd s_n$, this estimation is done by propagating the availability values of the resources $s_1, \ldots, s_n$ in the inverse direction of the corresponding dependency relationship expressed in the dependency graph, i. e. along the edges $(s_1, r), \ldots, (s_n, r)$, and then calculating $A(r)$ according to equations (1) and (2) with $A_I(r) = 1.0$. This can be efficiently done, for example, with a depth-first search algorithm, starting from the user vertices and terminating at vertices with $deg_{\text{out}}(r) = 0$.

As opposed to the dependency graph, the accessibility graph shows the actual current state of the network.

2.4. Response Selection

Once an attack has been detected, an appropriate reaction should be selected automatically. With the current dependency graph $\hat{\mathcal{G}}$ and the current availability graph $\mathcal{G}$, we define a *countermeasure* or *response measure* as a transformation

$$\theta : \hat{\mathcal{G}} \times \mathcal{G} \rightarrow \hat{\mathcal{G}} \times \mathcal{G}.$$

The dependency graph $\hat{G}' = (\hat{V}', \hat{E}')$ is obtained from $\hat{G}$ by adding or removing vertices or edges, and the accessibility graph $G' = (V', E')$ is derived from G by also adding or removing vertices or edges but, additionally, vertex availability values can be changed. Let the set of available countermeasures be denoted as Θ.

As response measures may be arbitrarily complex in nature, we assume that a single response measure θ can be divided into $N_\theta \subset \mathbb{N}^+$ successive atomic *response steps* $\theta^{(i)}$:

$$\theta = \theta^{(1)} \circ \ldots \circ \theta^{(N_\theta)}.$$

Each of these real-world response steps $\theta^{(i)}$ corresponds to one of the graph transformation primitives mentioned above (adding/removing vertices or edges, setting availability). The effects of such a response step can be either directly associated with the action (*explicit impact*) or a result of changes in the environment due to the response step application (*implicit impact*).

For the automatic response measure determination, each available countermeasure $\theta \in \Theta$ is now applied in parallel to the current accessibility graph. For each change in the accessibility graph, the propagation algorithm mentioned above needs to be executed. The resulting graphs (so-called *response graphs*) are then compared with respect to different metrics, e. g. *expected response success* or *expected response costs* [2], and the most appropriate response is then applied to the real-world network.

The resulting dependency graph of one such GrADAR cycle is used as the input for the subsequent iteration. Thus, the process constitutes a so-called *closed-loop control system*.

2.5. Current Limitations

During the validation of the GrADAR approach it became apparent that correctness and robustness could only be achieved if the precise effects of real-world countermeasures in the graph space could be accurately predicted. Because of the closed control loop structure of the approach, subsequent iterations of the loop would operate on incorrect input.

So far, the response measures for which effects were specified consisted only of blocking access to specific resources (e. g. closing a firewall port) or migrating resources to different locations (e. g. installing a new server to replace another). The effects of these operations on the graph could be specified fairly easily in terms of changed availability values or changing edges in the graph.

However, more complex behaviour of resources, such as the interactions between workload placed on a resource and its processing capacities and their effect on that resource's availability, are only expressible if fairly well known and specified in advance. This is because the effects are taken into account by propagating them through the accessibility graph according to the update algorithm. This imposes constraints in that effects can only be propagated in the reverse direction of the dependency relationship. Effects on resources in the other direction cannot currently be expressed. For example, blocking a DoS attack against a Web server at a firewall port would have the explicit impact of a 0.0 availability at the firewall port. This would be propagated to the (dependent) user node. However, a possible implicit impact could be an increased availability of the Web server. This is currently expressible only if the changes in availability are known beforehand and "hard-coded" into the graph transformation, something which is often impossible. It is desirable to predict such effects on resource availability dependent on the current network situation.

Furthermore, availability measurements are currently performed only through active probing, i. e. by sending requests to the respective resources and measuring and normalising the time required for an answer. This poses two problems. First, the measuring process itself produces workload for the network and the target resource, and second, it requires the measuring process to produce a traffic pattern which is representative for the specific resource.

3. Beyond GrADAR: Improving Availability Estimation

Due to the limitations recounted in section 2.5, we propose to enhance the GrADAR approach by incorporating the effects of workload and resource capacity into the graph-based model and avoiding active probing in favour of passive availability measurements. We believe that through the enhancements described in this section the correctness and robustness of the approach can be significantly improved.

3.1. Solution Idea: Propagation of Workload

To more accurately represent the graph-based equivalents of DoS attack response measures, we propose to incorporate the workload placed on the various resources into the GrADAR model. For this, the relationships between resource workload and its availability need to be determined, especially in the area of a resource's processing capacity. Additionally, dependency relations between the workload of different resources should be investigated. If such relationships exist, this could allow the inference of workload values for resources of which workload cannot be directly measured. Similar to availability estimation, this could be done by propagation according to the workload relationships; see figure 2 for an example. Here, two users (or user groups), `User_1` and `User_2`, communicate with an HTTP server, `HTTP_S1`, via two separate firewall ports, `HTTP_F1` and `HTTP_F2`. In figure 2(a), the workload generated during a DoS attack by `User_1` is propagated (in the direction of the dependency relationship) to the firewall port and the HTTP server. Because of this workload, the HTTP server has an availability of 0.0 which, in turn, is propagated (against the dependency relationships) to the resources dependent on it. An example response to such an attack, namely blocking the attacker at the firewall, is depicted in figure 2(b). As a result, the workload generated by `User_1` is no longer propagated to `HTTP_S1` which causes an increase in the server's availability. This, in turn, is propagated to `User_2`.

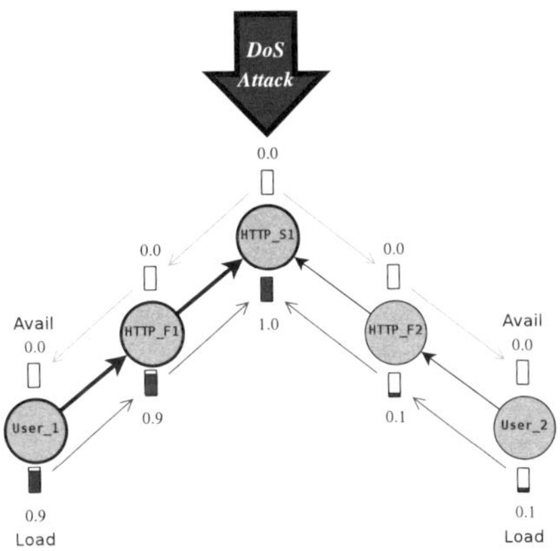

(a) Workload-based DoS attack

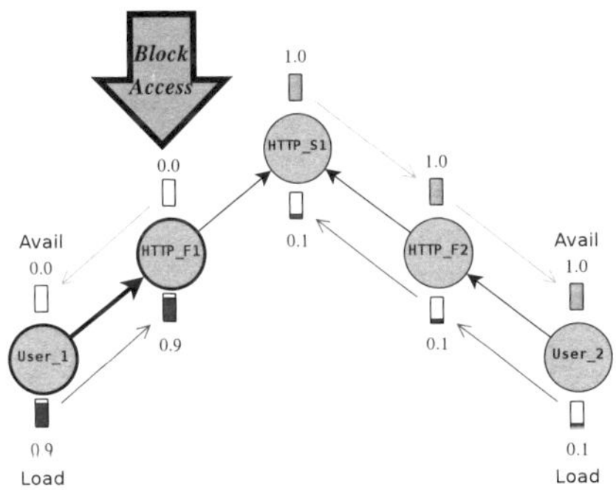

(b) Example response to DoS attack.

Figure 2. Example of workload propagation in the GrADAR model.

3.2. Workload Definition

Before the workload $L(r)$ of a resource $r \in \mathcal{R}$ can be effectively measured, it needs to be defined in a suitable fashion. The dictionary defines workload as "the amount of work assigned to, or done by, a worker or unit of workers in a given time period" (The American Heritage Dictionary, 2nd Edition). Similar to the definition of availability as the normalised duration of a resource-typical transaction (see e. g. [12,13]), a resource's workload can thus be defined as the number of typical transactions a resource needs to process per unit of time. Since a networked application scenario contains multiple types of resources, different types of "work" need to be considered as each resource has transactions that are typical for it; for example, the number of concurrent transactions a Web server needs to process. Table 1 contains a listing of possible resources along with the availability and workload definition for each of them.

Table 1. Workload definitions for selected resources.

Resource	Availability definition	Workload definition
IP stack	ICMP ping response time	IP packets/time
CMS	Delay for receiving backend content	Current active transactions
IRC server	Delay for connection, joining channel and sending a message	Current active transactions
DB server	Delay of query from Web server backend	Current no. of transactions
DNS server	Delay for result of lookup query	Requests/time
MAC layer	Interface up/down	Frames/time
CPU	Execution delay for application requiring CPU/memory/HDD	Average CPU load
Memory		Average memory consumption
HDD		Average consumed HDD capacity

To adequately compare the workload of different types of resources, workload values need to be normalised:

$$\tilde{L}(r) = \frac{L(r)}{L_{max}(r)}, \quad (3)$$

where $L_{max}(r)$ is the maximum workload which a resource can adequately process within a certain time frame. This is closely related to the definition of availability (c. f. [1,2]) where request-response delays are normalised with respect to a maximum acceptable time from a user's perspective.

3.3. Measurement Framework

The measurement of availability and workload is performed according to the framework outlined in the SDL [16] diagram depicted in figure 3.

Passive sensors at appropriate locations in the network (e. g. at central switches or a firewall) constantly observe passing traffic. Different *conversations* between consumers and providers of a service, and the *transactions* they comprise are identified by analysing packet headers and correlating certain fields, e. g. sequence numbers, IP addresses or port numbers. For each recognised transaction, various properties of the traffic such as average packet loss, transaction duration, average round-trip time, jitter, etc. can be used to evaluate the availability of a resource; c. f. [12] and [13] for details of how traffic properties can be used to make quantitative statements about the quality of a service. Thus, a resource's availability value can be updated after each completed transaction.

By logging the number of transactions for different services and/or protocols, these sensors can also establish a current workload for the observed resources. As already mentioned (c. f. table 1), the number of typical transactions within a specified period of time (or possibly the ratio of initiated vs. completed transactions) can serve as a workload metric.

The goal of our work is the identification of current threats to a network and the ability to react in near-real-time. Bearing this in mind, it seems advisable to consider the development of resource availability and workload over a customisable period of time rather than only the currently measured values. Using this as a base for decisions may reduce the likelihood of overreactions or false positives.

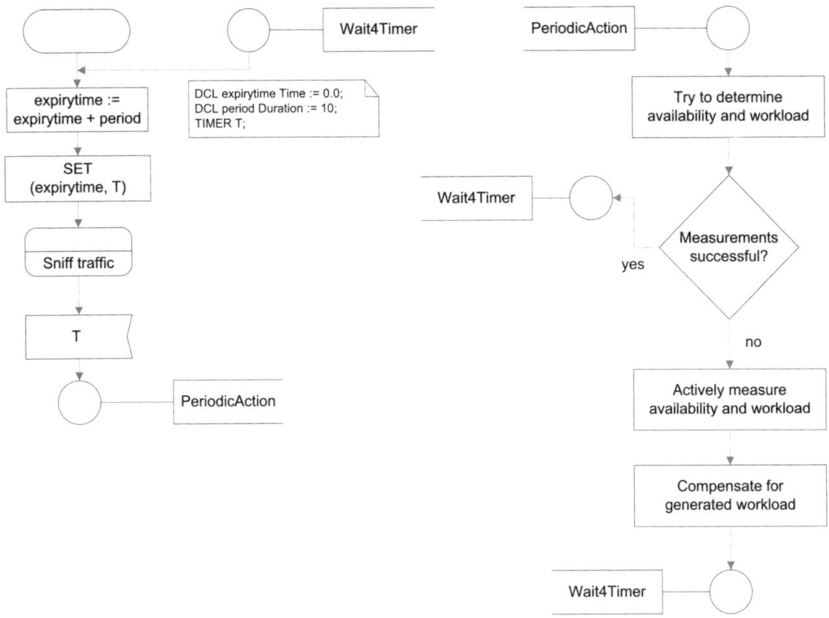

Figure 3. SDL diagram of the measurement framework.

The aforementioned description of passive measurements assumes that representative traffic for all relevant resources can always be observed. This may not be the case in real-world configurations, e. g. if clients are under EMCON in military settings. Thus, to always retain an overview of the current network status, active measurements need to be performed if no traffic was observed for a certain amount of time. In the case of availability, this is done via an active probe which consists of a (representative) request to the respective resource. The resulting availability is the normalised duration of this request. The resources' workload needs to be queried directly via appropriate interfaces at the resources themselves (e. g. via SNMP [17]), although this introduces a certain degree of dependency on specific applications and protocols. Note that, when performing active measurements, the load generated by these artificial requests needs to be taken into account and compensated when determining resource workload.

We are aware that both active and passive measurements have disadvantages. In the case of active probing this is the injection of additional traffic into the network resulting in increased workload for the targeted resources. On the other hand, when observing traffic passively, the volume of traffic to be dealt with is potentially prohibitively large. We try to avoid this by restricting ourselves to the analysis of only packet headers instead of entire packets.

4. Preliminary Results

We have performed first simple workload and availability measurements in a network topology depicted in figure 4. A client accesses a content management system (CMS) which, in turn, retrieves its data from a database server on a different host (scenario 1). In a second measurement scenario (scenario 2), both the CMS and the database are located on the same host. In both cases, the servers are separated from the client by a firewall host. The servers hosting the Web and database servers are older-model single core machines and are thus not able to process requests entirely in parallel.

The values shown in figures 5 and 6 were obtained by generating an increasing number of concurrent requests per time frame. For each number of concurrent requests, the measurement process was repeated 300 times. The mean request-response durations were used as the basis for the availability calculations.

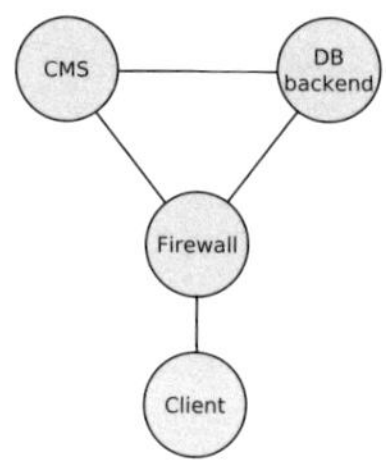

Figure 4. Server setup during measurements.

Figure 5 shows the measurement results for scenario 1. The normalised CMS workload and availability are plotted against an increasing number of concurrent requests directed at the Web server. At first, the CMS availability degrades linearly with the increase in workload. Beyond around 15 concurrent requests, an overload situation is entered, in which the availability remains at zero (shown as triangle-shaped data points in the diagram). At this point, the server cannot process requests within an acceptable time frame.

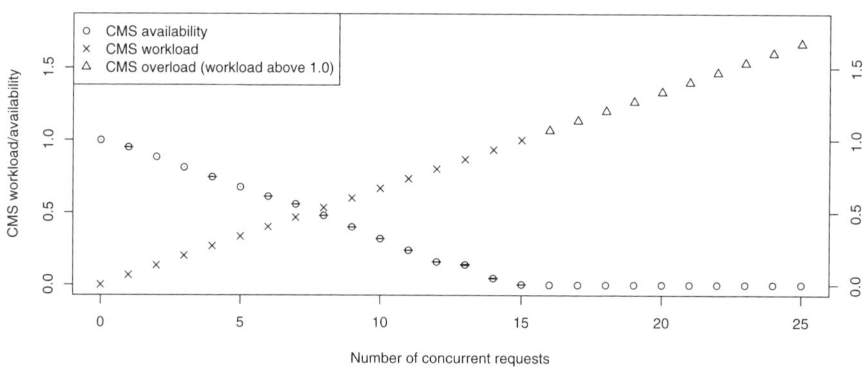

Figure 5. CMS workload and availability plotted against increasing number of concurrent requests; CMS and DB server on separate hosts.

In the second scenario, where the DB server is on the same host as the CMS server, the CMS availability degrades slightly more quickly (depicted in figure 6). This is most probably because both server processes share the same system's resources. Load processed by the CMS is partially transferred to the database, which, in turn, reduces the

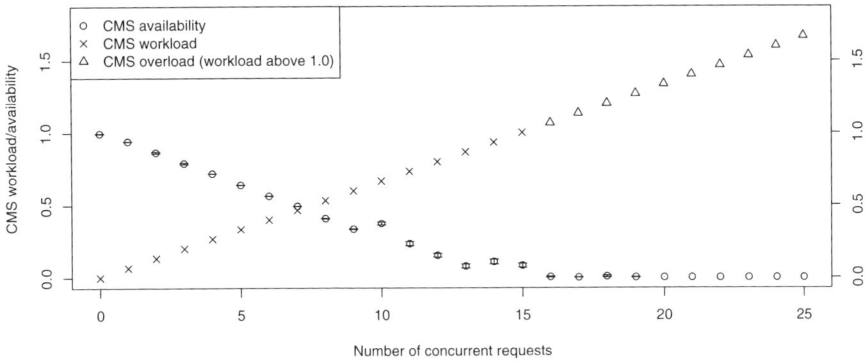

Figure 6. CMS workload and availability plotted against increasing number of concurrent requests; CMS and DB server on the same host.

available resources for the CMS. Also, the availability curve is less smooth. The small confidence intervals suggest that the reasons for the outliers are systematic. They could, for example, be caused by changes in scheduling policies. This is also true for the outlier observed in scenario 1.

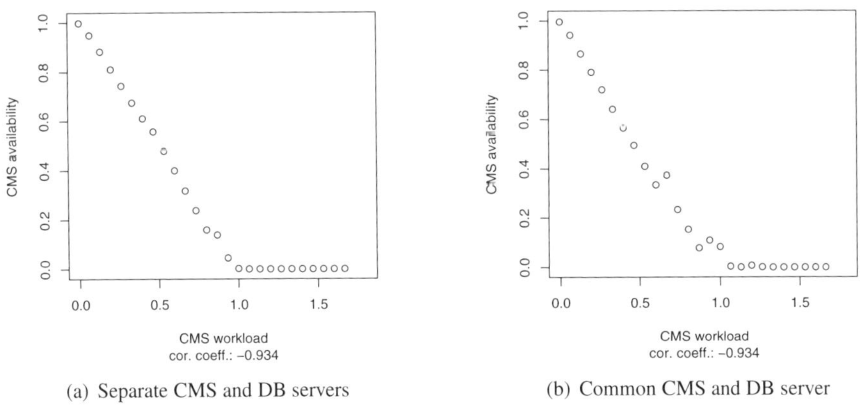

Figure 7. Relationship between resource workload and availability.

Figure 7 shows the relationship between workload and availability for both scenarios. In both cases, the correlation coefficient is very close to -1. This suggests the existence of a functional dependency between resource workload and availability. In this simple first example, we observe a linear dependency between the two values. However, more complex dependencies may exist, e. g. in the case of a multi-core system which is able to process multiple requests in parallel. Here, up to a certain workload, the availability should not be markedly impaired at all. Also, when considering other types of resources (e. g. the operating system kernel), binary relationships are possible, where a resource remains fully available up to a certain workload, after which it immediately drops to zero.

5. Summary and Further Work

This contribution has discussed an extension to GrADAR, an approach for automatically assessing the effects of denial-of-service countermeasures. Including the effects of resource workload into the GrADAR model permits the specification of complex countermeasure effects.

Workload and availability measurements were performed in a simple DMZ-like setup which included machines representing a client, a firewall host and two servers. They were conducted using a passive monitoring solution capable of calculating resource workload and availability from observed network traffic. The results of these measurements indicate a possible functional dependency between resource workload and availability which justifies the incorporation of workload measurements into the GrADAR approach.

The work done regarding the GrADAR extension is of a preliminary nature. There are numerous aspects which need to be considered in future work. So far, the generated traffic used for measuring workload and availability consisted only of a steadily increasing number of concurrent clients requesting the index page of an e-commerce Web site. Representative traffic for such a scenario needs to be generated for a more detailed evaluation; e. g. according to a formal customer state model as depicted in figure 8 with different states for each type of viewed page and appropriate state transition probabilities. Also, possible dependencies between the workload of different resources needs to be investigated, e. g. workload placed on the CMS and its backend database.

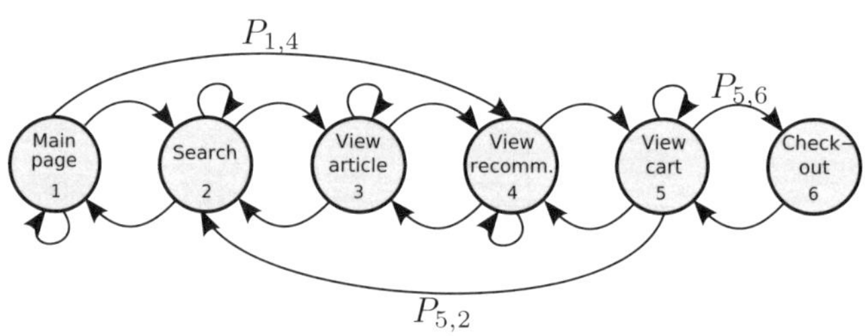

Figure 8. Possible state model underlying browsing by Web shop customers.

Where the passive analysis of traffic is concerned, problems may arise when only parts of conversations between resources can be observed, e. g. due to node movement in mobile ad hoc networks. In this case it might become necessary to harmonise the observations of multiple sensors distributed throughout the network.

Acknowledgements

The authors wish to thank Christian Thul for his excellent development work and his valuable feedback.

References

[1] M. Jahnke, C. Thul, and P. Martini. Graph based metrics for intrusion response measures in computer networks. In *Proc. of the 3rd LCN Workshop on Network Security. Held in conjunction with the 32nd IEEE Conference on Local Computer Networks*, Dublin, Ireland, October 2007.

[2] M. Jahnke, C. Thul, and P. Martini. Comparison and improvement of metrics for selecting intrusion response measures against DoS attacks. In A. Alkassar, editor, *Proc. of the Sicherheit2008 Conference*, Saarbrücken, Germany, April 2008.

[3] P. Barford and M. Crovella. Measuring Web performance in the wide area. *SIGMETRICS Performance Evaluation Review*, 27(2):37–48, September 1999.

[4] B. Eriksson, P. Barford, R. Nowak, and M. Crovella. Learning network structure from passive measurements. In *IMC '07: Proceedings of the 7th ACM SIGCOMM Conference on Internet Measurement*, pages 209–214, New York, NY, USA, 2007. ACM.

[5] S. Seshan, M. Stemm, and R. H. Katz. SPAND: Shared passive network performance discovery. Technical Report UCB/CSD-97-967, EECS Department, University of California, Berkeley, August 1997.

[6] B. B. Lowekamp. Combining active and passive network measurements to build scalable monitoring systems on the grid. *SIGMETRICS Performance Evaluation Review*, 30(4):19–26, 2003.

[7] B. Claise. Cisco Systems NetFlow Services Export Version 9. RFC 3954 (Informational), October 2004.

[8] N. Williams, S. Zander, and G. Armitage. Evaluating machine learning algorithms for automated network application identification. Technical Report 060410B, Centre for Advanced Internet Architectures (CAIA), March 2006.

[9] P. Barlet-Ros, V. Carela-Español, E. Codina, and J. Solé-Pareta. Identification of network applications based on machine learning techniques. In *TNC 2008: Proc. of the Terena Networking Conference*, 2008.

[10] Y. Himura, K. Fukuda, K. Cho, and H. Esaki. Characterization of host-based traffic with multi-scale gamma model. In *Proc. of the 2nd CAIDA/WIDE/CASFI Workshop*, Seoul, South Korea, April 2009.

[11] T. Toth and C. Kruegel. Evaluating the impact of automated intrusion response mechanisms. In *ACSAC '02: Proceedings of the 18th Annual Computer Security Applications Conference*, page 301, Washington, DC, USA, 2002. IEEE Computer Society.

[12] J. Mirkovic, P. Reiher, and A. Hussain. Measuring denial of service. In *Proc. of the ACM Workshop on Quality of Protection (QoP)*, pages 53–58. ACM Press, 2006.

[13] J. Mirkovic, A. Hussain, B. Wilson, S. Fahmy, P. Reiher, R. Thomas, W. Yao, and S. Schwab. Towards user-centric metrics for denial-of-service measurement. In *Proc. of the Workshop on Experimental Computer Science*, 2007.

[14] S. Bagchi, G. Kar, and J. Hellerstein. Dependency analysis in distributed systems using fault injection: Application to problem determination in an e-commerce environment. In *Proc. of the 12th Intl. Workshop on Distributed Systems: Operations & Management*, 2001.

[15] P. Bahl, P. Barham, R. Black, R. Chandra, M. Goldszmidt, R. Isaacs, S. Kandula, L. Li, J. MacCormick, D. Maltz, R. Mortier, M. Wawrzoniak, and M. Zhang. Discovering dependencies for network management. In *Proc. of the V HotNets Workshop*, 2006.

[16] ITU. *Specification and Description Language (SDL) (ITU-T Recommendation Z.100)*. International Telecommunications Union, August 2002.

[17] J. D. Case, M. Fedor, M. L. Schoffstall, and J. Davin. Simple network management protocol (SNMP). RFC 1157 (Historic), May 1990.

The Virtual Battlefield: Perspectives on Cyber Warfare
C. Czosseck and K. Geers (Eds.)
IOS Press, 2009
© 2009 The authors and IOS Press. All rights reserved.
doi:10.3233/978-1-60750-060-5-262

On nth Order Attacks

Daniel BILAR [a,1]

[a] *Department of Computer Science, University of New Orleans, USA*

Abstract. An nth order attack seeks to degrade, disable or subvert an end system indirectly by targeting one or more end mission-sustaining ancillary systems. We discuss the vulnerability etiology enabling such attacks. We illustrate the notion of these attacks with concrete historical, current and forward-looking examples; also in the context of cyberwar against advanced computerized societies. We sketch the challenges and requirements to detect and mitigate the effects of nth order attacks.

Keywords. nth order attack, Highly Optimized Tolerance, ancillary system, assumption violation, economic warfare, critical infrastructure

1. Introduction

The goal of nth order cyber-warfare is to induce instabilities in mission-sustaining ancillary systems that ultimately degrade, disable or subvert an end system. Such systems may be technical/algorithmic; however, societal, psychological, ideological, economic, biological and natural systems may be targets, as well. Ancillary systems include pars pro toto memory resource allocation, throughput control, hardware/software manufacturing, visualization environments, social welfare systems, human networks, power generation/transmission/distribution, voting systems, data and goods supply lines, reputation management, entropy externalization, business models and economic systems.

For example, a denial of service attack against a web server can be seen as a case of a 2nd order attack against the resource allocation subsystem of the TCP transport subsystem. Thompson's trojaned compiler in "Reflection on Trusting Trust" may be seen as a 3rd order attack against software manufacturing tools [1].

This paper defines and discusses this class of attacks and tries to explain their etiology via reference to Highly Optimized Tolerance (HOT) processes. HOT processes induce structured systems through optimization mechanisms that incorporate tradeoffs between objective functions and resource constraints in probabilistic environments. Pertinent to our discussion is the property that such optimization-generated systems are *robust towards common perturbations, but especially fragile towards rare events*, such as unanticipated changes in the environment. Inducing such 'rare events' in mission-sustaining ancillary systems is thus the goal of nth order attacks.

The rest of this paper is organized as follows: Sec. 2 explains the main concepts that motivate our subsequent discussion. Sec. 3 reviews related work. We give concrete examples of nth order attack in Sec. 4. Sec. 5 discusses analytical aspects of nth order

---

[1]Corresponding Author: Daniel Bilar, Department of Computer Science, University of New Orleans, 2000 Lakeshore Drive, New Orleans LA 70148, USA; Email: daniel@cs.uno.edu

attacks. Sec. 6 briefly sketches theoretical and practical remediation approaches. Sec. 7 gives final thoughts on the urgency of addressing the theme of the paper.

2. Overview

The following section serves to flesh out the nomenclature and concepts used throughout the paper. We shall start with the abstract notion of a 'system'; the definition of which varies across time and domains. For the purposes of this discussion, we adopt a recursive variant of biologist von Bertalanffy's seminal work on General Systems Theory [2]:

A system is a whole that functions by virtue of the interaction between constitutive components. As such, it is defined by these relationships. Components may be other systems.

For our purposes, the attractiveness of the definition lies in its emphasis on *openness* and the allowance for *structural similarities* across different domains with concomitant correspondence of governing behavior. For an short, readable, largely non-technical overview of competing system theories, the reader is referred to [3, ch. 2].

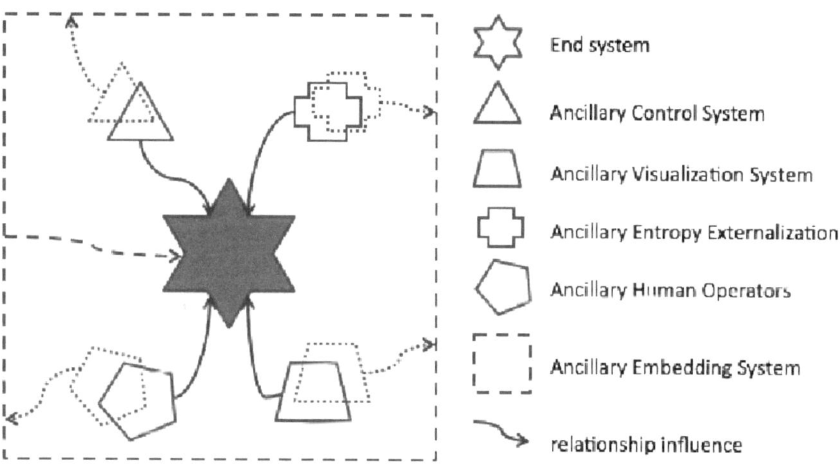

Figure 1. System view end system

Ancillary systems are responsible for control mechanisms, fault detection/resilience/recovery, energy/data flow, economic viability, human usability, data processing/structures, graceful startup/shutdown, reputation management, governance, social order and more. Such systems may be technical/scientific/algorithmic; however, societal, psychological, ideological, economic, biological and natural systems are included, as well.

Ancillary systems span different scales and varying orders of complexity. They may be embedded in or encompass the end system, and may in turn be composed of and influenced by other ancillary systems. Figs. 2(a) and 2(b) list an embedding (say a business model) and embedded ancillary system (in this example human operators) with reference to an end system (denoted by the center star) from Fig. 1.

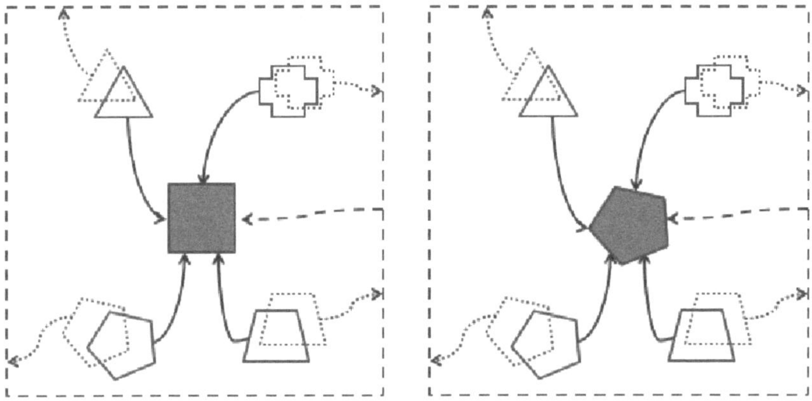

(a) Embedding System: Business Model (b) Embedded System: Human Operator

Figure 2. Examples components of embedding system (a) embedded system(b) of an end system

A Network Intrusion Detection System (NIDS) may serve as an illustrative end system example. Its ancillary control system negotiates the data and instruction interplay between sensors, analysis, database and decision/response engine. The ancillary visualization system displays the events and possible remediation options. The human operator subsystem must interpret the happenings and subsequently make the decisions that are not automated to the best of its reasoning ability. The entropy externalization subsystem is (among other things) responsible for cleaning out accumulation of dynamic data through sliding windows/logging and filtering out sensor noise. The end system itself is embedded in a business model that governs aspects of its design, implementation and activity: profit model, signature update cycles, customer support and more.

The ancillary systems of the NIDS end system have themselves subsystems: Human operators (Fig. 2(b)) field a visual subsystem subject to parameters (no UV sight, certain percentage of color-blindness, angular resolution etc). Their control system may be thought of as their reasoning strength and limitations (cognitive dissonance, herd instinct, unconscious intelligence [4] etc), as well as their physiological mechanisms (hormone secretions of the hypothalamus that regulate sleep, hunger, temperature etc). The human subsystem of human operators may be coworkers, friends, the fellow polity, family. Entropy externalization systems manifest themselves in physical (as in human waste product expulsion), as well as mental and psychological mechanisms (stress relief through exercise, keeping a diary, art, talking on the phone etc).

The business model (Fig. 2(a)) is embedded itself in an economic environment, say a free market economy, which influences its setup (tax codes, corporate structure, sales channels, liquidity parameters such as interest rates which determine acceptable debt ratios etc). The control subsystem may consists of corporate governance, union influence, mission statement, and legislative regulations. Its visualization subsystem may include accounting publication systems (standardized formats like IFRS with its own assumptions), dress codes, as well as marketing approaches (corporate image, advertisements etc). Human operator system may be stockholders, consultants writing the business plan, company workers, product consumers, company management, and competitors. Finally, the entropy externalization ancillary system of the business model may include mechanisms to off-set losses to subsidiaries, third-tier rebranding of products for steep sales discount, 'poison pills' to counter hostile takeovers, corporate fusion plans, and more.

2.1. n^{th} order attacks

An $\mathbf{n^{th}}$ order attack tries to indirectly degrade, disable or subvert an end system by targeting one or more ancillary systems.

With this qualitative definition in hand (which we shall pick up in Sec. 5), let us revisit the NIDS example in Fig. 1 with its control, human, entropy externalization, embedding and visualization ancillary systems. How would one go about perpetuating an nth order attack against an NIDS? One could take on the control system via a DoS attack against the response/decision engine, or try to supply fake/poisoned data to the analysis engine. Given biological, cognitive and psychological human parameters, enough false positives at 3am will make human operators tone down the sensitivity of the analysis component. One form of entropy electronic equipment produces is heat. The vast majority of Intel and AMD CPUs, for instance, reach critical heat at about 55-85°C[5, p. 5-13], which may cause the BIOS to shutdown to prevent damage: Hence, one attack against this entropy externalization system raises the ambient temperature of the building in which the NIDS components are deployed (say by low-tech clogging the climate intake vents). PNNL's Starlight [6] offers a comprehensive NIDS visualization system, replete with 2-D and 3-D multimedia visualizations supporting comparisons and emphasizing interrelationships. As can be intuited by the Starlight Network Intrusion Detection Graph[2] in Fig. 3, once data flow reaches a critical mass (by virtue of screen resolution and human limitations) visuals will degenerate into saturated pixel blobs, obviating their usefulness. The susceptibility of security visualization methods to intentional noise remains a serious concern, as described by [7].

Why do these attack work? Why does any attack, cyber- or otherwise, work? The answer we propose is surprisingly simple: *Attacks work because they violate assumptions.* Any finite system by design must incorporate implicit and explicit assumptions into its structure, functionality, and language. These systems are formulated with 'expected', 'typical' cases in mind and the assumptions reflect these expected use cases: A man-in-the-middle attack violates the assumption that you are talking to the party you expected; a race condition attack violates ordering assumptions; a buffer overflow attack violates an explicit resource assumption; BGP routing and DNS case poisoning attacks violate implicit trust assumption of non-malicious open architecture participants. Likewise, terroristic activities in open societies are easy to pull off because spaces are open, population freedom of movement not controlled - hence they violate implicit societal trust assumptions. Lastly, many democratic voting schemes assume 'honest' voters, and hence can be undermined by strategic voting [8]. There are scores of examples, in every domain.

We shall revisit the trust assumption in open societies in Sec. 7. Our next goal, however, is to gain some intuition about the etiology of the problem: We present a putative generative mechanism which crucially depends on assumptions to highlight the consequences of violating said assumptions.

[2]In the interest of fairness, it should be noted that this image is originally in color, not gray shades.

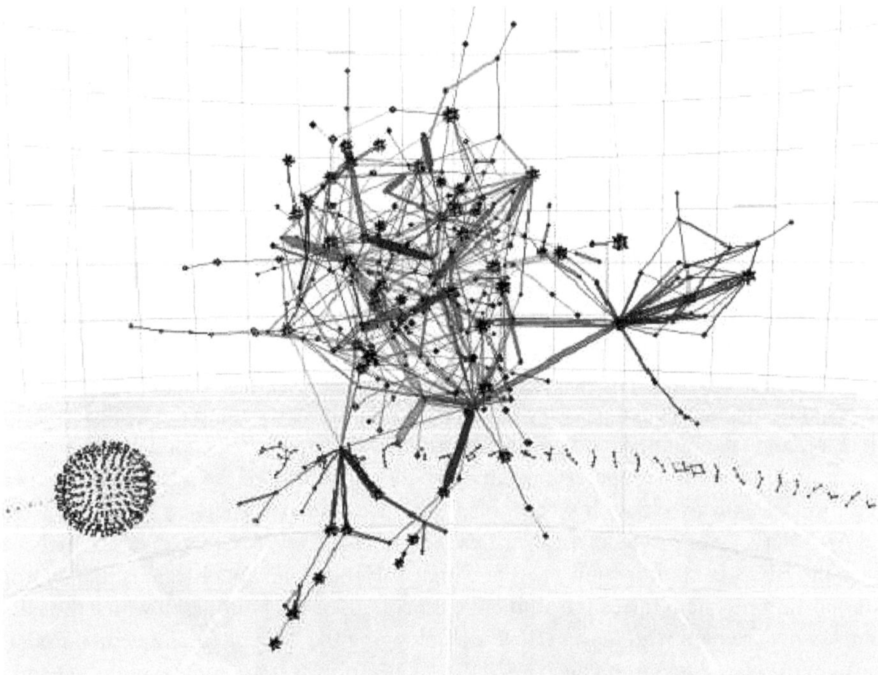

Figure 3. Starlight NIDS Graph

2.2. *Highly Optimized Tolerance*

Highly Optimized Tolerance (HOT) is a generative mechanism that seeks to explain the structure, statistics and resiliency of interconnected systems. Originally proposed to account for the ubiquity of so-called power laws in natural and engineered systems, it has been fruitfully applied to the study of forest ecosystems, router network robustness, internet traffic, power systems and immune systems. The strength of HOT models is four-fold: First, by virtue of its emphasis on evolved and engineered complexity through feedback, tradeoffs between objective functions and resource constraints in a probabilistic environment, it models the majority of real-life systems which are subjected to such pressures. Secondly, its features include high efficiency, performance, and robustness to designed-for uncertainties, i.e. 'average' cases. Thirdly, it conversely exhibits hypersensitivity to unanticipated perturbations, i.e. 'rare' cases. This too, is a feature of most systems, as we will see. Lastly, unlike rival generative mechanisms, the resulting structural configurations are domain-specific and non-generic [9]. For a discussion of power laws, a primer on HOT and a survey of generative mechanisms (including HOT), the reader is referred to [10,11,12], respectively.

2.3. *HOT example: Buffer overflow*

We shall proceed to present a first example to highlight a HOT process-induced vulnerability that can be subject to a 0th order attack.

Below we find an instantiation of a HOT model: A Probability, Loss, Resource (PLR) optimization problem [13]; a generalized restatement of Shannon source coding for data compression yielding the Shannon-Kolmogorov entropy for the objective function J. The reader is referred to [14] for details and more examples.

$$\min J \tag{1}$$

subject to

$$\sum r_i \leq R \tag{2}$$

where

$$J = \sum p_i l_i \tag{3}$$

$$l_i = f(r_i) \tag{4}$$

$$1 \leq i \leq M \tag{5}$$

We have a set of M events (Eq. 5) occurring iid with probability p_i incurring loss l_i (Eq. 3), the sum-product of which is our objective function to be minimized (Eq. 1). Resources r_i are hedged against losses l_i, with normalizing $f(r_i) = -\log r_i$ (Eq. 4), subject to resource bounds R (Eq. 2). We will demonstrate a mapping between this abstracted PLR model and the following short C program (adapted from [15]) which will be subjected to a buffer overflow.

```
#include <stdlib.h>
#include <stdio.h>
#include <string.h>

int provePequalsNP()
{
/* Next paper .. */
}
int bof()
{
char buffer[8]; /* an 8 byte character buffer */
strcpy(buffer, gets()); /* get input from the user */
/* may not return if buffer overflowed */
return 42;
}

int main(int argc, char **argv)
{
bof(); /* call bof() function */
/* execution may never reach
next function because of overflow */
provePequalsNP();
return 1000000; /* exit with Clay prize */
}
```

We shall assume here that the probabilistic environment is adequately represented by the user. She is asked for input in `gets()`, this represents the event. In the `C` code, the human designer specified an 8 byte buffer (`char buffer[8]`) and the compiler would dutifully allocate the minimum buffer needed for 8 bytes (this is the resource r). Hence, the constrained resources r is the variable `buffer`. The loss associated with the user input event is really a step function; as long as the user satisfies the assumption of the designer, the 'loss' is constant, and can be seen (simplified) as just the 'normal' loss incurred in proper continuation of control flow. Put differently, as long as user input is ≤ 8 bytes, the resource r is minimally sufficient to ensure normal control flow continuation. If, however, the user decides to input 'Honorificabilitudinitatibus' (this lengthy wink to the Bard was implicitly assumed to be an unlikely/impossible event by the human designer in the code declaration), the loss l functions takes a huge step jump: a catastrophic failure ensues since `strcpy(buffer,gets())` overflows `buffer`. The improbable event breaches the resource and now, control flow may be rerouted, the process crashed, shellcode executed via a stack overflow - or in our example, fame remains elusive.

How did this vulnerability come about? In keeping with our hypothesis, we may discern two distinct, domain-specific HOT (Highly Optimized Tolerance) optimization processes at play - one involving human designers and the other, code compilers - that had a hand in allocating the resource that was breached. The first domain-specific mechanism that induces a cost-optimized, resource-constrained structure on the executable program is the human element. Humans using best-practice software development techniques have to juggle at various stage of the design and coding stages: Evolvability vs specificity of the system, functionality vs code size, source readability vs development time, debugging time vs time-to-market, just to name a few conflicting objective function and resource constraints. The second domain-specific mechanism that induces a cost-optimized, resource-constrained structure on the executable is the compiler. The compiler functions as a HOT process. Cost function here include memory footprint, execution cycles, and power consumption minimization, whereas the constraints typically involve register and cache line allocation, opcode sequence selection, number/stages of pipelines, ALU and FPU utilization.

3. Background and Related Work

The issues of vulnerabilities in ancillary systems and their impact on end systems have been discussed in the popular press. Makansi issues a clarion call to action - part historical, current and future US survey, part Cassandra-cry [16] - on the sorry state of the US electricity grid. Pertinent to our discussion is his focus on the grid's transmission subsystem: Maintenance neglect of transmission lines, pylons and most importantly, the nearly-unguarded substations. It is the opinion of the author that the neglect of the ancillary transmission system viz. the grid system constitutes a prima facie example of constraint-based value optimization as suggested by HOT, given that the former accounts for less than 10% of the electricity asset value chain.

Within a more general framework of catastrophic societal scenarios, Clarke [17] raises awareness of seldom-mentioned ancillary systems. He stresses hidden but pervasive technological and social interdependence and subsequently calls for a more expansive definition of critical infrastructure. In the context of nth order attacks, he mentions the essentially defenseless railway system and abounding chemical plants (a devilish target, since chemicals are very often shipped on railways through population centers). His emphasizing near-blind spot subsystems like kindergarten teachers (in the US, around

20% of the population is in K-12 schools for about half the day) and morticians/undertakers[3]) remains a rare and meretricious exception.[4]

The modeling tools provided by complex network theory have been used to evaluate the susceptibility of critical infrastructure to both failure and attack. Network theory lends itself to the main concepts of this paper, in that network graphs can be used to represent influence diagrams, and system decomposition. In addition, through statistical link-node distribution analysis, one is able to define a variety of centrality (vulgo 'importance') metrics (see Newman [19] for a book-length academic primer). Static social network analysis was applied by Celebi [20, pp. 127-141] to network graphs of websites affiliated with the terrorist PKK. Using graph metrics such as geodesic distance, connectivity and principal component analysis, the goal was to identify the most influential websites (so-called hubs) order to break information connectivity; in other words, pinpointing neuralgic nodes for removal to impede the functioning of the network.

Saddling the horse from the other end- and as a cautionary tale of what can be learned in open societies built on trust - is the nigh unbelievable story[5] of the PhD thesis White House cybersecurity czar Richard Clarke wanted 'burned' in 2002. Sean Gorman, a geography PhD student at George Mason University, gathered data on the US's fiber optic cable network entirely from open sources. He managed to layer the fiber-optic infrastructure - the information backbone supporting much of the US's military, civilian, financial, air traffic, water, power and control critical infrastructure - onto business and industrial sectors. The resulting map, which he could mine algorithmically with network analysis methods for neuralgic points, was termed a 'terrorist treasure map'. In the end, he was allowed to publish a neutered version of his thesis [21].

From a dynamic modeling perspective in the context of TCP network/web server request adaptation mechanisms, the paper series by Guirguis and Bestravos serve as a good starting point [22,23]: They systematically investigate so-called Reduction of Quality (RoQ) attacks. RoQ attacks target adaptation mechanisms used in network protocols. They achieve their effectiveness by non-DoS, low-bandwidth traffic maliciously optimized against the admission controllers and load balancers, thereby continuously forcing the adaptive mechanism to oscillate between over-load and under-load conditions. Conceptually speaking, RoQ attacks may be viewed as a class of nth order attacks (1st or 2nd order degradation attacks). Fig. 4 shall help us understand the generalizable modus operandus of RoQ attacks.

Assume the system services requests at a high steady state rate x^*, thanks to its adaptation subsystem that seeks to optimize service rates. Malicious traffic in form of an RoQ attack (burst time t shaded) push the system from its steady state equilibrium; the system, through its adaptation mechanism, slowly convergences at rate ν to the new, lower steady state y^*. Since attacks have ceased, after some time, the system's adaptation mechanism is able to converge at a higher rate μ back to the the high steady state x^*. Optimized RoQ attacks would then begin anew, forcing the system to oscillate between x^* and y^* just when it has settled, thereby degrading performance of the end system.

[3]From [18]: " .. the most terrifying aspect of the epidemic was the piling up of bodies" and from historian Alfred Crosby as quoted in [17, p.166]: ".. the accumulation of corpses will, more than anything else, sap and even break the morale of a population"

[4]Clarke's epistemological mindset of possibilistic vs probabilistic thinking heeds poet's William Carlos Williams' admonition: *What would happen in a world, lit by the imagination?* If on nothing else, decision

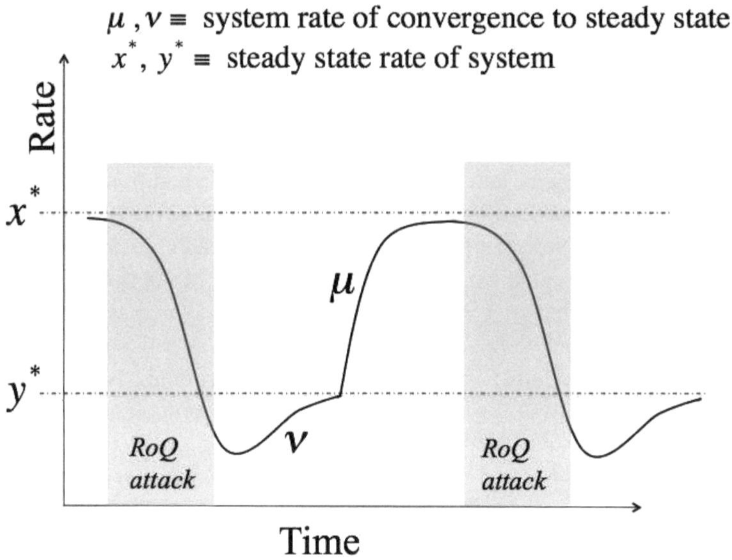

Figure 4. RoQ attacks force the adaptation mechanism with malicious traffic into dropping from a high system steady state rate x^* into to lower system steady state y^*. Picture adapted from [22, p. 3]

Putting it in the nomenclature used in this paper: The RoQ attack's δ requests per second for burst time t (grey shaded) repeated over period T constitutes the 'rare event' which the adaptation system was not expected to handle efficiently. Hence, the adaptation mechanism - as a HOT process designed for common perturbations, but fragile towards rare events - finds its internal assumptions (designed for normal traffic) violated. We now move on to concrete examples of nth order attacks.

4. Example of nth order attacks

4.1. Embedding Ancillary System

Estonia, after regaining independence in 1991, decided on a massive nation-wide 'cyberfication' program: Comprehensive Internet access together with a population registry for authentication/ID purposes would enable the Baltic nation to 'Tiger-Leap' into the 21th century. The result of this push was an extraordinarily far-reaching state information system consisting of (among other things) a PKI infrastructure, over 70 state information systems, financial institutions, state/private portals and associated data exchange layer subcomponents.

In April/May 2007, Estonia suffered a two-phased denial of service attack (predominantly ICMP and TCP SYN[6]). The first phase (04/27/07 - 04/29/07) knocked out government web servers and news sites, and included some semantic hacking such as web de-

makers are strongly urged to follow up on Clarke's works.

[5]See a 2003 Washington Post article at `http://tinyurl.com/zuyrv`

[6]Nazario offers insightful traffic analysis of Estonia (`http://tinyurl.com/2359fq`) and the more intense 2008 South Ossetia attack (`http://tinyurl.com/6psa6r`)

Attacks	Destination	Owner	Description
35	195.80.105.107/32	pol.ee (now politsei.ee)	Estonian police
7	195.80.106.72/32	www.riigikogu.ee	Estonian Parliament
36	195.80.109.158/32	www.riik.ee, www.valitsus.ee	State communication entry portal, Estonian Government
2	195.80.124.53/32	m53.envir.ee	Ministry of the Environment
4	213.184.50.6/32	Estonian CERT	
6	213.184.49.194/32	www.agri.ee	Ministry of Agriculture
35	213.184.50.69/32	www.fin.ee	Ministry of Finance
1	62.65.192.24/32	starman.ee	Private telecom provider

Table 1. Second phase, 128 DDoS attacks: ICMP (115), TCP SYN (4), generic (9). Most serious 10 attacks: 10+ hours at 90 Mb/s. Peak on May 9: Attack shut down 58 sites at once. Data from Nazario (Arbor Networks)

facements. The second phase (04/30/07-05/17/07), coordinating a botnet encompassing some 178 countries, was aimed at critical infrastructures: The two largest banks, neuralgic routers at the ISP level and some governmental portals which were unavailable for a couple of hours. As can be gleaned from Table 1, during the second phase of attacks, the police, government and state communication portals, as well as the Ministry of Finance bore the brunt of the traffic.

This case also highlights the question of perspective in classifying the level of indirection of an nth order attack. On one technical level, the attack could be classified as 2nd order degradation attack, since it consisted of relative primitive DoS traffic aimed at resource allocation mechanisms underlying electronic services. From the point of the individual, it may be classified as a 3rd or 4th order destabilization attack, since it, say, undermined the information infrastructure needed for data exchange between the supermarket and the banks that enable him/her to use credit cards to buy groceries. For a short description of Estonian development, a timeline of the two-phased cyber-attack that took place and subsequent reactions, the reader is referred to [20, pp. 93-103]. We would like to stress these cyberattacks went hand-in-hand with planned physical disruptions: SMS-coordinated flash mobs causing traffic jams, trade and tourism interruption by train and road blockades, physical attacks against parliament, and more. This synergistic *levée en masse* of the Russian ethnic minority to foment unrest on the ground, in conjunction with the cyberattacks against societal critical infrastructure (see Table 1) were aimed at destabilizing Estonian society. In its comprehensiveness and goals, these efforts constituted the rare event in our model; in terms of modern conflict, it heralds a new class of 'total war' (see Sec. 7).

4.2. *Business Model Ancillary System*

The email-born Bagle worm first appeared in January 2004 and still ranks - 5 years later - among the top 15 malware families found in the wild, with a prevalence of roughly 2%. It reached its apex in 2006/2007, ranking among the top four, with a prevalence of roughly 15%. For an incisive write-up, the reader is referred to [25].

What makes this worm noteworthy in our context is its 4th order attack m.o.: Through a clever blend of so-called server-side polymorphism and 'high variant-low instance' release, it managed to circumvent conventional pattern-based antivirus (AV) signature detection by *attacking the economic cost structure of the AV companies* itself. With server-side polymorphic malware, the mutation and encryption code transform engine that produce variants is not incorporated into the individual instances, but resides remotely on a server. This outsourcing make the job of traditional signature-based AV companies (who

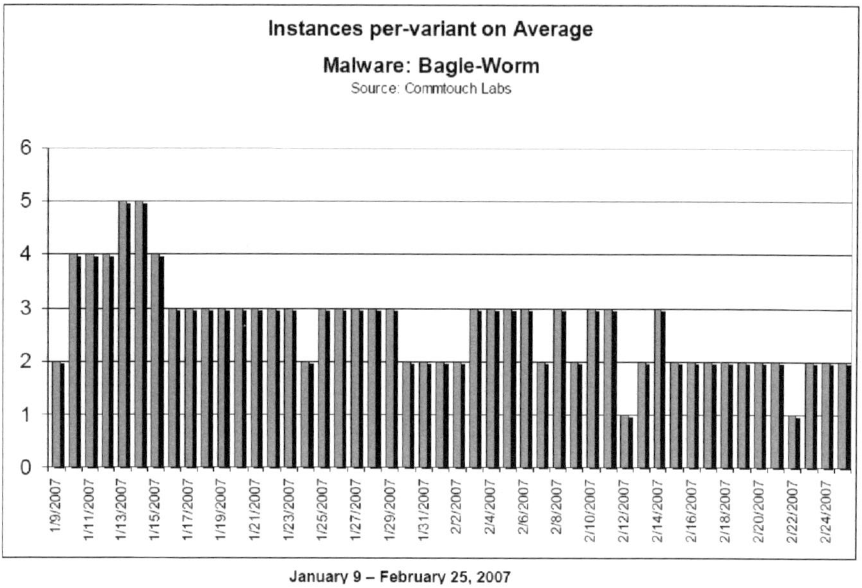

Figure 5. Small batches per variant. Picture from [24].

analyze the specimens) harder, since their analysts have less of a code base to work with. This in and of itself could have been dealt with: Bagle's true innovation was to sabotage the economic incentives of AV companies to distill such a signature by generating enormous number of variants in very small batches.

Fig. 5 illustrates the simple but highly effective distribution approach: It lists the average number of instances of the same code, per variant each day of the report period. We see that very small batches of the same code were released but a huge number of variants thereof (30'000 distinct variants server-side supplied in 2007 alone, an average of 625 new variants a day[24]). This constitutes arguably a 4th order attack, since this mechanism neither targeted a vulnerable program on the end system (0th order), nor disabled host or server-based AV services (1st order), nor targeted (say through denial of service or DNS rerouting) either the start or end points of the AV signature distribution system (2nd and 3rd order), but cleverly vitiated the economic incentives of the AV companies to develop signatures (4th order). With modern malware, it is simply not cost effective to invest even one day's worth of highly skilled analyst's time to develop signatures for rapidly mutating, low-count instances - exactly the type of rare event for which the business model was not designed.

4.3. Human Operator Ancillary System

Bond and Danezis invite the reader to entertain following Gedankenspiel [26], inspired by Faust's pact with Mephistoteles: Person W sends a program to person Z, accompanied by an email singing said program's praises. For it promises powers: The power to remotely browse X's hard disk, the power to read the emails between X and Y. Curiosity and maybe malice piqued, Z installs the program and lo, it does not deceive: It delivers on its promises, certainly, but surreptitiously keeps a log of Z's activities and rummages through Z's files. After a critical mass of incriminating evidence is gathered, the program now uses a combination of threats and bribes to get Z to propagate itself: From Data Destruction ("I'll delete all your files") to Revelation ("I'll tell Y you were spying on X and Y')' to Reporting ("I'll report your illegal downloads to the RIAA") to Access Denial ("I'll encrypt all your files") to Freebies ("You'll get tons of free software") and the promise of more powers ("You'll get the power to watch webcams").

The truly devious innovation of this SATAN virus consists of very elegantly leveraging the *psychological ancillary system of the human operator*: It appeals first to a mix of neutral (curiosity, risk) to base (greed, lust for power) psychological instincts. After a time of reward to re-enforce the risky behavior, it then brings the full gamut of shame, fear, cowardice and cognitive dissonance to bear in order to harness two additional subsystems of the human operator: His own human operator subsystem (select the next victim) and his rational subsystem (convince him/her to install me). The induced calculated betrayal of interpersonal trust (the rare event) seems particularly odious. You can almost see the friend exclaiming: "How could Z do this to me, as a friend?" As far as 1st or 2nd order subversion attacks against human operators are concerned, the conceptual SATAN virus is extraordinarily clever. [7]

5. Analysis

With reference to the schematic network graph given in Fig. 6, the US national end 'super' system of interdependent critical infrastructure ancillary systems, we outline some characteristics for a theoretical n^{th} order attack analysis framework.

1. We require first a notion of *evolving system state*, since we are dealing with dynamical systems.
2. Any model has to furthermore incorporate notions of *cross-dependencies*, since systems are open and coupled.
3. These dependencies must include *ties to assumption violations* (as denoted in Eq. 2 of the IIOT model in Sec. 2.2) to propagate effects between systems.
4. These propagated dependencies must have an *impact* on the system state that is quantitatively measurable.
5. Lastly, the modeling formalism has to be high-level enough that there be a reasonably direct correspondence between the system elements modeled and the formalism of the approach.[8]

We explain the rationale for these requirements with the help of Fig. 6. For instance, the communications infrastructure is powered primarily through the power infrastructure. If power delivery is disrupted, telecommunications may switch to backup generators which rely on fuel from the energy distribution infrastructure, delivered via the transportation infrastructure paid for through the financial infrastructure. Conversely, the communica-

[7] It is the author's opinion that this conceptual SATAN virus offers one more astounding innovation, namely symbiotic human-viral code. Even more extraordinary from the point of view of information complexity, the probably simpler viral code manages to induce the 'production' of the more complex human code (propagation module) *dynamically* by invoking evolutionarily and socially generated 'factory routines'.

[8] As an wished-for bonus (maybe there is a Santa Claus), model analysis should be tractable, i.e. any modeling approach used must try to avoid combinatorial state space explosions

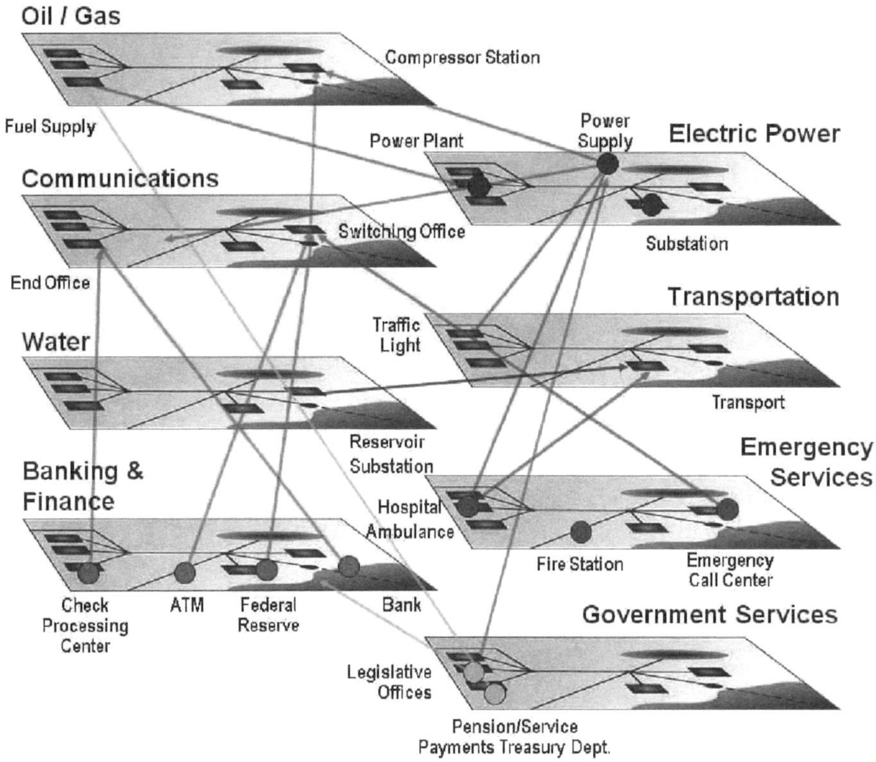

Figure 6. Network of Critical Infrastructure. Picture from Sandia as shown in [27, p.12]

tions infrastructure provides control to the power and transportation infrastructure and underlies much of the financial infrastructure. A fair question that a candidate framework should be able to answer: How much power will we lose for how long if we degrade the communications infrastructure's performance by 20%?

5.1. Theoretical framework

There is a wealth of research on static network graph analysis (see [28] for a practical overview); its main drawbacks remain the inadequate handling of evolving dynamic behavior and cross-dependencies/feedback loops. Since we are concerned with system failure/degradation/subversion, reliability theory formalisms and models suggest themselves.

A first stab system decomposition into constitutive subsystems can lend itself to a simple Fault Tree Analysis. FTA has been used for decades to model failure in multi-component systems. Invented in 1961 by Bell Labs to improve the reliability of the Minuteman Launch Control System, it has since then been extensively used for evaluating system safety in engineering disciplines as diverse as power, nuclear, electric, and source code analysis [29]. FTA investigates independent pathways between failures of components that lead to the fault tree's top-event. In our parlance, this would be affecting the the end system. Its representation takes the form of a logical diagram in which the top-event's occurrence depends on a specific combination of basic events, which are com-

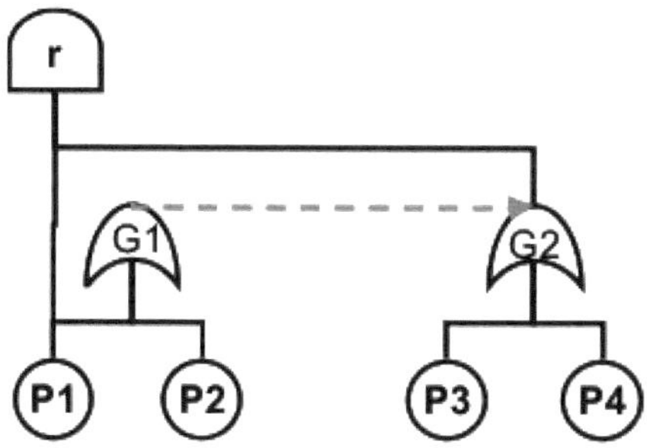

Figure 7. A BLDMP ($\mathcal{F}$, **r**, **T**, (P_i)) consists multi-top coherent fault tree $\mathcal{F}$, main top event r, set of triggers T, set of 'triggered' Markov processes P_i associated with leaves of $\mathcal{F}$ (denoted by the red dashed line), and two categories of state for P_i, normal and failure that are triggered via appropriate transfer functions. Picture from [31]

bined with two primary types of gates, AND and OR. Canonical FTAs have no notion of component dependencies, or conditional event sequence timing. As such, they do not meet our requirements; however, extensions such as ones offered in Dynamic FTA [30] do incorporate some, but not all, requirements delineated above.

We offer one modern approach (itself a generalization of Dynamic Fault Trees) that may be fruitfully applied for our purposes, subject to our requirements: Boolean Logic Driven Markov Processes (BLDMP) [32]. BLDMP combines low-level global Markovian state space evolution with a higher level FTA modeling approach. Each leaf is associated with a Markov process which can model the dynamic behavior of a system. Forms of cross-dependencies can be modeled by triggered Markov processes. This fault tree represents the 'structure function' of the system. This structure imposed on the Markov graph can be used to prune the state space, thereby avoiding combinatorial explosions and making analysis more tractable. It remains an open question, though, whether the quantitative impact of these propagated dependencies can be determined analytically, given the non-linear complex dynamics of the setup, or whether one has to resort to an mix of expert judgment, simulation results and historical empirical data.

There exists an alphabet soup's worth of alternative formalisms describing dynamical systems, each with their strength and weaknesses. We briefly mention so-called Dynamic Reliability Block Diagrams (DRBD), as developed by [33]. In contrast to BLMDP's hybrid state space/combinatorial formalism, DRBD is based on the single, high level formalism of RDB [34, Sec. 3.10]. Of interest to us is primarily its dynamic expressiveness, which derives from a technique to model at a low level simple dependencies. This basic dependency 'building block' can be combined with others to model any dynamic behavior (see [35, sec. 3]). The topic space is by no means exhausted: For an extensive reference list of methodologies/formalisms and a reference work, the reader is referred to [33, Sec. 6][34], respectively.

In terms of developed models, the Vulnerability of Information System (VIS) project [36] very recently took a stab at answering questions similar to the critical infrastructure one posed at the beginning of Sec. 5. VIS attempts to quantitatively measure the impact of unexpected Information Communication Technology (ICT) breakdown on economic sectors deemed potentially most ICT-vulnerable. Fondazione Formit (VIS' project lead) singled out five countries (Ireland, Italy, Luxemburg, Romania and Spain) and identified key ICT-impacted economic sectors within each: Among these, we find public administration, sewage disposal, telecommunication, finance, water and electricity supply - sectors which qualify as critical infrastructure. They subsequently selected representative companies within these sectors for micro-analysis to study ICT breakdown effects, existing recovery strategies and costs. On a macro-level (and more interesting in the context of our discussion), a sophisticated econometric partial equilibrium model taking EU sectorial interdependences as well as cascade effects into account was developed. The model, which allowed for free variables such as ICT breakdown intensity, breakdown and recovery time length, measured the effects of reduced ICT performance on output and value loss, employment and price change, as well as social welfare loss with a time horizon of one day to three months. The validity of some model output was also assessed by means of expert judgment impact analysis in the case study companies and subsequent country-specific micro-simulations.

As of the time of this writing (July 2009), the final report is in the works; perusing available material, the numbers seem overly optimistic: Cumulative Spanish output loss after one month (assuming 10% ICT instantaneous loss, with 50% recovery in five days) hovers below 1% across all economic sectors. One might take issue with the strong equilibria assumptions in the econometric model, yet the crux lies with the recovery assumption: 10% ICT loss with 50% recovery within 5 days may be realistic in terms of accidents or technological glitches, but very likely unrealizable in the face of intentional attacks (in fairness, intentional attacks were puzzlingly not in the project scope's risk space). Thus we stress again, albeit in a different context, the pitfalls of strong assumptions, as well as the dangerous allure of fantasy recovery ('error handling') documents, a topic which we shall return to below.

5.2. *Practical framework*

Since analytical modeling proves to be non-trivial in its requirements, perhaps an approach along the lines of a simulation offers an alternative. Indeed; given a controlled, instrumented environment in which the end system can be situated, actual n^{th} order attacks against ancillary systems and their concomitant effects can be observed and then evaluated. Such is the case with software application running on a single machine, where destabilization efforts can be effected through an embedding ancillary system acting as mediating OS middleware. We list Holodeck[9]; a fault injection framework that allows Windows programs to run in simulated hostile environments [37]. Its functionality includes the ability to create resource starvation situations affecting ancillary systems such as memory, hard disk, network bandwidth; as well as error handling ancillary system in the form of data poisoning such as corrupted resource files/network streams, unexpected API return values, and a gamut of explicit fault injections.

Empirical evidence collected over two decades support Holodeck's emphasis on *error handling ancillary systems*. Miller subjected Unix, Windows and OS X utilities in the simplest case to random (not malicious) keyboard input, and reported end system crash failure rates of 25%-40%, 24%, and 7%, respectively [38][39]. Sociological and organizational case studies by Clarke [40], analyzing what he terms 'fantasy documents'

[9]Commercially available at `http://www.securityinnovation.com/holodeck/`

(disaster contingency plans[10]), corroborate the brittleness of error handling subsystems, as well.

6. Remediation

In our view, remediation efforts must either address the assumption violations underlying the vulnerabilities, or devise a control mechanism to keep the system in a stable state, should it come under attack. We crystallized thusly: Since we posited that the etiology of nth order attacks (any attack) lay in the HOT-induced violations of assumptions, is there a way of dynamically mutating those assumptions? If not, can we prevent malicious parties from learning of these assumptions? Lastly, if we cannot prevent a violation, can we return a system back to a stable state?

An effective, protocol compliant, but rarely used TCP feature in Linux kernels exists which prevents some forms of degradation attacks against the TCP resource allocation mechanism: SYN Cookies. The server outsources the state of a half-open connection (kept normally on the server) in the form of a cryptographic challenge (the cookie) back to the client[41]. This is an example of an assumption mutation. Internet cognoscenti have heard of the 'Slashdot' effect - when legitimate connection requests overwhelm the server because of popularity of content. This problem was tackled early on in 2001 by Akamai [42] in the form of dynamic load balancing, which constitutes a runtime assumption mutation.

Keeping parties from learning about exact resource boundaries (and subsequent exploitation) may be able to borrow methods from thwarting so-called side channel attacks. Side channel attacks try to infer a system process' state by means of (sometime inadvertently, sometimes unavoidably) leaked observables like time to completion, EM radiation, sound, protocol return values generated in course of the system's evolution. These attacks range the gamut from ingenious timing analysis on B-tree lookup operations and data structure rebalancing (which lead to the release of database privileged information [43]), to differential power analysis where current used in switching reveal activities that can be mapped to processes [44], to CPU operation inferences through characteristic acoustic spectral signatures [45]. In all these instances, processes leaked information. It may be possible to design and operate systems in such a way that the leaking of resource boundaries (the assumptions an attacker wants to violate) is minimized. We hypothesize that designs that incorporate the insights of Maximum Entropy Principles (for an introduction see [46]) are a step in the right direction.

For state control, Ott's [47] work on controlling chaotic systems may yield some fruitful insights, since the interdependent, nested systems under consideration in this paper are more than likely to exhibit non-linear, complex, chaotic behavior due to feedback relationships. In a nutshell, Ott's OGY method injects tiny perturbations into the system when it threatens to veer off towards an unstable state. These perturbations (a control vector based on the system state's Jacobian eigenvectors) 'nudge' the chaotic system back towards a fixed point and into a stable state. For a beginner's primer on non-linear systems, the reader is referred to [48].

[10]A classic remains LILCO's ill-fated February 13th 1986 Shoreham evacuation plan. The aim of this exercise was to demonstrate the evacuation plan feasibility. It failed at step 1: The bus drivers (a logistical and psychological vital link; tasked among other things to evacuate children) failed [40, pp. 26-30]

7. Epilogue

In a worthwhile comparative study [49], Fukuyama of 'End of History' fame discusses the notion of societal trust as a gateway to prosperity. He maintains that members of 'high-trust' societies (like the United States) can leverage wide-circle (beyond family ties) trust to form efficient, optimized civic and economic organizations. It is hard to overstate how deeply this trust subsystem permeates every facet of open societies, how much it lowers tangible and intangible transaction costs between individuals, corporations and the state, and how easy an assumption it is to violate for malicious actors - with disastrous effects on the end system.

This realization was not lost on Bin Laden and his fellow strategists. In a 2004 broadcast, he boasted (quoting research from Chatham House [50]) that the 9/11 attacks had cost al-Qaeda only $500,000 while inflicting at least $500 billion of economic losses on America. Accordingly, the Islamist supremacists' playbook calls for beating the US by systematically attacking the US economy's vulnerabilities. The most accessible vulnerabilities in open societies are induced by deeply ingrained trust assumptions these societies have developed over decades and take for granted: that freedom of movement, freedom of speech, freedom of religious assembly, assistance from the social welfare state, immigration policy will not be used to subvert society from within; that a participant in mass transit, a shopper at the mall, a fertilizer buyer, a student reading nuclear engineering, a worshipper at a house of prayer will not commit mass murder. The chilling passage (excerpted from [51]) is worth quoting at length (italics are ours):

> The Islamic nation has entered through al-Qa'ida's war with America a new period that is different from all the other periods experienced by Muslims against their enemies. This period is based on economic war due to the peculiar nature of the adversary in this ferocious battle. Usually, wars are based on military strength and victory belongs to those who are militarily superior on the battlefield...But our war with America is fundamentally different, for the first priority is defeating it economically. For that, anything that negatively affects its economy is considered for us a step in the right direction on the path to victory. Military defeats do not greatly effect how we measure total victory, but these defeats indirectly affect the economy which can be demonstrated by the breaching of the confidence of capitalists and investors in this nation's ability to safeguard their various trade and dealings [..] *Any operation targeting an area of infrastructure in a new country that does not have a history of countering these operations is considered as bleeding (exhausting) to the greater enemy America and the targeted nation itself. It is so because these nations will be required to protect all similar potential targets which results in economic exhaustion (bleeding)... For example, if a hotel that caters to western tourists in Indonesia is targeted, the enemy will be required to protect all hotels that cater to western tourists in all countries which may become a target of similar attacks. You can say the same thing about living residences, economic establishments, embassies [..]*

Similarly, the PRC People Liberation Army's emphasis on asymmetric warfare and ongoing push to develop modern "Assassin's Mace" weapons within the doctrine of "The Inferior Defeats the Superiors"[11] should give some pause. The Director of Foreign Military Studies at the Academy of Military Sciences in Beijing, Major General Pan Junfeng, offered following tidbits reminiscent of nth order warfare (presumably against the US) in a 1996 issue of China Military Science (as cited in [53, p.12]):

[11] Philosophic outlines of said doctrine are already found in Sun Tzu, the modern incarnation can be traced to Mao, implementation to the 1980s, and open discussions among specialized scholars abound since the early 1990s [52]

> We can make the enemy's command centers not work by changing their data system. We can cause the enemy's headquarters to make incorrect judgments by sending disinformation. We can dominate the enemy's banking system and even its entire social order.

The interested reader is invited to peruse the some of the PLA's official and unofficial takes on future warfare in [54,53].

We would be remiss in our discussion if we were not to mention an nth order attack against the ultimate ancillary system: Electromagnetic pulse attacks against the electricity grid. An April 2008 report to the US House Armed Services Committee [27] outlined the effects on critical civilian infrastructure, should a nuclear weapon[12] be detonated 200-400 miles over Kansas (italics are ours):

> The functioning of society and the economy is critically dependent upon the availability of electricity. Essentially every aspect of American society requires electrical power to function. Contemporary U.S. society is not structured, nor does it have the means, to provide for the needs of nearly 300 million Americans without electricity. Continued electrical supply is necessary for sustaining water supplies, production and distribution of food, fuel, communications, and everything else that is a part of our economy. [..] *No infrastructure other than electric power has the potential for nearly complete collapse in the event of a sufficiently robust EMP attack* [..] Large-scale load losses in excess of 10 percent are likely at EMP threat levels. Instantaneous unanticipated loss of load, by itself, can cause system collapse. This is possible at 1 percent loss, and is very likely above 10 percent [..] Should the electrical power system be lost for any substantial period of time, the Commission believes that the consequences are likely to be catastrophic to civilian society. Machines will stop; transportation and communication will be severely restricted; heating, cooling, and lighting will cease; food and water supplies will be interrupted; and many people may die.

We therefore close on a somber note: The issues touched upon in this paper are not merely of academic or scientific interest. In practical terms, they go to the very heart of how future conflicts between open societies and their enemies will be waged - and are waged as we speak.

References

[1] K. Thompson, "Reflections on Trusting Trust," *CACM*, vol. 27, pp. 761–764, August 1984.

[2] L. Von Bertalanffy, "An Outline of General System Theory," *British Journal for the Philosophy of Science*, pp. 134–165, 1950.

[3] P. Érdi, *Complexity Explained*. Springer, November 2007.

[4] G. Gigerenzer, *Gut feelings: The Intelligence of the Unconscious*. Viking Books, 2007.

[5] G. Torres and C. Lima, "Maximum CPU Temperature." http://tinyurl.com/oytsnv, October 2007.

[6] US DOE, "Starlight Information System." Pacific Northwest National Lab, 2003.

[7] G. Conti, "Attacking Information Visualization System Usability Overloading and Deceiving the Human," in *Proceedings of the 2005 Symposium on Usable Privacy and Security*, pp. 89–100, ACM, 2005.

[8] W. Poundstone, *Gaming the Vote: Why Elections Aren't Fair*. Hill and Wang, 2008.

[9] J. M. Carlson and J. Doyle, "Highly Optimized Tolerance: A Mechanism for Power Laws in Designed Systems," *Physical Review E*, vol. 60, no. 2, pp. 1412+, 1999.

[10] A. Clauset, C. R. Shalizi, and M. Newman, "Power-Law Distributions in Empirical Data," *SIAM Reviews*, June 2007.

[11] J. Carlson and J. Doyle, "Highly Optimized Tolerance: Robustness and Design in Complex Systems," *Physical Review Letters*, vol. 84, pp. 2529+, March 2000.

[12] The exact yield necessary is presumably classified and/or undeterminable through simulation, expert physicist estimates range from 10-1500 kilotons.

[12] M. Newman, "Power Laws, Pareto Distributions and Zipf's Law," *Contemporary Physics*, vol. 46, pp. 323–351, September 2005.

[13] L. Manning, J. Carlson, and J. Doyle, "Highly Optimized Tolerance and Power Laws in Dense and Sparse Resource Regimes," *Physical Review E*, vol. 72, pp. 16108+, July 2005.

[14] J. Doyle and J. Carlson, "Power Laws, Highly Optimized Tolerance, and Generalized Source Coding," *Physical Review Letters*, vol. 84, p. 5656:5659, June 2000.

[15] J. C. Foster, V. Osipov, N. Bhalla, and N. Heinen, *Buffer Overflow Attacks*. Syngress, 2005.

[16] J. Makansi, *Lights Out: The Electricity Crisis, the Global Economy, and What it Means to You*. Wiley, 2007.

[17] L. Clarke, *Worst Cases: Terror and Catastrophe in the Popular Imagination*. University of Chicago, 2006.

[18] J. Barry, *The Great Influenza: The Epic Story of the Deadliest Plague in History*. Penguin, 2005.

[19] M. Newman, A.-L. Barabasi, and D. J. Watts, *The Structure and Dynamics of Networks: (Princeton Studies in Complexity)*. Princeton University Press, April 2006.

[20] Centre of Excellence Defence Against Terrorism, ed., *Responses to Cyber Terrorism*, vol. 34 of *NATO Science for Peace and Security Series E*. IOS Press, 2008.

[21] S. Gorman, *Networks, Security and Complexity: The Role of Public Policy in Critical Infrastructure Protection*. Edward Elgar Publishing, 2005.

[22] M. Guirguis and A. Bestavros, "Reduction of Quality (RoQ) Attacks on Internet End-Systems," in *2005 Proceedings IEEE INFOCOM*, vol. 2, March 2005.

[23] M. Guirguis and A. Bestavros, "Adversarial Exploits of End-Systems Adaptation Dynamics," *Journal of Parallel and Distributed Computing*, vol. 67, no. 3, pp. 318–335, 2007.

[24] Commtouch, "Server-Side Polymorphic Viruses Surge Past AV Defenses." http://tinyurl.com/2vewz8, May 2007.

[25] Commtouch, "Malware Outbreak Trend Report: Bagle-Worm." http://tinyurl.com/39gnz4, March 2007.

[26] M. Bond and G. Danezis, "A Pact with the Devil," in *Proceedings of the 2006 Workshop on New Security Paradigms*, pp. 77–82, ACM, 2006.

[27] W. Graham, "Report of the Commission to Assess the Threat to the United States from EMP Attack: Critical National Infrastructures," tech. rep., Congressional Report, April 2008.

[28] W. De Nooy, *Exploratory Social Network Analysis with Pajek*. Cambridge University, 2004.

[29] C. Ericson, "Fault Tree Analysis – A History," in *Proceedings of the 17th International System Safety Conference*, 1999.

[30] J. Dugan and S. Bavuso, "Dynamic Fault-Tree Models for Fault-Tolerant Computer Systems," *IEEE Transactions on Reliability*, vol. 41, no. 3, pp. 363–377, 1992.

[31] M. Bouissou, "A Generalization of Dynamic Fault Trees through Boolean logic Driven Markov Processes (BDMP)®," in *Proceedings of the safety and reliability conference (ESREL07)*, 2007.

[32] M. Bouissou and J. Bon, "A New Formalism that Combines Advantages of Fault-Trees and Markov Models: Boolean Logic Driven Markov Processes," *Reliability Engineering and System Safety*, vol. 82, no. 2, pp. 149–163, 2003.

[33] S. Distefano and A. Puliafito, "Dependability Evaluation with Dynamic Reliability Block Diagrams and Dynamic Fault Trees," *IEEE Transactions on Dependable and Secure Computing*, vol. 6, no. 1, pp. 4–17, 2009.

[34] M. Rausand and A. Hoyland, *System Reliability Theory: Models, Statistical Methods, and Applications*. Wiley-Interscience, 2004.

[35] S. DiStefano, "How to Capture Dynamic Behaviours of Dependable Systems," *International Journal of Parallel Emergent Distributed Systems*, vol. 24, no. 2, pp. 127–150, 2009.

[36] Formit, "VIS - the Vulnerability of Information System and its Inter-Sectorial, Economic and Social Impacts." http://www.formit.org/vis/, May 2009.

[37] J. Whittaker, *How to Break Software Security: Effective Techniques for Security Testing*. Addison Wesley, 2004.

[38] B. Miller, L. Fredriksen, and B. So, "An Empirical Study of the Reliability of UNIX Utilities," *CACM*, vol. 33, no. 12, pp. 32–44, 1990.

[39] B. Miller, G. Cooksey, and F. Moore, "An Empirical Study of the Robustness of MacOS Applications Using Random Testing," in *Proceedings of the 1st International Workshop on Random Testing*, pp. 46–54, ACM, 2006.

[40] L. Clarke, *Mission Improbable*. University of Chicago, 1999.

[41] A. Zuquete, "Improving the Functionality of SYN Cookies," in *Proceedings of 6th IFIP Communications and Multimedia Security Conference*, pp. 57–77, 2002.
[42] T. Leighton, "The Akamai Approach to Achieving Performance and Reliability on the Internet," in *Proceedings of the 26th ACM Symposium on Principles of Distributed Computing*, ACM, 2007.
[43] A. Futoransky, D. Saura, and A. Waissbein, "The ND2DB Attack: Database Content Extraction Using Timing Attacks on the Indexing Algorithms," in *Proceedings of the 1st USENIX Workshop on Offensive Technologies*, pp. 1–9, USENIX, 2007.
[44] T. Popp, S. Mangard, and E. Oswald, "Power Analysis Attacks and Countermeasures," *IEEE Design & Test of Computers - Design and Test of ICs for Secure Embedded Computing*, vol. 24, no. 6, pp. 535–543, 2007.
[45] A. Shamir and E. Tromer, "Acoustic Cryptanalysis. On Nosy People and Noisy Machines," tech. rep., RSA and MIT CSAIL, 2004.
[46] H. Kesavan and J. Kapur, "The Generalized Maximum Entropy Principle," *Systems, Man and Cybernetics, IEEE Transactions on*, vol. 19, pp. 1042–1052, Sep/Oct 1989.
[47] E. Ott and M. Spano, "Controlling Chaos," *Physics Today*, vol. 48, no. 5, pp. 34–40, 1995.
[48] L. Lam, *Nonlinear Physics for Beginners: Fractals, Chaos, Solitons, Pattern Formation, Cellular Automata and Complex Systems*. World Scientific Press, 1998.
[49] F. Fukuyama, *Trust: The Social Virtues and the Creation of Prosperity*. Free Press, 1996.
[50] B. Hoffman and G. Weimann, "Econo-jihad," May 2009.
[51] S. Salama, "Unraveling Al-Qa'ida's Target Selection Calculus," *Terrorism and Political Islam*, p. 41:44, 2007.
[52] J. R. Lilley and D. L. Shambaugh, eds., *China's Military Faces the Future*, ch. 3-4. Studies on Contemporary China, M.E. Sharpe, 1999.
[53] M. Pillsbury, "China's Military Strategy Towards the United States: A View from Open Sources," tech. rep., US-China Economic and Security Review Commission, November 2001.
[54] M. Pillsbury, *Chinese Views of Future Warfare*. National Defense University Press, September 1998.

The Virtual Battlefield: Perspectives on Cyber Warfare
C. Czosseck and K. Geers (Eds.)
IOS Press, 2009
© 2009 The authors and IOS Press. All rights reserved.
doi:10.3233/978-1-60750-060-5-282

Business and Social Evaluation of Denial of Service Attacks in View of Scaling Economic Counter-measures

Louis-Francois PAU[a1]
[a]*Copenhagen Business School and Rotterdam School of Management*

Abstract. This paper gives an analytical method to determine the economic and indirect implications of denial of service and distributed denial of service attacks. It is based on time preference dynamics applied to the monetary mass for the restoration of capabilities, on long term investments to rebuild capabilities, and of the usability level of the capabilities after an attack. A simple illustrative example is provided for a denial of service on a corporate data centre. The needed data collection methodologies are categorized by classes of targets. The use of the method is explained in the context of legal or policy driven dissuasive, retaliation or compensation/ restoration actions. A concrete set of deployment cases in the communications service and transport industries is discussed. The conclusion includes policy recommendations as well as information exchange requirements.

Keywords. Cyberwar, Economics, Business impact, Social impact, Time preference dynamics, Mobile communications, Transport, Denial of service, Restoration costs

Introduction

This work in progress aims at addressing two strategic aspects of cyber-warfare mostly via communications networks and IT applications: a) first to take a total economic and social view in the assessment of evaluating damages of a cyber-warfare attacks on a society or business target; b) scaling a trade, economic, or legal retaliation or dissuasion for decision makers. It is assumed that the target of the attack does not in general have itself any or sufficient defence or attack means, so that a corporate or national level may decide ex-ante (dissuasion) or ex-post (retaliation, compensation) to scale a business defence affecting the economic sphere of the attacker. Such an approach is also relevant sometimes when attacker cannot be identified and localized precisely, so that the economic sphere of the attacker is restricted to business networks to which the attacker belongs.

Traditionally the damage assessment has been considered "binary" and limited in time, in that the target was considered to be rendered totally dysfunctional until full restoration only of its information and communication capabilities. Lessons learnt tell us that other organizational, physical, human and social capabilities are to be counted

[1] Correponding Author: Louis-Francois Pau, Prof. Mobile business; Email: lfp.inf@cbs.dk

as representing often larger collateral damage of the attacks; their restoration eventually takes quite some time, especially if the surrounding society does not have enough civil defence means and skills in place. Vice-versa, sometimes, the replacements made to infrastructure damaged by the attack will be less obsolete leading to better future robustness. To address this issue, the approach is to capitalize on the ability of cost-benefit analysis to bundle into the internal rate of return both tangible and some intangible effects .The internal rate of return expresses the time preference on tangible and intangible assets ,old and new, which gives a break even net present value over the long term. It is then proposed to treat short term dynamics of this internal rate of return , when exposed to a Brownian shock linked to an attack affecting the command and control node for the society or business target which have their normal long term equilibrium return rates.

Assuming the dynamic time preference resulting from a cyber-attack, it becomes possible to estimate all of the following :a) the incremental monetary mass needed short term for restoration of equilibrium business and social capabilities; b) long term investment over a given pay-back horizon needed over time to restore and improve capabilities to get back to the equilibrium rate; c) the value of the assets degraded by the cyber-attack as short term and long term restoration measures impact the target.

Apart from relevance in a national or corporate budgeting process, such a three-dimensional scaling of compensation, retaliation or dissuasion gives decision makers a way to communicate efficiently around them and to implement such counter measures against the attacker's economic sphere while referring eventually to a game theoretical equilibrium required by legal/treaty provisions.

As a conclusion, the proposed methodology empowers decision makers to scale eventual economic counter-measures or threats against attackers, the efficiency of which cannot be guaranteed as economic-social effects may not always impact attackers but surely their surroundings, and as the resolution of decision makers may also vary. It will be up to the reader to assess relevance in her/his own context, while this project has assessed some concrete cases. This project has also been motivated by specific concerns and abilities of wireless communications operators.

1. SURVEY

The cyber attacks considered in this paper (denial of service DOS , and distributed denial of service DDOS) are those damaging information, capabilities, and sometimes network and infrastructure elements owned or operated by a target, with resulting damages not only to the target but also to third parties dependent on this information, or those networks and infrastructure [1, 14, 16]. Damage assessment is considered difficult, as the intrusions and attacks cannot always be detected short term [2, 15, 17]. Nevertheless, large economic and social impact is felt, reaching from a unit in an organization to whole sectors; have been carried out as part of earlier work: descriptive assessments of the impact from surveys with input-output analysis of effect from outages and propagation models (e.g. [3, 7]), evaluations of incentives and investments to protect the information infrastructure (e.g. [4, 6, 8]), and evaluations of cyber-insurance premiums in relation to security procedures [5, 20]. Very few papers deal with models for damage assessment, which would allow a company to qualitatively and quantitatively estimate possible financial losses due to partial or complete interruption of connectivity; in [9] a systems engineering approach is taken, while in the present one an economic and business approach is taken and a simple numerical example is given in Section 4.

Also we will address in Section 5 the use of damage assessment estimates on legal grounds for retaliation or compensation [18, 19]. A distributed Denial of Service attack aims to deprive legitimate users of a resource or service provided by a system, by overloading the system with a flood of data packets, thus preventing it from processing legitimate requests. Therefore it is necessary as in [10] to take into account the doctrines governing the allocation of liability among key players in a distributed denial of service attack. Such doctrines are well established and based on common law tort principles and policy considerations.

Regarding related types of attack, such as malware, viruses, identity theft, exploiting vulnerabilities in control software / management functions/ protocols (such as DNS and BGP errors, lack of authentification of users, services or flows, payment solutions vulnerabilities), some studies like those of Ferris Research and Gartner Research have shown the huge business impact thereof as well as the very high handling plus restoration costs. But such estimates are at best interview based, and lack an analytical framework.

2. THEORETICAL BASIS

Comparison with economic theory

One way of looking at the economic consequences of a denial-of-service is to consider that the target has a diversity of assets included in a portfolio , each with varying life-cycles, and that any attack affects the overall value and sustainability of the portfolio. Whereas in economics and finance the typical research question is one of asset allocation in view of returning some performance goals [13], the cyber-warfare economics question is one of asset preservation over time. Another difference with economics and finance is that in these fields' risks and returns are usually mutualised across populations of owners or users via legal contracts, in cyber-warfare economics the target normally stands alone at the time of attack with all risks and must have made all required preventive investments. The only subfield of economics where some common features can be identified, is the area of pension economics where the retired person wants to maintain over time a purchasing power level, although here again assets are a mix of own assets and mutualised assets.

Regarding the definition of capabilities exposed to an attack, they are defined at any time as the net difference between a normal time-dependent operational capability profile of the attacked entity, and the complete or partial combined effect of the attack and of restoration measures on normal capabilities due to the nature of the attack and restoration processes. Consequently, dynamics play an important role, and the proposed methodology encompasses situations with a net reduction in capabilities. If the attack on one target involves reduced capabilities of other asset owners (like in the case of a "netbot", or the halving of transmission rate capability by the TCP protocol in case of a transmission error / congestion) one can either take a systemic view or the view of the target alone.

Regarding the description of the stochastics of attack processes, only attack specific process specifications with related methods would allow to model them closely, but macro-level approximations by known or tailored distributions already provide a good basis.

Regarding the restoration process, it is also to have its specific dynamics. However, restoration is supposed to be possible, at a cost, but not impossible, thus implying that data protection, integrity and security must be in place. In the case of data loss prevention (DLP) , the Ponemon Institute has estimated from commercial cases the cost of data loss to 100 k Euro- 5000 k Euro of which 36 % due to commercial losses and lost customers, and 36 % from loss of portable data storage. Although VOIP content is vulnerable, repeated calls remain possible in general.

Proposed methodology: time preference dynamics

The proposed methodology is to assume that the target applies different time preferences to the assets in its portfolio, where the time preference profiles express the urgency at which restoration of capabilities must be carried out in view of a time distributed attack (including a shock) degrading suddenly specific assets in the portfolio. In economics, **time preference** (or "discounting") pertains to how large a premium a user will place on usage nearer in time over more remote usage.

Taking one class of assets, assume that the time preference rate r(t) fluctuates around an equilibrium level r(eq) while subject to a Brownian point process W(t) .The short term dynamics are modelled by [12]:

$$dr(t) = a(r(eq) - r(t))dt - V.dW(t) \qquad (1)$$

where :

- r(t) is the short term time spot preference at time t for a given asset, $0<r(t)<1$
- a is the intensity of the feedback force towards the equilibrium time preference r(eq)
- r(eq) is the equilibrium time preference for that given asset
- V is the volatility of the time preference fluctuations
- W(t) is the stochastic Brownian point process driving the attack diffusion process

Monetary mass requirements for restoration of capabilities

The incremental monetary mass dM (t) needed short term for restoration of equilibrium capabilities of the asset can then be determined. Assuming for simplification purposes the short term time preference rate r (t) to drive short term interest rate dynamics by near a constant rM:

$$dM(t) = M(t-dt). (r(t)+rM).dt \qquad M(0)=M0$$

where :

- M(t) is the monetary mass used short term to invest in rebuilding the asset capability to its levels just before t=0 where monetary mass represented by the asset value was M0
- rM is the fixed increment to the short term time preference producing the short term interest rate payable to finance the rebuild of the capabilities

Long term investments to rebuild capabilities

The long term investment K (t) over a given horizon TK needed long term to restore and improve the assets capabilities to get back to the equilibrium time preference rate r (eq) can be expressed as follows:

dK(t) = K(t) [r(t).dt + V.BetaK.(dW(t)+Lambda.dt)]
BetaK = (1-exp(-a.TK))/a

where :

- K(t) is the long term bond-like investment needed over the horizon TK to restore asset's capabilities
- K(0) is the initial annuity value of the assets capability value over the horizon TK
- TK is the time horizon to rebuild and possibly improve on the asset's capabilities; this parameter is essential in all practical cases
- Lambda is the premium by unit risk needed by the market to support the randomness over the real time preference
- BetaK s a constant

Usability of the capabilities over time after an attack

The value of the assets degraded by the cyber-attack as short term and long term restoration measures impact the target, is linked to a specific usability risk characterization WA (t) of the asset's capabilities. The change in the degree of usability A (t) of this asset, bounded between 0 and 1 is:

dA(t) = A(t). [r(t).dt +V (dW(t)+Lambda.dt) +VA (dWA(t)+LambdA.dt)]

where:

- the first term in the parenthesis is the effect of short term restoration via the monetary mass investment
- the second term in the parenthesis is the contribution from long term fixed horizon asset capability rebuild
- the third term is the reduction in recovery speed linked to the volatility and risk in the asset's specific capabilities as they impact its degree of use
- A(t) is the effective degree of usability of the asset , A(eq)=1
- VA is the volatility of the asset's capabilities usability risk
- WA(t) is the Brownian motion of the usability risk characterization of the asset's capabilities
- LambdA is the premium by time unit in unit usability risk needed by asset users to support the randomness over the asset's usability risk.

The unique property of this model is that all time preference variations are subject to the short term time preference and that the risk exposure, which is here the investment needed to restore the asset's capabilities, is by one bond-like financing the duration of which determines the size. The usability of the asset is a Brownian movement correlated with the time preference rates over time. Another characteristic of this model is that it is decoupled from the initial asset valuation , which can be tailored to specific cases and rely on data pre-existing to an attack (see Sections 4 and 5).

3. A SIMPLE NUMERICAL EXAMPLE

Scope

This very simple example does not allow to show and exploit all the dynamic effects taking place, but to show how a concrete situation can lead to estimations of short term and long term financing needs tied to the time preference expressed. It also shows that, even if financial means are made available to rebuild capabilities, the actual restoration time of usable capabilities is very much subject to the stochastic distribution properties and to the quality of actual means for capabilities restoration. It also leads in Section 4 to further data collection methodology considerations.

Description

The numerical example pertains to a data centre in a company, with a scrap value of 10 MEuros, running services to support company operations. The equilibrium state is one where all services operate 100 % to support all divisions and operations with a company turnover of 500 MEuros/year; furthermore client capabilities are dependent on the company's operations being supplied to them for another 500 MEuros /year (treated as contingent liabilities). The equilibrium time preference r (eq) is equivalent to the company's net operational profit margin from operations r (eq) = 50 %/year, approximated as 0.5/ (365x24) = 5.7E-05 /hour. The short term monetary interest rates are only about 10 %/year, so that rM= -4.76E-05/hour. A full instantaneous attack W (0) =1 on the data centre at time t=0 reduces services usability to A (0) =0 with a minimum nominal restoration time of TK= 3 months for all resulting services and operations to internal divisions and third parties after such a disruption. The attack lasts dt= 1 hour , taken also as time increment, creating a shift in the time preference to a very high spot time preference value ; the maximum which can be chosen is r(1 hour)=1, meaning the target wants perceptually all measures to be taken for immediate recovery of the data centre . With a maximum volatility in time preference fluctuations of V=1 /hour, the needed reactivity becomes: a ~ 5.8E-05. Post attack, the short term time preference grows tremendously leading to a strong rise in perceived short term monetary flows for restoration dM(1) of slightly under 10 MEuros/hour ; this expresses the perception that the data centre must be restored at once . The total capability value of the assets over TK=3 months is 250 MEuros with an hourly annuity of 115 740 Euros. With a risk premium Lambda= 0.2, the initial long term investments dK (1) needed to recoup lost supplies to customers, and to rebuild the capability, can be estimated at about 235 M Euros. For the usability risk WA (t) a simplified linear decreasing profile can be taken over the restoration period TK, that is WA (t) =1-(t/TK); we also assume LambdA=0. However, the quality and efficiency of the restoration are highly volatile especially in downstream supply chains from the company ; this leads to the usability of the target's capabilities only increasing again (dA(t)/dt >0) , despite a high time preference, if the volatility VA is less than 1,2*TK . Half of the overall capability is only restored at time 0,5/ (1,2-VA/TK) which can be longer than TK= 3 months for some values of VA.

4. APPLICATION AREAS AND DATA COLLECTION METHODOLOGIES

This paper cannot give cases or fictive examples for all the application areas for which economic and social impact of denial of service need to be quantified. This Section only serves to survey such areas by categories and to give when known established approaches to assess relevant data to be fed into the calculations.

4.1. Public services

The denial of service of public services on a national basis or on an agency basis (administrative services, social services, water, air traffic, waste management, financial payments), have wide ranging consequences where the indirect impact encompasses prejudice caused to citizens (in their ability to act, to get benefits or to contribute tax etc) measured in time lost, benefits / contributions lost, and of qualitative damage (health, safety, administrative registrations etc). In this field, traditional cost-benefit analysis of tangible and intangible services applies. As to the setting of the time preference rates, they should be high for those public services where public authorities by law have obligations of service continuity, while they would be less and derived from minimal service obligations in other cases.

4.2. Company products and services

In this case, the applicable methodology to the data collection is the one used for corporate liability insurance assessment. This includes loss of capabilities (physical, raw material and service related) with their replacement, loss of revenue due to non delivery in time, physical loss of output such as manufacturing with associated logistic and CRM overheads , indemnification of human resources if work or life is jeopardized, and indirect loss and damage to clients. As to the setting of time preference rates, in-company rates should correspond to the average return on assets or operational margin (whichever is largest) within the sector in which the company was denied services, while the same would apply for the clients in their respective sectors.

4.3. Loss of shared infrastructure

There is no established methodology to cover loss of shared infrastructure, "critical" or not (such as communication or transportation networks, denial of service of a satellite by jamming, etc). However the normal approach would be to make the inventory of the lost capabilities (physical and service related) by infrastructure operator, of lost revenues by infrastructure operator including claims payable to customers under contract terms, of verifiable loss and damage by individual and institutional users, and moreover of social costs to the same. As to the setting of time preference rates, this is a difficult issue as infrastructure suppliers quite often do not have contractual quality of service obligations. On the contrary, suppliers of "critical" infrastructure whose control systems may have been compromised, bear a responsibility beyond just service provisioning, and there recovery processes may be longer. Judgment would have to be applied to the time preference of the infrastructure operator (normally very high but not coupled to financial rates of returns) and to the users taking diversity into account. For users the principle of setting the time preference could be based on the tolerable postponement of the access and use of the shared infrastructure to next normal period (such as shift by e.g. one day, or to next available equivalent infrastructure provider).

4.4. Technology providers

Some well known technology providers in such areas as communications, software, control systems, transport technologies, biomedical devices, etc.., may be liable to claims by their customers for vulnerabilities in their products, although third parties are those exploiting them. While the "customer cum users" would know the attack profiles, while not always knowing the technical roots for the vulnerabilities, technology providers may benefit from the proposed framework for risk assessment if they share attack profiles with their customers. The risk assessment method in turn allows them to quantify reasonable levels of investments in improving the technologies and their distribution mechanisms.

5. DENIAL OF SERVICE IMPACT ANALYSIS USAGE PROCESS

The concept is to use the damage assessment methodology of Section 2 , with its different time scales, to specific data collected by established methodologies moderated by neutral judgement (like best practices or eventually arbitration courts) (see Section 4) , to calculate estimates of the set of damages . Such assessments must be transparent and done by neutral parties.

The assessed damages can then be used by executive authorities for a spectrum of actions:

- Dissuasive process: preemptively to a denial of service, by policy makers or companies, to announce that these claims would be raised if an attack occurs. The policy makers or companies may not have evidence yet or from past cases to identify the attackers, but may communicate to make such a categorization of attackers credible and visible to attackers .Also, subject to proper later judicial tracing and identification of the attackers, the policy makers or companies would communicate that they intend to recover the amounts of the claims by all legal means in case of an attack. As the average cost to attackers of a cyber-attack is usually small, dissuasion followed by retaliation or recovery may be of some concern to attackers or their backers.
- Retaliation process: if the attackers are traced and identified by technical and/or judicial means, or if strong assumptions and partial evidence exist (e.g. from IP addresses, software code structure, software forensics, etc…), legal or forceful retaliation would be done for the same size of claims against direct or indirect interests of the attackers. One obvious instance of this would be to seize quarantine or destroy the physical and communications assets used by the attackers, or assets owned controlled by them. This may happen in a judicial framework (with fines and penal measures) or an international treaty framework, but may be replaced by policy maker coercitive decisions including offensive means.
- Compensation / Recovery process: if the attackers are traced and identified by judicial means, and can be put on trial, this process would use the damage assessments as normally done in a judicial court procedure. In this case

however the data collection methodology and data would be subject to a contradictory evaluation, there may be issues of sovereignty leading to inability of enforcement/ extradition, and the delays involved are normally quite long.

- "Keep silent" process: There is of course a fourth process, which is to ignore attacks, keep silent, report nothing, and not to sue, often for "image» reasons. It is unfortunately very common that banks, communications and infrastructure operators so far do not report attacks and even figure out other reasons vis-à-vis their users.

It is conjectured that the main practical relevance of the proposed method is for dissuasive and retaliation processes, resting ultimately on the ability of the asset owner / target to carry out and update his own exposure valuations based on estimates related to user and client damages (tangible and intangible).

This same conjecture is obviously reinforced by the consideration that the tracing and identification of the true attackers may not always be possible, or may take so much time, that the strategy to use a recovery process may not work while a dissuasive or retaliation process may have effects when used together.

Likewise, if attackers are using innocent identifiable resources, a recovery process would take time establishing that they are not responsible, while giving time to the responsible attackers.

It should not be forgotten that cyber-attacks against corporate assets often are initiated from inside the company or past employees, which too opens up for a combination of dissuasive, retaliation and partial recovery processes.

Finally, as some types of defensive measures (such as anti-virus) have fast deployable get-around's known to attackers, dissuasion and retaliation processes may in some cases be the only way forward.

6. APPLICATION CASES

This research has found its way into a number of deployment cases summarized below spanning all categories identified in Section 4:

Public services

Case: minimal public transport service under employee strikes (Western Europe)
Contribution: the proposed method allowed determining the public damage- number of employees on strike curve, allowing for the union and the employer to settle on a minimum service level.

Company products and services

Case: corporate liability insurance estimate for a Scandinavian CRM provider
Contribution: the customer relationship management (CRM) company's services were outsourced by several operators in the communications and credit card fields. The contracts between these operators and the CRM service supplier stipulated damage claims should the CRM supplier not be operational. The method allowed the CRM

provider to determine the liability insurance amount it had to get cover for vis-à-vis cyber attacks to compensate its customers.

Loss of shared communications infrastructure

Case: attack on 3G operator BSC with partial recovery via other operator(s)
Contribution: The wireless 2G and 3G base system controller manages the connectivity to and from radio base systems (RBS). Due to bad network management or practices, some BSC are not totally immune from certain types of attacks. When redundancy and restoration procedures have failed, radio coverage and connectivity may be lost unless back-up is activated from other operator's BSC (when feasible). Such operators have to be compensated, as well as possibly some wireless service users under contractual terms, and total damage assessment with/without insurance may be necessary.

Mobile networks not only provide great benefits to their users but they also introduce inherent security issues. With respect to security, the emerging risks of denial of service (DOS and DDOS) attacks will evolve into a critical danger as the availability of mobile networks becomes more and more important for the modern information society. There are ways to mitigate the attacks by adding minimal authentication to the radio channel assignment protocol, but this too has business implications and requires risk assessment. At the same time, via subscriber management, interoperable management and signalling / control networks, they carry the potential for tracing and retaliation measures, besides lawful interception in support of legal procedures. In particular is highlighted the retaliation process which international inter-carrier settlements allow for, as such agreements reach out worldwide.

Finally, it has been brought to the attention of the author, that other applications exist, e.g. in the case of water distribution protection, where attacks have wide reaching implications, and where physical-chemical forensic evidence may be collected. In this case the attack has both time-based as well as spatial distributions.

7. CONCLUSION

While law and jurisprudence regarding denial of service and other cyber-attacks is making slow progress in both national and international arenas, this paper presents a quantitative approach respecting attack and restoration dynamics likely to be used in dissuasion as well as in retaliatory processes, in the hope that ultimately attackers will feel a largely missing retroaction. It may also allow institutions and companies to determine by self-analysis in the presence of a given threat profile, which assets to protect in priority on economic, business and social grounds.

In the event international organizations like GATT, European Space Policy Institute, OECD or the European Parliament ("Declaration on the reinforcement of international security", 25 March 2003 and report to the Council of 11 December 2008) also embark on putting an economic and social measure to cyber-attacks, supplemented by constraining legal measures, instead stating of political / cultural or defence values only, this research may give elements of the analysis.

Specific policy recommendations linked to the above research and the deployed cases would be the following:

- in international commercial contract law, allow for compensation and information exchange clauses whereby attackers using one party's facilities or services to mount an attack on the other party, may retaliate against the attackers on the basis of damage assessment and evidence provided by the other party; an example of this are international communications operators inter-operator settlement procedures;
- enhance auditing procedures, to verify the basis for insurance or damage claims in the case of cyber-attacks;
- mandate reporting and information exchange about attacks to designated governmental bodies, for sharing of attack profiles and partial evidence (like envisaged by the EU).

Just as technical vulnerability reduction demands collaborative efforts between users, technology providers and operators, the business and social impact assessments also demand such collaboration and information exchange, besides internal due diligence. The issue is which governments, players and sectors, like the communications industry, will take concrete steps in this direction. One reason why this is an issue is that "patches" and additional costly imperfect technologies are too often preferred to demanding and longer lasting technical, legal, architectural and economic measures. It is in this context that humanities, economic and social disciplines can clarify the way towards peace in cyber conflicts [21].

What this research does not allow to do is to account for interdependencies between targets and attackers, or proxies to the attackers, due to cross-ownership, exclusive agreements, shared infrastructure (buildings, communications, transport, and energy), geo-economics and political / cultural / social influence.

REFERENCES

[1] Lech J. Janczewski, Andrew M. Colarik (Editors) (2007), *Cyber Warfare and Cyber Terrorism*,
[2] Boca Raton: Idea Group Inc (IGI), ISBN 1591409918, 9781591409915
[3] O. Sami Saydjari (2004), *Cyber defence: art to science*, Communications of the ACM,
[4] Vol 47, Issue 3 (March), 52-57
[5] Scott Dynes, Eva Andrijcic, M. Eric Johnson (2006), *Costs to the U.S. Economy of Information Infrastructure Failures: Estimates from Field Studies and Economic Data*, Proceedings of the Fifth Workshop on the Economics of Information Security, Cambridge: Cambridge University
[6] Lawrence D. Bodin, Lawrence A. Gordon , Martin P. Loeb (2005), *Evaluating information security investments using the analytic hierarchy process*, Communications of the ACM, Volume 48 , Issue 2 (February) , 78 - 83
[7] Jay P. Kesan_ Rupterto P. Majuca, William J. Yurcik (2004), *The Economic Case for Cyber insurance*, University of Illinois College of Law and Economics Working Papers, Paper no 2, http://law.bepress.com/uiuclwps/papers/art2
[8] Huseyin Cavusoglu (2008), *Economics of information security*, in: L. Jean Camp and Stephen Lewis (Editors), Advances in Information Security, Vol.12, Springer, US, 978-1-4020-8089-0 (Print) 978-1-4020-8090-6 (Online)
[9] Marco Benini & Sabrina Sicaria (2008), Risk *assessment in practice: A real case study*,
[10] Computer communications, Vol 31, no 15, 3691-3699
[11] Bruce H. Kobayashi (2005), *An Economic Analysis of the Private and Social Costs of the Provision of Cyber security and other Public Security Goods*, George Mason University School of Law, Working Paper Series, Paper no 26, http://law.bepress.com/gmulwps/gmule/art26
[12] 9. T. Dubendorfer, A. Wagner & B. Plattner (2004), *An economic damage model for large-scale Internet attacks*, in Proceedings 13th IEEE International Workshops on Enabling Technologies: Infrastructure for Collaborative Enterprises (WET ICE) , 14-16 June, 223- 228, ISBN: 0-7695-2183-5

[13] Meiring de Villiers (2007), *Distributed Denial of Service: Law, Technology & Policy*, Sydney: University of New South Wales Faculty of Law Research Series, Paper no 3, http://law.bepress.com/unswwps/flrps/art3

[14] Valer Bocan & Vladimir Cretu (2005), *Mitigating Denial of Service Threats in GSM Networks, The First International Conference on Availability in GSM networks*, Department of Computer Science and Engineering, Politechnica University of Timisoara, http://www.dataman.ro

[15] O. Vasicek (1977), *An equilibrium characterization of the term structure*, Journal of financial economics, Vol 5, no 2, 177-188

[16] K.C. Butler, D.L. Domian (1991), *Risk, diversification and investment horizon*, Journal of portfolio management, Vol 17 ,Spring, 41-47

[17] D. Ventre (2007), La guerre de l'information, Paris : Lavoisier, ISBN: 978-2-7462-1883-3

[18] J. Kizza, F. Migga Kizza (2008), *Securing the information infrastructure*, Cyber Tech Publishing, ISBN: 978-1-59904-379-1

[19] T. Shimeall, P. Williams, C. Dunlevy (2001), *Countering cyber-war*, NATO Review, Vol 49, no 4, Winter, 16-28

[20] Jason Fritz (2008), *How China will use cyber warfare to leapfrog in military competitiveness* , Culture Mandala, Vol. 8, No. 1, October, 28-80

[21] H. Axlerod & D. Jay (1999), *Crime and punishment in cyberspace: Dealing with law enforcement and the courts*, Paper presented at the SIGUCCS Conference, Denver Colorado

[22] M. Erbschloe (2001), *Information warfare: How to survive cyber attacks*, Berkeley, California: Osborne/McGraw Hill.

[23] Shari Lawrence Pfleeger, Daniela Golinelli, Robin Beckman, Sarah K. Cotton, Robert H. Anderson, Anil Bamezai, Christopher R. Corey, Megan Zander-Cotugno, John L. Adams, Ronald Euller, Paul Steinberg, Rachel Rue, Martin C. Libicki & Michael Webber (2008), *Cyber security Economic Issues : Corporate Approaches and Challenges to Decision making*, Rand Corp. Research brief RB-9365-1

[24] M. Godelier & H. Dawod (Editors) (2004), *Guerre et paix*, Special issue, Thema, Paris: CNRS, March, www.cnrs.fr/presse

The Virtual Battlefield: Perspectives on Cyber Warfare
C. Czosseck and K. Geers (Eds.)
IOS Press, 2009
© 2009 The authors and IOS Press. All rights reserved.
doi:10.3233/978-1-60750-060-5-294

Virtual Plots, Real Revolution

Roelof TEMMINGH[a] and Kenneth GEERS[b]
[a] *CEO, Paterva, Pretoria, South Africa*
[b] *Scientist, Cooperative Cyber Defence Centre of Excellence, Naval Criminal Investigative Service (NCIS), Tallinn, Estonia*

Abstract. It is increasingly difficult to separate 'cyberspace' from what we think of as the 'real world'. Human beings respond to stimuli from both. Threats to persons, organizations, and governments require timely and accurate evaluation, but cyber attackers can exploit the imperfect and maze-like architecture of the Internet to make threat evaluation difficult. In cyberspace, it is possible to create fraudulent online identities – potentially millions of them – that could programmatically support any personal, political, or military agenda. In the future, computer botnets may evolve from spam and Distributed Denial of Service (DDoS) generators to semantic creatures that can post opinions, arguments and threats on the Internet. Counterfeit identities on the World Wide Web (WWW), complete with randomized or stolen biographies, pictures, and multi-year histories of Internet activity, will be difficult to separate from real human beings because there is no quick way to determine whether a virtual person really exists.

Keywords. Artificial Intelligence (AI), botnet, cyber threat, cyber warfare, identity theft, Information Operations (IO)

Introduction: a 'semantic botnet'

Cyber attackers exploit the relative anonymity of Internet communications to send unwanted data, including spam, malicious code and Denial of Service (DoS) attacks around the world with near impunity. In the future, computer 'botnets' – networks of compromised and organized computers within a common Command and Control (C2) infrastructure [1] – will evolve to encompass virtual populations of randomly-generated and/or stolen identities, which could be used to support any personal, political, or military agenda. Each fabricated virtual identity will have a 'life' of its own, whose credibility grows over time as the number and variety of its Web postings proliferate.

In 1950, Alan Turing wrote that even the "dullest" human could outperform a computer in a conversation with another human. Turing believed it was inconceivable that a machine could provide a "meaningful answer" to a truly wide variety of questions [2]. In 2009, that may still be the case, but there is a big difference between the formal test that Turing proposed and posting a comment to a blog. Even with adequate time for analysis, there is simply not enough content to evaluate whether it was posted by a man or a machine.

On the Internet, the best way to separate real people from artificial people – without time-consuming, in-depth and unlikely cross examination – is with statistics.

However, for every mathematical defense strategy, there seems to be an effective response for the attacker.

1. The search for intelligent life on the Web

Until recently, most Internet conversations were conducted via email, in relatively interactive, one-on-one correspondence (Internet Relay Chat (IRC) does not count as it was never mainstream). Today, new technologies such as YouTube, Facebook and Twitter allow each of us to be a prolific producer of digital information. The current model is frequently not one-to-one, but one-to-many or many-to-one.

1:n/n:1 communications are popular because they cover a lot of ground very quickly; for example, Sophie can now update her whole family at once instead of each member individually. However, the trade-off here is a loss of interactivity. To pass or fail the Turing Test, some level of cross-examination is required. In short, banal, or low-interaction conversations, a cyber attacker (in the form of an artificial, 'proxy' personality) will find it easier to push information to the world and have it accepted as is, i.e. with no follow-up question and answer time.

Further, it is reasonable to assume that most Internet traffic consists of trivial, everyday information. Serious political and philosophical discussions normally do not take place in the 1:n/n:1 environment. Even when it does, Natural Language Processing (computer analysis of human languages) is unproven technology, and requires significant human oversight to be effective.

Thus, this paper assumes that most of the information found on the Internet is not unique, and can be stolen and repackaged for nefarious purposes. Even among unique photos of a common sight such as the Eifel Tower, the photographers themselves might be hard-pressed to find their own picture among others. Effective authentication technologies such as digital signatures exist, but they are rarely used for common communications, which remain open to theft and malicious manipulation.

2. Internet-enabled intelligence

With an unrestricted Internet connection, the average Web user now has access to more information than her head-of-state did just five years ago. All of Wikipedia, for example, can fit on one hard drive. The data points are now all there; the best strategy is to choose one's sources carefully and to discriminate between good and bad analysis. This is the art of Open Source Intelligence (OSINT).

Computer hackers conduct OSINT just like everyone else. In fact, they also begin their search for information at a target's homepage. Good OSINT can quickly lead an attacker from a name to a date-of-birth, address, education, medical records, and much more. Via social networking sites, the attacker may soon discover intimate details of a person's life, from the clubs she frequents to where she might physically be at any given moment. Eventually, a web of connections to other people, places and things can be constructed.

Good OSINT researchers – and hackers – master both the semantic side of the Web (e.g. the content of a webpage), and the technical side, like the Domain Name Service (DNS), or the 'phone book' of the Internet. The DNS registry catalogues who owns a given website, and often provides a point-of-contact for them in the real world.

Hackers 'enumerate', or conduct in-depth technical reconnaissance, against cyber targets. Technical information, including barely-hidden 'metadata' such as an Internet Protocol (IP) address or a timestamp, is analyzed for anything that can be exploited in the real world. Common applications like webmail are frequently targeted. Sooner or later, hackers normally find an open, misconfigured, or vulnerable Internet access point, which is analogous to a thief finding or forcing open a door or window in the physical world.

The real magic of an effective cyber attack lies in combining technical data with real-world information. Likewise, threat actors can be divided into those who have 'reach' into the real world, and those who do not.

3. It's good to be the king

The combination of Internet monotony and hacker creativity described above can make for a volatile mixture. The average computer programmer could never pass the Turing Test, but she can write a program that updates the world via Twitter on how a bogus Web user is spending his day, or what a bogus Web user thinks about how you are spending yours. And if it is theoretically possible to create one false Web identity, perhaps millions of them already exist.

A large virtual population, scattered all over the world and encompassing different socioeconomic backgrounds, could be programmed to support any personal, social, business, political, military, or terrorist agenda. The nature of an attack could be limited only by the attacker's imagination. For example, in the week before an election, what if both left and right-wing blogs were seeded with false but credible information about one of the candidates? It could tip the balance in a close race to determine the winner.

Via Internet-enabled OSINT, targets can be meticulously profiled by an attacker to learn personal, organizational, or national sensitivities and vulnerabilities. For example, if the target were a multinational corporation (MNC) engaged in oil exploration, OSINT might reveal a wide range of attack vectors: disgruntled employees, friction with indigenous populations, whistleblowers, and/or ongoing lawsuits. A zombie army could be used to target any or all of the above – including judges and jury – by manipulating industry blogs, commenting on news articles, sending targeted email, etc. The MNC, of course, could have its own botnet army pushing its side of the story.

In the impersonal world of cyberspace, who can say for sure whether a message was sent by a real person? Even highly idiomatic language can be stolen by a robot and used (perhaps incorrectly) in another context. It is beside the point to say that one could *eventually* authenticate the information. Propagandists seek first and foremost to bring attention to their cause; ethical considerations are secondary. And the attacker may simply need for the effect to be temporary. If a certain momentum toward the desired goal is achieved – that is, if real people begin to follow the robots – then the attacker can begin to 'plug out' the artificial intelligence. The robots could then be reprogrammed for their next assignment.

Over time, if fake users cannot be distinguished from human users on the Internet, the latter will be forced into a situation not unlike Harrison Ford in *Blade Runner*. The difference will be that there is insufficient interactivity with the robot to spot the fake.

4. The technical details

Today, botnets spam the world, perform DDoS attacks, and hack other computers. Tomorrow, they could be used by ideologues to sway public opinion.

Programmatically, a complex, copy-and-paste algorithm can steal biographical information from web pages, news reports, blogs, and other Internet resources. These in turn can be reconstructed to form the skeleton of an artificial personality. Details from popular news and current events will put meat on the bones. Once created, these artificial 'people' will be instructed to begin interacting with the Web in multiple ways. In due course, they will assume a 'life' of their own, and might even make a few human friends in the process.

The following steps have been field-tested with good results:

Her name is Violet:

- Visit the Census database (http://www.census.gov/genealogy/names/names_files.html)
- Select random first name
- Select random last name

She looks real:

- Select random but common first name/lastname combination
- Search Google (Images) for "fname lname @ Facebook" inurl:profile, medium size with face recognition
- Select random image after page 1

She has a real job:

- Mine the LinkedIn Directory (http://www.linkedin.com/pub/dir/fname/lname)
- Mine the ZoomInfo Developer API (http://developer.zoominfo.com)
- Pick random data and combine creatively

She said what?

- Violet opens a social networking account
- She befriends people
- She posts to their site

5. Evaluating the credibility of a cyber threat

In chess, it is often said that a threat is mightier than its execution. The very existence of a threat – regardless of whether it can be realized – tends to have a harmful effect on the victim, which may behave differently or even begin to act in a way that undermines its long-term security.

OSINT can yield enough information about a target to make even an empty threat seem credible. It is always difficult to quickly and accurately evaluate newly-discovered information, but cyber threats are especially complicated due to the power of modern OSINT and the relative anonymity behind which cyber attackers can hide. For example, phishing attacks are successful even though they normally employ only one layer of deceit: the website itself. Intelligent attackers can weave a much more intricate web of deception than that; an entire organization could successfully be faked if the time were taken to invest in enough third-party references.

In cyber terminology, the classic 'I know where you live' can be articulated as 'I know your Oracle server runs on 10.7.0.33, its administrator is Bob, and Bob likes passwords that relate to Manchester United'. OSINT specialists, especially those with some knowledge of computer hacking, could quickly develop the following threat: 'You have an appointment today with Dr. Livingstone at the Olympic Hotel … if I were you, I would cancel it'. Business leaders, military officers, and even heads-of-state have personal lives that can be targeted.

Botnet armies could be used to amplify a threat or to artificially enhance its credibility. If an attacker threatened a corporation or a government with strikes or civil unrest, a barrage of hard-to-verify complaints on Web fora could augment the threat, especially if the attacker had been seeding the fora for some time. The challenge for the attacker is to make the fabricated 'evidence' seem real while making verification a complex and time-consuming challenge.

When evaluating a cyber threat, it is important to remember that what makes a cyber attack easy – the power, ubiquity, vulnerability and anonymity of the Web – can also lessen its credibility. Good OSINT can lead to a significant bluff. In fact, the problem of attribution is the most complicating factor in cyber threat analysis. If the attacker is careless and leaves a large digital footprint (e.g. his home IP address), law enforcement may be able to take quick action. If the cyber attacker is smart, and covers his digital tracks, then deterrence, evidence collection, and prosecution become major challenges.

In almost all cases, computer log files alone do not suffice. Unmasking a cyber attacker requires the fusion of cyber and non-cyber data points. Investigators must enter the real world if they want to arrest a computer hacker. There will always be clues: if the goal is extortion, where is the money to be paid, and is there a point-of-contact? If the threat is Denial of Service, the target could ask for a proof of capability. The point is to generate a level of interactivity with the cyber threat actor that might be used against it. Further, cross-checking suspect information against trusted sources is always one of the best defenses.

From a technical perspective, solutions to the attribution problem exist. They include the increased use of Public Key Infrastructure (PKI), Internet Protocol version 6 (IPv6), and biometrics. Neural networks have also played a considerable role in reducing credit card fraud [3], and their ability to locate suspicious patterns in voluminous network traffic could be helpful outside the financial sector in the future. However, wide-scale deployment and proper implementation of such technologies are still years away. The widespread use of anonymous email services to support criminal activity, for example, has convinced some that an international convention is needed to regulate its use [4].

In the short term, one inexpensive counter to the threat posed by fake online identities is the simple use of a live video feed. As in *Blade Runner*, before you can really trust someone, it may be necessary to look her in the eye.

6. Attacking zombie armies with mathematics

Cyberspace mirrors the real world, and as such, it is complex and highly dynamic. Nonetheless, security analysts must find signals within the noise, or a targeted attack in a sea of normal network traffic. By way of example, let us examine an attempt to hack a simple, online poll.

The Internet Movie Database (IMDB) ranks Sergio Leone's *Il buono, il brutto, il cattivo* as the top-rated 'Western' film of all time, with an average user-determined score of 8.9 on a scale of 1 to 10 [5]. High IMDB rankings are lucrative in DVD sales, so a rival production company might try to raise the value of its own, low-ranked Western *Five Bloody Graves* by artificially increasing its number of high votes.

The IMDB, and the copyright holders of *Il buono*, must defend their turf. A sound strategy could consist of a two-step process:

1. the discovery of statistics that distinguish humans from computer programs as they vote in an online poll, and
2. using these statistics to support traffic analysis and database integrity.

Is it possible to separate human voters from robotic voters in a given data set? The trick is to keep sorting the data until identifiable fault lines appear. The goal of an attacker is to skew the poll result without being discovered; the goal of an IMDB security analyst is to identify the artificial votes and discard them. In concrete terms, the analyst should try to isolate portions of the data set that look different than those created by humans. While human beings are occasionally irrational, their behavior on the whole can be qualified and quantified as human. For example, when asked to vote on a scale from 1 to 10, human results normally lie within a 'bell curve': some are high, some low, but most votes fall somewhere in the middle.

Statistical analysis should reveal characteristics that distinguish humans from robots throughout the entire voting process. For example, if a computer program were to rate films in a purely random fashion, there would be no qualitative bell curve at all (instead, an equal number of 1s, 2s, 3s, etc). In terms of voting frequency, humans may typically cast their ballots over lunch or before bedtime; computers do not share the same requirements for nourishment and rest, so any serious divergence on vote frequency may be a sign of bot infestation. Humans are also prone to some highly subjective choices: top-ranked *Il buono* has received over 100,000 votes, while fourth-ranked *The Wind* has barely 2,000 to its credit. *The Wind* thus may be a 'hidden gem'; qualitative distinctions such as current popularity and off-beat taste may be difficult to program accurately.

On the technical side, it is possible to analyze the Internet traffic that brought the vote from the remote computer to IMDB in the first place. The 'source' Internet Protocol (IP) address can be geo-located on the Earth with the help of DNS. A good security analyst brings some knowledge of culture and politics to her analysis, and understands that there should not be too large of a discrepancy between what she expects to find and what she does find in the data.

Think of an IP address as a car. Not every parking space should be occupied by a red, 1989 Fiat Uno, just as not every entry in a computer log file should contain the same IP address. At the other extreme, randomizing IP addresses also does not work; one might then see just as many Maseratis in the lot as there are Hondas. To make his cyber attack credible, a hacker needs to make the final distribution of his source IP addresses mirror real Internet traffic patterns, which would require a large and sophisticated botnet.

Internet browser activity also offers computer network defenders valuable data points for analysis. When a human accesses a webpage, she typically waits for images, forms, and advertisements to load in the browser. Computers lack the curiosity and patience of a human. Robotic voters may move mechanically from one data request to the next; all such regimented Web requests should be investigated for other non-human properties.

Finally, cyber defense against virtual army attacks should involve a statistical analysis of the alleged identities themselves. The basic strategy is similar to a game of 'twenty questions'. Is the user male or female? Young or old? In entertainment or politics? Strange patterns and sudden ratio changes should be investigated. Advanced analysis might consist of an algorithm that combines first name, last name, country of origin, IP address and vote to known or expected baselines. Attackers can never be completely sure of what a security analyst expects to see, so their attack will always require some guesswork and likely entail some miscalculations.

7. Conclusion

In 2009, hackers steal data, send spam, and deny service to other computers. In the future, they may also control virtual armies, in the form of millions of artificial identities that could support any personal, business, political, military, or terrorist agenda. This attack vector exists because humans now communicate via ubiquitous software that is by nature impersonal and non-interactive. Further, given the pure amplification power of the Internet, it is not necessary that every target fall for the scam. And it may not matter if the ruse is eventually discovered, because the attacker may desire to sway public opinion only for a short period of time, such as prior to an election [6], business deal [7], or military operation [8].

Technologies exist, such as PKI, IPv6, and biometrics, to mitigate this threat. Smart system administrators, through network traffic analysis and rigorous database oversight, can also theoretically ensure a high level of data integrity. And if an attacker tried to fly 'under the radar' by using an insignificant number of bots for an attack, there would likely be a correspondingly insignificant impact on the target data set to merit the effort.

Unfortunately, the widespread use of good defensive tactics and technologies is not on the horizon. Most system administrators do not have the time, expertise, or staff to undertake a sophisticated analysis of their own data. Furthermore, clever programming can obfuscate many common signatures: if IP addresses and browser settings are scattered within the attack in a realistic way, the bar for cyber defenders is raised considerably.

For the foreseeable future, individual Web users must improve their own ability to evaluate threats emanating from cyberspace [9]. In most cases, the key is credibility. Illustrations from the Turing Test and *Blade Runner* suggest that sufficient interactivity with a computer should reveal that it is not human. But in the 1:n/n:1 computing environment in which we now live, the danger is that adequate dialogue is becoming rarer all the time.

References

[1] Freiling, Felix C., Holz. Thorsten, and Wicherski, Georg. "Botnet Tracking: Exploring a Root-Cause Methodology to Prevent Distributed Denial-of-Service Attacks". S. De Capitani di Vimercati et al. (Eds.): ESORICS 2005, LNCS 3679, pp. 319–335, 2005.
[2] Oppy, Graham and Dowe, David. "The Turing Test". *The Stanford Encyclopedia of Philosophy (SEP)*, http://plato.stanford.edu/entries/turing-test/, 2008.
[3] Rowland, Jan B. "The role of automated detection in reducing cyber fraud." *The Journal of Equipment Lease Financing*; Spring 2002; 20, 1; pg. 2.
[4] Mostyn, Michael M. "The need for regulating anonymous remailers". *International Review of Law*, Computers & Technology; Mar 2000; 14, 1; pg. 79.
[5] Top Rated "Western" Titles, The Internet Movie Database, www.imdb.com/chart/western.
[6] Consider the enormous impact of the 2004 Madrid train bombings on Spain's national elections three days later: "Europe: An election bombshell; Spain, a week on." *The Economist*. London: Mar 20, 2004. Vol. 370, Iss. 8367; pg. 41.
[7] Financial institutions often take the loss when their clients are defrauded: Patterson, Aubrey B. "Fighting hackers, fraud and wrong perceptions." *American Bankers Association. ABA Banking Journal*; Apr 2003; 95, 4; pg. 14. However, the court case of *Ahlo, Inc. vs. Bank of America*, in which malicious code on the company's computer was likely used to steal almost $100,000 from its bank account, demonstrated that coverage is not absolute: Cocheo, Steve. "Privacy rumblings grow louder." *American Bankers Association. ABA Banking Journal*; Jun 2005; 97, 6; pg. 56.
[8] All political and military conflicts now have a cyber dimension. The conflict between Russia and separatists in Chechnya has clearly demonstrated the power of well-timed Internet propaganda: Geers, Kenneth. "Cyberspace and the changing nature of warfare." *SC Magazine*, http://www.scmagazineus.com/Cyberspace-and-the-changing-nature-of-warfare/article/115929/, August 27, 2008.
[9] In 2006, identity theft was already the fastest-growing crime in the United States, affecting almost 20,000 persons per day: Ramaswamy, Vinita M. "Identity-Theft Toolkit." *The CPA Journal*; Oct 2006; 76, 10; pg. 66. Nearly a third of all adults in the U.S. reported that security fears had compelled them to shop online less or not at all: Acoca, Brigitte. "Online identity theft." *Organisation for Economic Cooperation and Development. The OECD Observer;* Jul 2008; 268; pg. 12.

THE INTERNATIONAL INSTITUTE
FOR STRATEGIC STUDIES
ARUNDEL HOUSE
13-15 ARUNDEL STREET
LONDON WC2R 3DX

© 2009 The authors and IOS Press. All rights reserved.

Subject Index

The Virtual Battlefield: Perspectives on Cyber Warfare
C. Czosseck and K. Geers (Eds.)
IOS Press, 2009

© 2009 The authors and IOS Press. All rights reserved.

Author Index